Writing and Speaking in the Technology Professions

Writing and Speaking in the Technology Professions
A Practical Guide

Edited by

David F. Beer
Director of Technical Communication
Department of Electrical and Computer Engineering
The University of Texas at Austin

A Selected Reprint Volume
IEEE Professional Communication Society, *Sponsor*

The Institute of Electrical and Electronics Engineers, Inc., New York

Printed in the United States of America

10   9   8   7   6   5

ISBN 0-87942-284-X

IEEE Order Number : PC0278-2

Library of Congress Cataloging-in-Publication Data

Writing and speaking in the technology professions : a practical guide
 / edited by David F. Beer.
  p. cm.
 ''A selected reprint volume, IEEE Professional Communication
Society, sponsor.''
 Includes bibliographical references and index.
 ISBN 0-87942-284-X
 1. Technical writing—Handbooks, manuals, etc. 2. Communication
of technical information—Handbooks, manuals, etc. I. Beer, David
F. II. IEEE Professional Communication Society.
T11.W75 1992
808′.0666—dc20
           91-21418

Contents

Preface

THIS ANTHOLOGY provides engineers, engineering students, and others working in technical and administrative fields with material that will help them become skillful and effective communicators. People have needed to communicate on technical matters at least since the building of Stonehenge and the Pyramids, and the amount of technical communication taking place in the last decade of the twentieth century is staggering. Whether they like it or not, working engineers and other technical people must communicate on a daily basis, often both in writing and in speech. With the modern development of advanced technologies, this demand has become even more critical—and complicated. Nowadays there is no question of the integral nature of efficient communication within the technology professions.

Rough estimates range from 20 to 50 percent of an engineer's time being taken up in communicating with others. You may have heard how engineers spend their first five years on the job wishing they had taken more math courses in college, and the second five wishing they had taken more English. Perhaps this indicates how essential communication skills are to the professional who wishes to rise beyond an entry-level position. In fact, plenty of evidence from industry confirms that communication skills become even more critical as you advance in your career. These skills might be needed for anything from writing a memo, an inspection report, a set of procedures, a proposal, or a manual, to spoken communication in brief reports or full-length technical presentations. Additionally, there are the myriad situations where conversational communication is important, whether it takes place over the water fountain or in a meeting or seminar.

You may well ask why this is an anthology rather than a textbook. The main reason is that I want to provide guidelines written by engineers and others who are involved in communicating in their everyday work. Thus you will find articles in this book written by people well acquainted with industry, whether in science, technology, or a branch of engineering. Many authors are working engineers themselves. The articles are brief and to the point. They not only contain insightful and practical information, but are themselves models of clarity and coherent organization. The information within each article is accessible. It is easily retrieved by a busy reader looking for information on how to communicate in a specific way or on how to improve some aspect of writing, speaking, listening, or organizing. In summary, the articles are practical and "user friendly," rather than theoretical or scholarly additions to the publish-or-perish pool.

Another reason this is not a textbook is that there is no shortage of textbooks directed at students who need to learn technical writing skills applicable to a very wide range of fields and careers. Most of these texts assume the reader is training to be a technical writer. Few are directed specifically at engineers or technicians and their workplace. Some provide a chapter on oral communication, but only one or two that I know of also give information on organizing and speaking in meetings—or on that most neglected, but vital, part of communication: effective listening.

The selections in this anthology are separated into eight parts. The first six parts deal with written communication and follow a logical sequence. Thus, Part 1 contains information and help on the initial stages of the writing process. The material here is as applicable to a one-page memo as to a book-length proposal. These opening articles will give you practical advice on how to write clearly and to the point while efficiently conveying your message to your selected audience.

Part 2 deals with some formatting and visual aspects of technical writing. Not only will you find information here on such basics as margins, fonts, and headings, but also on choosing appropriate graphics and incorporating them into your text when needed. Other visual/verbal combinations, namely lists, abbreviations, and symbols, are also covered. From this section you will get some excellent ideas on how careful use of these tools can improve the economy, appearance, and success of your documents.

Once the basics of clear and economical writing are considered, we are ready to look at some specific modes of writing, and so Parts 3, 4, and 5 discuss some of the documents you may be called upon to write on the job. Part 3 includes several articles on fairly specific writing tasks, including introductions to reports, some reports themselves, plus job application letters and recommendations. The problem of writing persuasively is also considered here, followed by a look at how to produce newsletters and technical articles for publication.

Parts 4 and 5 contain practical advice on how to write manuals and proposals. The usefulness of many products depends on how well written their accompanying manuals are. Moreover, a company's success may depend on how many winning proposals it submits. Although a proposal can be as brief as a one-page suggestion, many manuals and proposals are lengthy documents. Most technical professionals are likely at some time in their career to be part of a team working on such a document. Being aware of the "larger picture" will make you a much more useful member of that team. Thus Parts 4 and 5 provide many insights into how to put together effective manuals and proposals, whether you are writing alone or as part of a team. You will find your questions on how to focus, organize, and design these documents answered here.

It has been said there is no such thing as a good writer—only good *re*writers. This may well be true. At any rate, few of us would be foolhardy enough to write something to be read by others, whether a one-paragraph letter, five-page report, or five-hundred page proposal, without first carefully reviewing and testing it to ascertain its quality. The articles in Part 6 will enable you to approach the task of editing your drafts methodically and confidently. The first four articles in this section deal with revising and editing your writing in general. The remaining three focus on some interesting possibilities for document improvement, such as collaborating with others, using an outline to reorganize a completed piece, or starting a report at the beginning of the project it deals with and developing it as the project progresses.

Because many people in engineering and the sciences find they have to communicate verbally as much as in writing, Part 7 contains a collection of articles showing how you can give effective oral presentations. These selections, like all others in this book, are practical rather than theoretical. From them you will get help in successfully planning, organizing, and delivering both short, informal talks and longer, highly technical presentations. You will also find guidelines on how to handle an audience unfavorably disposed to what you have to say; the concluding article in effect becomes a checklist of everything you need to consider before getting up in front of an audience.

The final section of the book will be useful when you have to organize, conduct, or spend time in meetings. The articles in Part 8 enable you to prevent a meeting from deteriorating into the lengthy, ineffective exercise in frustration and boredom that we are all familiar with. You will also find guidelines on how to organize a technical seminar, and will learn when and how to argue, should it become unavoidable. The final two articles in the collection deal with listening, and it is worth noting that the word *effective* appears in both titles. We all have to do a lot of listening, but, as these articles reveal, much of our listening is inefficient and does not lead to real understanding of what we are hearing. If we can eliminate the barriers to efficient listening, we will become much better communicators, while enabling others to do the same.

The articles in this collection were published in the past ten years, and consist of hands-on and practical advice from people who are communicators in the technical workplace. It is unlikely that the need for this kind of communication will diminish in the future, and all indications are that the need will increase. Thus, to be a successful engineer, scientist, or manager, it is inevitable that you will have to successfully communicate through the spoken and written word. Technology might have replaced the typewriter with the keyboard, and the pen and ruler with computer graphics, but the need for clear, economical writing and coherent and effective speaking will always be with us. The articles in this anthology will help readers in any technical field to significantly improve their writing, speaking, and listening skills, thus increasing the efficiency and effectiveness of communication taking place in the technical world as we move from the twentieth to the twenty-first century.

Many people have contributed directly or indirectly to this anthology. First, of course, are the authors of the articles themselves. Next, I must acknowledge all the engineers I have worked with at Texas Instruments and elsewhere. They have enabled me to get a clear picture of the kinds of communication skills people working in various technical fields need, and thus have helped me decide which articles to select. My friends and colleagues in the Electrical and Computer Engineering Department at the University of Texas at Austin have contributed to the book in the same way, as have my students throughout the years. Specifically, I would like to thank Leo Little, Ray Neubauer, and Candy Young of the University of Texas at Austin for the assistance they have kindly provided in the production of this anthology. Also, Ron Blicq, Kimberly Manthy, Herb Michaelson, and Cheryl Reimold of the IEEE Professional Communication Society have contributed valuable advice, as have the staff of the IEEE Press. My wife Ruth and our daughter Natasha have, as always, been a source of encouragement—and have helped me keep my priorities straight.

Part 1
Initial Approaches
Writing Your First Drafts

"CAN ENGINEERS WRITE?" Joan Knapp asks in the title of this section's opening article. The answer is, of course—provided engineers are given the training and practice anyone needs to do a complex task well. Thus the articles in this section are chosen to help engineers and others understand some of the most important basics of technical writing.

Starting Out

In response to her own question, Knapp takes you through ten stages of writing an engineering report. She describes five kinds of people most likely to read your reports and shows how your writing must, to a great extent, be determined by who you are writing for. Her emphasis is on research reports, operation manuals, proposals, and feasibility studies, but what she says applies to many other kinds of technical writing as well.

Who Is Your Reader?

In the second article Ruth Savakinas develops some of Knapp's ideas and gives pointers on finding out who your audience really is. The payoff for careful audience analysis prior to writing is a reduction in the need to rewrite. This article's title is a reflection on how we tend to rush into written work too quickly, without allowing time for thoughtful "prewriting"—and hence fall into the trap of writing first and afterwards doing a lot of re-aiming and rewriting.

If you're not yet convinced that one of the most important parts of writing is identifying your readers *before* you write, then M. C. Cosgrove's analysis of writing for practical people such as designers should win you over. This article, with its subdivisions of Problem, Solutions, and Recommendations, is full of practical advice on determining what your readers most need from your report, and then writing accordingly. Communication is the equivalent of technology transfer, and its efficiency can be greatly improved by an awareness of its "real-world" context.

Breaking the Rules

Perhaps you are still haunted by so-called rules of English learned in high school or later. Don Bush's article helps dispel those fears, showing how some rules are obsolete, some are just pet peeves of "pop grammarians," and some should be broken whenever necessary. Language is a tool engineers use to convey information, so you need not worry about violating a rule if following it would prevent effective communication. A good practice is to try to write the way you would speak if you were verbally giving information.

Brevity Saves Money

You may reap economic benefits by writing clearly and concisely, as Yvonne Day illustrates in the next article. Succinct writing can indeed be an "inflation fighter" when it costs $7–$12 to produce one typed page. Day provides plenty of evidence to support her thesis, and her numerous illustrations of "dog puppies" and other clutter are instructive and entertaining. Following Day, Richard Nailen shows how to avoid verbal inflation. He presents humorous examples of obscure language prevalent in much engineering writing, including wrong, superfluous, and unneeded words, and shows you how to avoid such gobbledygook. His quotes from wide-ranging authorities, such as Mark Twain and a British engineering editor, are much to the point.

Douglas Mueller's article is valuable for its examples and an instructive and entertaining quiz (with answers). Like Day, Mueller illustrates how "foggy" writing costs industry and government millions of dollars each year. Getting to the point right away is one of the prime requirements of efficient technical communication, and there are plenty of ways to avoid written waste when you write to convey information rather than to impress readers with your vocabulary.

Style Makes the Message

In the next-to-last article in this section, Daniel Plung explores the subtle question of style in technical writing. If this seems a rather literary approach to technical communication, there is still no denying that style (which Plung defines) helps shape and determine the effectiveness of your communication. His presentation of ten stylistic tools is illuminating, especially since he calls on many well-known authorities to illustrate his points.

Time for Writing

It's all very well to carefully prepare and produce an effective piece of written communication. The first eight selections give you ample advice on how to do this. But where do you get the time? We all have other things to do besides write. In the final article, however, J. K. Borchardt offers 11 concrete suggestions on how to allocate chunks of time for writing and—we hope—for practicing the recommendations found in the other articles in this section. Engineers can indeed write, and can come to write well. As with everything else worthwhile, it just takes some time, care, and practice.

Can Engineers Write?

JOAN KNAPP

Abstract—Writing skill is an important element in engineering success. To supplement engineering curricula that provide little help in developing writing ability, this article describes ten steps in report writing that apply to research reports, operation manuals, proposals, and feasibility studies. The steps are (1) analyze your audience; (2) classify the report; (3) design the report; (4) do the research; (5) write a rough draft of the body of the report; (6) write a conclusions section; (7) write an introduction; (8) write an executive summary or abstract; (9) revise the report; and (10) add missing elements.

CONGRATULATIONS, graduates! Engineering school has taught you all you need to know to guarantee a successful career, right? Maybe not. You haven't learned how to write a report. However, the fact that you got through engineering school means that you can surely handle such trivialities as writing assignments, doesn't it? Perhaps. But recent studies, such as the major study of graduate adequacy conducted by Colorado School of Mines in 1978, suggest that many graduating engineers are deficient in communication skills, particularly writing. The studies also conclude that those skills are necessary for a successful career.

"My son never writes home—he's an engineer."

The experience of practicing engineers bears out the correlation between writing skills and successful careers. Engineers who write well advance in rank, eventually reaching management levels. Those who cannot write tend to be passed over for promotions and remain in routine jobs. Junior engineers may find writing requirements minimal during the first year or two of practice, but as they gain expertise, they are increasingly required to communicate

Reprinted with permission from *Colorado Engineer*, Summer 1983, vol. 79(4), pp. 4–8; copyright 1983 by Colorado Engineer, Boulder, Colorado.

The author is an Information Developer with the Field Engineering Division, IBM Corporation, 3131 28th Street, Boulder, CO 80301; (303) 441-2367.

that expertise. Those who can do so are promoted to more challenging, better paying positions.

Unfortunately, the usual engineering curriculum offers little or no help in developing students' writing abilities. Since grades depend on tests involving calculations only, students have no incentive to improve their writing. Further, constant emphasis on mathematical rather than verbal material may actually decrease ability to communicate verbally over a four- or five-year period. And, finally, many students adopt a defeatist attitude toward writing; because they don't write, they assume they can't.

Yet most engineers have the potential to write well. Indeed, because they have been trained to think logically, engineering students often have the capability of becoming better writers than humanities students. What they need is instruction and practice.

Although this article cannot substitute for the supervised practice and detailed instruction offered in a writing class, it is addressed to senior students with the intent of providing a method and some guidelines for report writing. To simplify the process of planning and writing a report, I have divided that process into ten steps and I discuss each step as it commonly applies to engineering reports.

1. Analyze your audience. Nobody gets up in the morning, looks out the window, and says, "What a nice day! I think I'll write a report." Poems may be written on such occasions; reports are not. Whether it be a handwritten memorandum or a handsomely bound proposal, a report is a communication needed and directed by an audience.

Thus, when you are asked to write a report, your first step should be to ask three questions: Who will read the report? For what purpose are they reading it? Are they engineers? Based on the answers to those questions, readers usually fall into one of five audience categories:

a. Expert	Other engineers, preferably in the same field, who read the report for information relating to their own projects
b. Executive	Managers, usually lacking an engineering background, who read the report to make executive decisions
c. Technician	People usually lacking an engineering background, who read the report for direction in using products and systems designed by engineers

Reprinted from *IEEE Trans. Prof. Comm.*, vol. PC-27, no. 1, pp. 10–13, March 1984.

d. Lay   People lacking an engineering background, such as special-interest groups, who read the report for non-engineering reasons

e. Combined   A group such as a government agency, comprising engineers and non-engineers, who read the report to make decisions

Placing your reader in one of these categories helps you structure the report to fulfill readers' information needs and also helps you select the level of language that will communicate best. For instance, if you're reporting to the Sierra Club on the cost/benefit ratio of a water project, you should avoid highly technical language where possible and insert simple definitions for the terms you have to use. If, on the other hand, you're reporting to an engineering firm on thermodynamic analysis of an engine, nontechnical language would be both inefficient and insulting.

2. Classify the report. The foregoing examples suggest the second step you should take in writing a report. Not only must you analyze the audience and their purposes, but you also must classify the report and its purposes to decide what format you will use. The more common reports engineers write are physical research reports, operation manuals, proposals, and feasibility studies. Although in-depth directions for writing these reports cannot be given here, a brief description of the purpose and format for each type follows.

• *Physical research reports* are written to describe research projects, such as stress tests on metals. They follow a rigid format: introduction, problem statement; materials; methods; results; discussion; summary; and conclusions.

Students who have written lab reports are familiar with this format, but a report written in a work situation differs from that in the academic situation in four ways. First, the practicing engineer is not writing for a professor who knows the answers but for people who don't know the answers and are primarily interested in answers. Second, if new testing methods are used or if a physical system is modeled via equations, those features must not only be described but also justified. Third, as in all professional reports, a title page, table of contents, lists of illustrations and symbols, and other reader aids are required. And fourth, different language levels are needed in different sections of the report. These distinctions are illustrated in the discussion of feasibility reports.

• *Operation manuals* for water and sewer systems, for power plants and other industrial complexes, and for specialized tools and instruments are written to tell the user how to operate and maintain these systems and products. A manual must include a brief introduction, stating its purpose; a set of performance directions; and a set of debugging procedures. Theoretical sections presenting principles of design and descriptions of mechanisms may be included, but they should be firmly separated from sets of instructions.

When writing a manual, use clear, simple language; keep sentences short; supplement the text with plenty of drawings; and separate individual instructions by using numbers and white space. Current research at American Bell suggests that effective use of white space can increase speed of comprehension by as much as 50 percent.

• *Proposals* are written to propose that a project be undertaken. Projects may range from practical (better parking facilities) through R&D (research and development of a space shuttle) to pure research (a method of soils testing). Because of this diversity of projects, no firm directives for format or language levels can be given. But since proposals are the most audience-oriented of all reports, great care should be taken to analyze the audience and to present a report as persuasive and attractive as possible for that audience.

The proposal begins with a statement of the problem (good parking facilities, a space shuttle, or adequate methods of soils testing do not exist) and then presents the immediate background to the problem, the benefits that will come from solving it, and the feasibility of the solution. It describes the methods to be used, the facilities available, the tasks to be done, and a schedule for doing them. The competence of the persons or organizations doing the work must also be demonstrated: What are their qualifications? How much previous experience have they had? What references can they supply? Finally, costs and method of payment must be specified.

More than any other type of report, a proposal must be visually attractive and inviting to the reader. Use headings and white space generously. Supply helpful, attractive graphics. Indicate different sections of the report clearly (index tabs at the outside edge are often helpful). And, finally, select a binding that is both durable and convenient for distributing sections of the report to various audiences.

As a junior engineer, you will probably not be asked to write a proposal. But you may be asked to write a section of a proposal, and you will benefit from knowing how a proposal works. The kind of report you will most commonly be asked to write is a feasibility study.

• *Feasibility studies* are written to answer three questions: Is a given project physically practical? Is it economically practical? Is it suitable from the viewpoint of those who will be affected by it? These questions require definite answers supported by factual evidence. This requirement, in turn, demands that the writer reverse the research process, beginning with conclusions and organizing facts to support them.

To illustrate, suppose you are asked to report to the Boulder City Council on the feasibility of a walkway over the intersection of Broadway and College Avenue. You would have to research the rationale for the walkway and various walkway designs to decide on a practical solution; your report would then present that decision supported by data. The most common error made by beginning writers of feasibility reports arises from confusion with physical research reports: Writers report what they have done rather than what they have discovered.

Because the ability to write a feasbility report is important to your career, this hypothetical example is used to illustrate the remaining steps in writing a report. You have defined your audience: the Boulder City Council. Although they aren't engineers, they will give the report to the city engineer or an outside expert. You are therefore writing for a combined audience. You have classified the report as a feasibility study. What is your next step?

3. Design the report. Most report-writing textbooks suggest that you begin research at this point. But much research and writing can be avoided if you first brainstorm the subject, examining possible approaches and arriving at the most practical, and then discuss these approaches with your clients to find out what they want to know and in what detail. Perhaps the city council already has a design in mind and is partly committed to it. If it's a good design, your report can support it with the modifications you suggest; if not, your report must show why another design is superior. You should also find out what dollar figure they have in mind: If they're thinking $200,000 and you're thinking $500,000, your report will not be useful. Third, you should find out what kinds of information they're most interested in. If they're not much interested in environmental impact, you can limit your research accordingly. And, finally, you should find out how massive a report they are expecting.

On the basis of this information, design the report. If the council is expecting a 25-page report, decide what proportion should be given to structures, what proportion to materials, and what proportion to environmental impact. Within the structures section, decide how many designs should be presented, what graphics are needed, what major and minor factors should be considered, and in what detail they should be described. You now have a format that prevents you from gathering facts willy-nilly and trying to fit them together to produce a report.

4. Do the research. Since research is a separate activity, only two suggestions for doing it are presented here.

First, begin with the obvious. Measure traffic flow to see whether the project is justified. Measure street width, maximum vehicle heights, and walkway grades to determine whether users of the walkway would feel as though they were climbing Mt. Everest; if so, the street may have to be lowered. Draw up preliminary specifications (final specifications will be made in the final report) and calculate costs. Determine who will be affected by the structure and how their views can be sampled.

Second, if library research is involved, take notes on cards that can be arranged according to topics and then rearranged within topics to provide supporting information for conclusions. You may find a note-card system convenient for recording and storing all information obtained through research. It provides a place to file bits of information and also helps you draw up an outline for each topic.

5. Write a rough draft of the body of the report. Since you have already designed the report, you can write any section when it has been researched. If you find the original report design inadequate, modify it, but don't let research alter the basic structure you designed to respond to your audience's needs.

Guide your reader through each section of the report. Suppose you've finished researching a section on design and have identified the most practical one. Do you describe the unsatisfactory possibilities and then, with great flourish, present your solution? Quite the opposite. Begin with the solution and support it through comparison and contrast with other possibilities. If it has disadvantages, admit them, but remember that you are the authority and that your audience is interested in answers.

At the beginning of each section, tell the reader what points you discuss. Listing these points displays them clearly and guides your writing so that each point can be the subject of a separate paragraph.

Paragraphs are the basic building blocks of writing. Each must be restricted to a single topic summarized in a topic sentence, usually the first sentence in the paragraph. The remainder of the paragraph develops the topic through factual evidence, explanation, or examples. Logical ordering of this material and transitions between ideas are necessary so that the reader experiences each paragraph as a structured body of information.

The sentences that compose paragraphs are also structured entities. Sentences usually contain a single thought or two related thoughts; in the latter case, the writer supplies the connection through punctuation and connecting words. Engineers tend to use too many short, unconnected sentences. Combining them improves the reader's logical grasp of the material and also improves the flow of the paragraph.

Two final suggestions are offered for writing rough drafts: First, don't be afraid of including too much detail. Extraneous material can always be edited out, but if your original draft doesn't contain the logical connections needed by your reader, you may find it difficult to supply them in later drafts. And, second, double-space your writing, use only one side of the page, and leave generous margins. These practices allow you to cut and paste to move subsections and to insert additional material and transitions without rewriting when you revise the report.

6. Write a conclusions section. Although individual

sections of your report present conclusions for that section, you need to gather those conclusions into a cohesive whole. Doing so requires formulating general statements from particular conclusions and arranging these in a hierarchy, with the most important (or logically prior) conclusions first. For example, state that the walkway is structurally feasible and why before stating that design A is the most practical design.

Since the conclusions section may be read by people who do not read the entire report, it must be self-sufficient and fairly detailed. It may be placed at either the beginning or the end of the report.

7. Write an introduction. Now that you have finished the report, you can write an introduction stating subject, purpose, scope, and plan of development. An introduction is not an overview of the subject of the report; rather, it is an overview of the report itself. Thus, conclusions, historical background, and literature reviews do not belong here.

Introductions frequently begin with a problem statement: "Two people, four students, and six dogs were injured at the intersection of Broadway and College Avenue in 1982." The remainder of the introduction explains how the report addresses the problem.

Beginning writers sometimes attempt to write the introduction before they have written the report—a virtually impossible task. Once the report is complete, however, writing the introduction should present no problem.

8. Write an executive summary or abstract. An executive summary presents the substance of the report in abbreviated form to an executive audience; an abstract presents this information to an expert audience. Abstracts and summaries are usually limited to a page or slightly more, but no rules about length or amount of detail apply to all reports. The best plan is to read well-received reports on similar topics and structure the abstract or summary accordingly.

Like the conclusions section, the summary or abstract must be self-sufficient and must be written at the language level appropriate for the targeted audience. Although you may have been writing for an engineer throughout the body of the walkway report, an executive summary for the city council demands a shift to language appropriate for a lay audience. But beware that a shift in level isn't accompanied by a shift in tone. Don't "talk down" to this audience; just use different terms.

9. Revise the report. Revision is such a complex process that specific directions for it cannot be given. A practicing engineer I consulted in writing this article offered this suggestion: "Put the report down for a couple of days and then come back and analyze it for clarity and concentration of ideas. Among other activities, this analysis may result in adding, deleting, and moving material; adding concluding statements and transitions between sections and subsections; checking topic sentences of paragraphs to see that they present a logical line of reasoning; adding transitions within paragraphs; checking sentences for subject-verb agreement, noun-pronoun agreement, and parallel structure; correcting punctuation and spelling; and substituting active for passive voice where appropriate.

10. Add missing elements. To add missing elements, consult a report-writing textbook and past reports in the company files. The elements to be added include letter of transmittal, title page, table of contents, list of illustrations, glossary, list of symbols, appendixes, list of references cited, and bibliography. Rules for documentation of sources depend on the firm or agency you are working for. All these additions take time but they are minor considerations compared with the conceptually demanding task of writing the report.

Finally, select an attractive binding, sign your name, and prepare to bask in the satisfaction of a job well done.

Ready, Aim—Write!

RUTH C. SAVAKINAS

Abstract—This paper details a methodical audience identification approach called a 'pre-write,' which will greatly reduce the need for rewriting a document. To pre-write, the writer must write a statement of the purpose of the document and a statement identifying both the intended audience and the implications of writing to that audience. Five questions the writer must answer about the audience are given, and information on audience identification and the needs of particular audiences is included.

[Author's note: To educators, pre-writing is the first of the six steps used to teach the writing process and includes group discussions, reading of models, and practice in research and organizational methods. In technical communication, the phase known as pre-writing encompasses all activities that occur before the first draft is begun—for example, performing research, taking notes, and preparing outlines. The pre-write method I present in this paper, which focuses on writing about technical subjects, may be applied by anyone preparing to write. This pre-write method involves identifying the document's purpose and intended audience.]

WHEN WE WRITE, we want to feel confident that our readers understand—perhaps even enjoy—what we have written for them.

If we know our readers, we may interview them to discover what they want in a document. When we cannot communicate directly with our audience, or when our audience is not clearly defined, we can use other methods.

Many test the readability of their writing by applying a formula such as Gunning's Fog Index. A technical writer or editor or a colleague may be asked to review a document for accuracy and clarity; this is also a readability analysis. A nontechnical person may also provide valuable feedback as to whether the material is understandable. Instructional or procedural materials can be tested for usability by a representative sample of the intended audience. However, since these methods are applied after writing, considerable rewriting may be necessary. It is obviously preferable to do the best possible work the first time.

PRE-WRITE VERSUS REWRITE

Complex technical writing is likely to be very difficult to read. Readability further decreases when the writer does not define major ideas for the reader and when the written

Ruth Savakinas is a senior technical writer with the Applied Technology Division of Computer Sciences Corporation in San Diego, California.

document is not relevant to the reader's experiences and interests. These two impediments can be eliminated if you clearly define your purpose and your audience; this definition is what I call a *pre-write*.

You can dramatically increase the clarity and luster of your writing, and reduce the need for tedious rewriting, by following this simple pre-writing approach.

WRITE YOUR PURPOSE

Whether you write a memo or a book, always consider your reasons for writing before you begin:

- What is the subject of your document? Will you introduce a theory, propose an improvement, explain a technique, describe a process, or report the results of your research?
- Where will your document appear? You may be writing a report that will be circulated within your company, an article that will appear in a commercial publication, a research paper that will be published in a professional journal, or a procedure that will be used in training.
- Why are you writing? Is your purpose to instruct, inform, persuade, or inspire?

The subject and purpose of your document must be clear to you if you hope to make it clear to your audience. As to the subject, think about the breadth (which topics) and the depth (how much detail) of coverage needed. Doubtless you will slant the material in one way if your purpose is informative, in quite another if your purpose is persuasive. You must also consider your goals and aims in writing the document.

With your goals and aims in mind, write your purpose clearly in a statement of a few short sentences. Keep this statement nearby and refer to it often as you write. This reference ensures that you communicate the document's purpose to your reader and that you maintain your focus on the purpose of your writing.

IDENTIFY YOUR AUDIENCE

When you write to a friend, the letter is easy to write because you know the person you're addressing. Also, you usually write the way you speak. Writing or speaking to someone you don't know isn't as easy, and you can't al-

Reprinted from *IEEE Trans. Prof. Comm.*, vol. 31, no. 1, pp. 5–7, March 1988.

ways be sure that you're communicating clearly. If your audience does not understand what they have read, or becomes so lost or bored that they don't finish reading what you have written, it may be that you did not truly write it to and for them. To write to and for your audience, you must get to know them.

To identify your audience, there are five questions you must answer:

- Who will be reading your document?
- What prior knowledge do these readers have about the subject?
- What do these readers need to know?
- Why will these readers read your document?
- How will these readers use the information you provide?

Who the Readers Are

Your readers may be administrators and executives, or your supervisors and colleagues. They may be professionals from various fields of science and engineering, or professionals from a single field of science or engineering. They may be nontechnically trained individuals, or the general public (a mixture of these categories of readers).

You may have direct access to your audience. Or you may have access to audience profiles maintained by your company's market research or personnel departments, or your professional society. If not, consult library sources to learn some things about the profession(s) and to become familiar with the educational level of your readership.

If most of your readers are in a specific profession, you can consult the Occupational Outlook Handbook published by the U.S. Department of Labor's Bureau of Labor Statistics to become acquainted with the profession. This handbook describes the nature of the work, the working conditions, training and other qualifications, opportunities for professional advancement, average earnings in the profession, and related occupations. If, for example, you are writing about an automotive electronic modification to an audience of automobile mechanics, this handbook reveals that although knowledge of electronics was quite a narrow specialization in the past, today's mechanics must be familiar with basic electronic principles.

Another helpful source is the Statistical Abstract of the United States, published annually by the Bureau of the Census of the U.S. Department of Commerce. Among the many subjects about which this book provides data is educational attainment by profession. Continuing our study of automobile mechanics, we learn that more than half (52 percent) graduated from high school and that 24 percent attended college.

The average educational level is an important factor both in the choice of vocabulary and the sentence structure, and in the psychological approach of your writing. Knowing the educational level will help you answer the questions pertaining to what readers already know and what they need to know.

What the Readers Know

Note the educational level of your anticipated readers and the specific educational background required in the profession. (Amount of experience also affects the readers' knowledge; however, such data may not be available to you.) Compare this information with the subject and purpose of your document to determine the relationship between the two. For example, the subject may be of a general nature given the readers' background or it may concern a highly advanced topic within a narrow, specialized branch of the profession.

This comparison will also help you to assume a specific level of experience at which to aim your document: you must decide if you will write to the beginner, the reader with intermediate-level experience, or the expert.

Let's assume you are introducing a new plating process. The new process is similar to a plating process that has been used by your company's production technicians for several years. Your technicians' familiarity with the existing process eliminates the need to explain steps common to both processes; instead, your document can refer to these similarities and focus on the differences.

What the Readers Need to Know

Now that you have an idea of who your readers are and what they know, you can decide how much background information to include. You can also determine what level of detail is necessary to ensure that your readers will understand your document.

If you are preparing a report for administrators and executives, you should condense the subject into an exact statement of the purpose and the outcome; they want the gist of the subject, not the minute details. Be logical, assume an appropriate tone and attitude, avoid overly technical language, and omit opinion. In writing for your supervisors and colleagues you should also remain factual and brief, but for this audience jargon and implication are usually acceptable.

Professionals from different fields of science and engineering want precise details; they want to know about procedures and measurements used, applications of results, and any sources of error. Though a high-level vocabulary can be used, field-specific terminology should be avoided. With readers from one profession of science or engineering, field-specific vocabulary is appropriate. These readers are interested in precise details; however, they also want you to locate the problem and indicate its importance in the field.

Readers in the nontechnical audience will require full

background information and explanation of any uncommon terms and concepts. Be diplomatic, clarify goals, and motivate your audience.

When writing for a general audience, you should decide what your readers need to know based on the level of experience at which you are aiming. Beginners require extensive coverage of topics and a great amount of detail; assume no prior knowledge on the part of these readers, and introduce all basic concepts and terminology to them. Those on the intermediate level also need detail, but they want to know about the unique aspects of the subject and also want a good reference section. Experts require detail only on the extremely technical aspects of the subject and prefer an extensive reference section. No matter which audience you are addressing, all readers want and need a carefully organized document. Give them concrete information and clear examples. Vary your vocabulary, sentence length, and sentence structure to maintain their interest.

Why the Readers Are Reading

Two people may read the same document for entirely different reasons: an engineering technician may read about computer architecture to gain an understanding of it; an engineer may read about it to find ways to improve its efficiency.

Few people read technical documents for pleasure or relaxation. More commonly, they read such documents to learn. They may be reading to increase their knowledge of the field or to improve their job performance. Some are motivated to read by financial reasons, such as those seeking promotion within their occupation or seeking to change occupations. Many simply need to learn how to operate a new piece of equipment.

Motivating psychological needs and desires may also differ. Some readers may be motivated by the need for personal fulfillment or discovery, or by the desire for achievement or prestige. Others may be motivated by the need for security or preservation, or by the desire for status or freedom from pressure. Considering the likely or possible motivating factors may cause you to alter your writing approach.

The subject and purpose of your document, and your knowledge of your intended audience, will all help you decide why readers are reading and how they will be using the information you present.

How the Readers Will Use the Information

When readers are seeking to increase their knowledge (whether in an effort to gain a promotion, to change occupations, or to learn for the pleasure of learning), be sure to show how your document fits into the literature in that field. If a reader is reading to improve job performance, examine the information and omit anything that is irrelevant.

When you are writing instructions for operating a piece of equipment or for performing a task, remember that your reader will want these instructions logically grouped and presented in 'bite-size' chunks (procedure steps). Decisions for grouping and presenting such instructions must also include environmental considerations, that is, the work area in which the document will be used.

WRITE ABOUT YOUR AUDIENCE

The last step in pre-writing is to write about your readers. It is important to actually write down your answers to the five questions above, on paper. The result will be a fairly clear picture of your audience. As you answer the questions about your readers, you should also write about the implications those answers will have for your writing approach; write about how you will focus or slant the material to meet your readers' needs.

USE YOUR 'PRE-WRITE'

Your pre-written statements of purpose and audience identification will help keep you 'on target' as you write, and help your readers understand—perhaps even enjoy—what you have written.

So, ready, aim—write!

SUGGESTED ADDITIONAL READING

1. Gowers, E., *Plain Words*, New York: Knopf, 1954.
2. Gunning, R., *The Technique of Clear Writing*, Rev. ed., New York: McGraw-Hill, 1968.
3. Klare, G. R., and Buck, B., *Know Your Reader: The Scientific Approach to Readability*, New York: American Book, 1954.
4. Leonard, D. C., and McGuire, P. J., ed. *Readings in Technical Writing*, New York: Macmillan, 1983.
5. McGehee, B., *The Complete Guide to Writing Software User Manuals*, Cincinnati: Writer's Digest Books, 1984.
6. Sides, C. H., *How to Write Papers and Reports About Computer Technology*, Philadelphia: ISI Press, 1984.

Writing for "The Audience"

M. C. COSGROVE

Abstract—Design technology transfer can be greatly facilitated by studying the audience expected to receive and act on the information. Before writing begins, knowledge of the audience should influence design of reports in the various choices made concerning technical level, vocabulary, organization, etc. During writing, audience awareness facilitates the writing of simple and direct prose. When writing is completed, it can be checked by reference to several mathematical formulas that can indicate its readability for a given audience.

IN the foreword to *Design Technology Transfer* [1], editor A. H. Soni expressed concern about the ever-increasing gap between practice and research in machine design. Several conferences were then organized to promote and accelerate design technology transfer. These efforts have continued to the present time.

Nevertheless, much remains to be done if the communication gap is to be significantly narrowed. This paper examines some possible causes of the communication gap and offers suggestions for bridging it, especially by catering to the needs of the audience.

PROBLEM

The complaint one often hears is that engineers, particularly those with advanced degrees, show a pronounced tendency to write their monographs, reports, etc. as if their readers were mirror images of themselves, with the same education and experience. The result is that their reports cannot be understood by those who need them most. It is likely that these engineers' uncommunicative writing habits can be traced to the following causes:

1. Poor training in writing Most graduate engineers were required, sometimes against their wills, to take one or more courses in college English. There, they wrote for an audience (the teacher) who knew more about the subject than they did. The principal purpose in writing was often simply to complete the assignment; no positive action (such as a publication) resulted from the writing. Thus, most engineers were not trained to write for the professional world.

Later, having gone to work, graduate engineers begin writing for a diverse audience, with perhaps less knowledge than they about the subject of their reports. They write for a variety of purposes, and action may be taken on what they have written. Many engineers do not recognize this reversal in their writing situation. They have a newly acquired vocabulary of technical jargon, a "style" learned to please the tastes of a Ph.D. English teacher, and a wish to display their erudition.

2. Poor examples in industry A second problem in communication derives from the poor examples of writing available to engineers once they are at work. In the course of "keeping up with the literature," the engineer soon picks up the pattern of long sentences, passive voice, jargon, stereotyped phrases, etc. that make reports inaccessible to anyone not on the same technical level. The engineer also quickly perceives that writing should be not only objective but impersonal. While keeping himself or herself firmly out of his writing, the engineer extends the idea unconsciously to keeping the reader out as well. He or she learns to set down hypotheses and, later, facts in a world seemingly without people.

3. Poor motivation to communicate The third problem in communication is more complex than the other two. Graduate engineers are poorly motivated to write well for several reasons:

- Their interest is in design, not writing. By the time they have conceived a design and seen it through numerous travails en route to becoming a reality, they have spent most of their creative energy. They are called on to begin writing about their projects at the moment their interest in them begins to wane.
- They have little or no "vested" interest in writing. If what they write is a piece of convoluted prose totally inaccessible to the rank-and-file designer, it is unlikely this will be immediately criticized or even noticed by their superiors.
- They may, on the other hand, have a "vested" interest in writing the results of research in prose nearly impenetrable even by superiors. Unfortunately, there is some tend-

Reprinted with permission from *Journal of Mechanical Design (Transactions of the ASME),* April 1981, vol. 103(2), pp. 342–345; Copyright 1981 by the American Society of Mechanical Engineers, New York.

The author is an Associate Professor, Dept. of Language Arts, University of Cincinnati, Cincinnati, OH 45221; (513) 475-2543.

Reprinted from *IEEE Trans. Prof. Comm.,* vol. PC-27, no. 1, pp. 38–41, March 1984.

ency in industry (as well as in academia) to gauge the sophistication of research by the difficulty in comprehending its description. If even the research director can't understand it, the work must be outstanding.

Given this combination of poor training, poor examples, and poor motivation, it is not surprising that so little of what is written by a graduate engineer is accessible to the working designer. Still, it is unfortunate.

An instructive contrast to this worrisome situation in design technology transfer can be found in advertising, a field that recognizes it is dealing with *people*. The people whose task is to write advertising copy take pains to learn everything they can about the so-called target group. They know age, sex, occupation, education, affiliations, and everything else available about the audience. Indeed, advertising writers may take the equivalent of a sabbatical to mingle with members of the target group. Moreover, in some agencies, they write copy while sitting before a picture of the typical consumer they hope to reach. Why do they take so much trouble? Because they wish to continue in the life style to which they are accustomed.

A second contrasting example can be found in free-lance writing. Free-lancers consult trade journals to learn where the markets are for their writing, and heed as if it were gospel the advice they always find there: "Never attempt to place an article or story in any magazine until you have steeped yourself in several issues of it. You will then know the special 'slant' of that publication and *how to tailor what you write to its readers*." Do free-lancers heed this advice? They do, because free-lancing is a precarious business at best, and failure to write for a specific audience may leave the writer hungry.

None of this is intended to suggest that engineers are either lazy or fat. Rather, they are victims of a system that has provided neither good training, good examples, nor even strong incentives to encourage their writing for an audience they know something about. Fortunately, there are remedies for the problem.

SOLUTIONS

The lack of incentives for engineers to write well might be remedied by research directors and others at supervisory level if they were made aware of the crucial communication gap between research and practice. The other two deficiencies—poor training and poor examples—are now being addressed in universities and colleges, where courses in technical writing are among the fastest-growing offerings on campus. If there is a single, common element in all technical writing courses, it is the oft-repeated advice that technical reports must always be written with a specific audience in mind. Of course, it will be some time before graduates of such courses can make a significant contribution to narrowing the communication gap in design tech-

nology. What can be done in the interim?

Special summer institutes in technical writing, generally week-long courses, are offered for working technical writers at several universities. The University of Michigan, Massachusetts Institute of Technology, and Rensselaer Polytechnic Institute offer the best known of these.

Short of attendence at one of these, there is always the possibility of on-the-job training, in the form of in-house courses, workshops, etc. At the very least, research engineers can be provided with several texts which can be kept ready-to-hand on their desks. Three technical writing texts that are particularly strong in their coverage of audiences and how to optimize one's communication with them are

- *Technical Writing*, 4th ed., by G. H. Mills and J. A. Walter; New York: Holt, Rinehart and Winston; 1978
- *Designing Technical Reports: Writing for Audiences in Organizations* by J. C. Mathes and D. W. Stevenson; Indianapolis: Bobbs-Merrill; 1976
- *Technical Report Writing*, 2nd ed., by J. W. Souther; New York: John Wiley & Sons; 1977.

These texts have been published within the last several years. The first is a good all-around text, a classic in the field; the second is particularly strong in writing for audiences within organizations; and the third includes a study, widely copied, of the reading preferences of managers, technical staff, and others in a large industrial organization. Not so new, but very useful, is an article by Joseph Racker, "Selecting and Writing to the Proper Level," which appeared in the *IRE Transactions on Engineering Writing and Speech*, EWS-2(1), January 1959, pp. 16–21.

Perusal of such sources would do much to develop an awareness in research engineers that much of the world needs their writing expressed in simpler terms. This new awareness could then be sharpened by recourse to a study conducted by A. H. Soni to survey the state of the art in machine design technology at practice level [2]. Approximately 2000 top-level engineering administrators, all ASME members, were contacted to assist in procuring answers to Soni's extensive questionnaire. The data collected are from a sample of 660 designers (out of about 100,000 nation-wide) representing 60 major industries. For every researcher who communicates with practicing designers, the study would make excellent required reading. Its breakdown of designers by education does much to explain the present communication gap within the industry:

Education Level	Percent
High school	23.0
A.S. (2-year college)	17.5
B.S.	44.5
M.S.	13.4
Ph.D.	1.5
Unknown	2.0

One of Soni's observations at the end of the report is particularly pertinent to a discussion of the need to write for a specific audience: "The quality of the technical material that is presented in the journals and transactions is far beyond the level of understanding of 85 percent of the designers."

It is to be hoped that such statistics will move research engineers to alter their writing style. The alterations needed would affect all elements of their reports, from the choice of words and the construction of sentences to the organization of paragraphs and the makeup of the overall work. These matters can be considered in chronological order:

Planning

1. Choice of words As far as possible, reports should be written in language that is simple, familiar, and concrete. Unfamiliar concepts can be explained by familiar analogies, and visuals can be used to clarify a topic still further. There is virtually no danger that the writer who follows this practice will be accused of patronizing the reader; technical writers and editors assert that in many years of experience they have yet to hear a single complaint that a report was too simple.

Adjusting one's vocabulary to the limitations of an audience is scarcely innovative. Years ago, Caterpillar Tractor Co. developed an 800-word vocabulary for use in correspondence with overseas offices and found it highly effective in minimizing confusion. More recently, National Cash Register in Dayton developed a dictionary of about 1500 words plus glossary for use in preparation of their manuals. The vocabulary was developed partly by computer and partly by trial and error. According to Charles T. Brusaw, who was in charge of the project, a writer with some practice in using this vocabulary can handle it so skillfully that the reader is aware only that it is easy to read. Doubtless, design technology transfer could be greatly aided if this practice became widespread. Brusaw has expressed his willingness to advise other organizations in developing their own simplified vocabulary.

2. Construction of sentences It is important in writing for audiences at the design-practice level to avoid overly long, complex sentence structure. Studies have shown that comprehension drops when long, involved sentences become the rule rather than the exception. This is not to suggest, however, that all sentences must be short and simple. A succession of such sentences reduces writing to the "Dick and Jane" level. In truth, it is hard to say how long is too long in sentence construction. Reading sentences aloud can help to detect those that are over-extended.

3. Organization of paragraphs Comprehension is easier for audiences at all levels if each paragraph begins with a topic sentence containing the main idea. All other sentences in the paragraph are then elaborations on that main idea. Writers can help readers still more if they use transitions between paragraphs and also with them, for transitions do much to show the relationship between sentences.

4. Organization of the report as a whole Readers are less likely to feel lost in the jargon jungle of reports if efforts are made to tell them where they have been, where the writer is taking them, and how he or she proposes to get them there. For this reason, the old "saw" that seems to appear annually in a hundred textbooks remains good advice:

- Tell the audience what you are going to tell them.
- Tell them.
- Tell them what you have told them.

The technical level at which an audience is told these things sometimes varies (acceptably) in different parts of the report. Current practice in many organizations is to write the abstract, conclusions, and recommendations of reports at quite a simple technical level, the discussion at a mid-technical level, and to relegate the highly technical material to the appendix. Also, a help for those readers likely to find the report "heavy going" is a detailed table of contents to match a liberal supply of headings. This enables readers to confine reading to only those parts they need to know.

The research engineer who plans a report with simple vocabulary and sentence construction, with topic-sentence paragraphs and easy-to-follow organization will have done much to make it accessible to practice-level designers. But there are two more things to do to make the report more comprehensible to them: One is the technique of the freelancer (point 5) and the other is borrowed from the advertising copywriter (6). Both should be used just before actually writing the report.

5. "Lingo" and slant If researchers would "go the second mile" to communicate with practice-level designers, they would do well to put aside briefly the kind of reading encountered while keeping up with the literature. Instead, they should read publications of the sort likely to be read by the audience they plan to address. Let them pore over the catalogs, as the Soni report shows the designers do. If they read trade magazines, the writers should skim two or three of those. They should steep themselves in the kind of writing with which the practicing designer is likely to be comfortable. They need not slavishly imitate such writing, but, likely, they will subconsciously pick up this style of writing for the time it takes to write their reports.

6. Readers in the room To copy the technique of the advertising writer is asking a good deal of research engineers; it's not the kind of thing they're accustomed to. Yet the technical writing texts strongly recommend it: *All technical writing should be done for specific readers, real or imagined.* What is suggested, then, is that the engineers *mentally* invite to the writing session a sampling of the audience for whom they are writing. If they are people within the organization, they can be conjured up by name; otherwise the writers call on their imagination to furnish the necessary people. They might picture a typical manager, technically trained but now "rusty"; a female slightly belligerent about some gaps in her education; a youngster fresh from high school and eager to learn, but seriously deficient in math; a couple of degree holders with expertise in a different field; these are the people for whom the researchers will write. For each, such a cast of characters would be different, drawn as they are from memory and acquaintance. But for purposes of discussion, one might imagine a motley crew not unlike those assembled in television's *M*A*S*H*. Colonel Potter can represent the senior designer of long experience but no recent training; Charles Winchester III will view with jaundiced eye whatever the writer attempts; Radar will serve for the high school youngster deficient in math; Margaret can represent all minorities with defensive postures; and BJ and Hawkeye can serve as technically trained individuals distracted by other responsibilities. With the audience assembled, the writer can begin.

Writing

With materials gathered and all planning done, resolution firmly fixed to write simple language in short sentences, with paragraphs and report organization set up, and with audience assembled, the writer begins to tell them what they need to know. One author [3] suggests beginning with "I want to tell you that . . ." (a phrase which can be edited out in revision) to facilitate the flow of a conversational style. During the writing, the author ought not to consciously choose short words, short sentences, etc. He or she need not even worry about grammar. At this point, simply tell the audience what they need to know, in their own language. Where engineers are "writing down" to practice-level designers, it is better to write conversationally and formalize it somewhat in revision than to write the usual complex prose and attempt to "humanize" it later.

Revision

The report or monograph should be allowed to "cool" for a day or so before attempting to revise it. This enables the writer to approach it freshly, in a situation more nearly like that of a reader. Once the cleanup of minor errors has taken place, there are two objective tests that can be used to determine whether the material is likely to be accessible to the audience for which it was intended. Both can be easily and quickly applied.

Rudolf Flesch began in the '40s to develop a readability test which he has refined over the years in a dozen or more popular books that exhort people to write the way they talk. His latest, *How to Write Plain English: A Book for Lawyers and Consumers* [4], contains a readability chart that will enable research engineers to determine whether they have written in the plain English needed by an audience like the one described in Soni's study. A second readability test, also first developed in the '40s, is Robert Gunning's Fog Index [5]. This index enables writers to determine how many years of schooling are required to read a given report with ease.

Some computerized "fog indexes" have now been developed, but to date none has achieved the wide popularity of the Flesch and Gunning tests.

If, according to the readability tests, the research engineer's material is still too difficult, it should be revised, with simpler sentence construction and vocabulary. The temptation to protest that the concepts being reported are too complex to be explained simply must be resisted. Albert Einstein once wrote an explanation of $E = mc^2$ for *Science Illustrated* that scored only 16 on the Gunning Fog Index.

Finally, when the formulas indicate that the level of reading is acceptable, the authors should consider whether their imaginary audience is likely to be satisfied. As far as possible, they should be their own "devil's advocate." Would Col. Potter say, "Son, you've got to remember it's a little while since I was in school"? Would Hawkeye say, "This is great, but could you cut it down a bit? We're a little busy here"? It is to be hoped, instead, that the engineer has produced a report with no communication gap, so that Radar could say, "Boy oh boy! I never thought I could 'get' this stuff!"

There is no danger that such writing would prevent Charles Winchester III from conceding that "This report might have emanated from Boston." Simplicity and clarity are appreciated at all levels.

RECOMMENDATIONS

Research engineers can narrow the communication gap between themselves and practicing designers if they take the following steps:

1. Scan several technical writing texts to develop general audience awareness.
2. Study Soni's report [2] to pinpoint a specific audience.
3. Plan reports couched in simple words, short sentences, single-idea paragraphs, and obvious organizational order.
4. Enter the designers' world by reading what they read.
5. Write as if talking to specific designers sitting in the room, so that instinctive habits of tailoring comments to different individuals will be called into play.
6. Revise by recourse to objective tests, combined with a final bit of creative imagination.

REFERENCES

1. Soni, A. H. Ed. *Design Technology Transfer*. New York: American Society of Mechanical Engineers; 1974.
2. Soni, A. H.; Torfason, L. "State of the Art in Machine Design Technology at Practice Level." In Ref. 1: 93–127.
3. Blicq, Ronald S. *Technically-Write!* 2nd ed. New York: Prentice-Hall; 1981.
4. Flesch, Rudolf. *How to Write Plain English: A Book for Lawyers and Consumers*. New York: Barnes & Noble; 1981.
5. Gunning, Robert. *The Technique of Clear Writing*. rev. ed. New York: McGraw-Hill; 1968.

Correctness vs. Communication

Don Bush

IN DECIDING THE "CORRECTNESS" OF ENGLISH EXPRESSIONS, *we must consider not only consistency but also history and current idiom, says this author, who slyly warns against falling into "grammatical traps" and offers a couple of remedies for the inconsistencies in our language.*

I have long been interested in language. I was introduced to grammatical rules very early, by my parents. I codified these rules in the sixth grade:

I - we - he - she - they - and - who - are-subject - pronouns.
Is - are - was - were - be - am - and - been - never - take - an - object.

I drilled on diagramming sentences. Miss McGillvray taught us the difference between good and bad, right and wrong. It was all so orderly, like "practical arithmetic," and bank statements, and batting averages.

This attitude was reinforced by my education in high school and college, and later on my first job, as a reporter for the Kansas City *Star*. It may surprise some of you that the press is one of the last strongholds of rigid grammatical consistency. There I learned that there was no such thing as a *raise* in pay—it was a *rise*, analogous to the rise in the river. To say a *raise* in pay would be—yes—inconsistent.

A few years after that, however, I took two graduate courses in advanced grammar at the University of Tulsa. There I learned a central principle, a career-changing principle: I learned that two usages can coexist: a *raise* in pay is just as correct as a *rise* in the river. The brass *shined*, and it also *shone*. A theory can be *proven*, and

also *proved*. *Technical Writing* is sometimes capitalized, and sometimes it is not. And the clincher: A tremendous number of words can be spelled in two ways, and either one is right.

The goal of consistency, of course, is too often the goal of technical writers. Somehow, it seems logical to police uniformity. Uniformity, however, is not a cure for bad writing. It is like taking aspirin to cure dandruff. Our aim is communication.

Even grammar books are not uniform. They do not always agree. In fact, one of my favorite books is Goold Brown's *Grammar of English Grammars*, which was written in 1859 and contains, in 1,000 pages of very small type, all of the differences the author found among contemporary grammar books.

Style books are even worse. One of the axioms of our business is that style books violate their own rules. This is true even for the good ones. Style just gets too complicated.

So what do we do?

I myself cannot disregard the rules. I am now an editor, and also a teacher. I subject my students to the same sixth grade rules. I am not against them.

However, I have learned to classify them in three ways. I use two criteria: history and idiom. There are rules that are both historical and idiomatic, non-rules that are neither historical nor idiomatic, and obsolete rules that are historical but no longer idiomatic.

HISTORICAL, IDIOMATIC RULES

Some rules are part of the language

historically and are also consistent with idiom. The most important of these is that subject and verb agree in number. Another concerns the use of subject and object pronouns, even *who* and *whom*. Some grammarians wishfully suggest that *whom* is disappearing. This is not so, of course. We are stuck with it. It is part of our history and our idiom.

I also believe that indeed we should usually avoid splitting an infinitive and ending a sentence with a preposition. These are good rules, not bad ones. The split infinitive is generally awkward and the ultimate preposition is often weak. The rhetoricians of the 18th century proclaimed that sentences should not be ended with any "inconsiderable word"[1], and that is a rule that I still try to follow.

But I will indeed—or *shall* indeed—accept violations of even these historic rules if the violations fit our idiom. I often say "We'd better watch out," or "We need to really hurry." These seeming violations have a historical basis in themselves and a strong foundation in English idiom.

NON-RULES

The second class of rules includes those that have a foundation neither in history nor in idiom. Some of them are imposed on us by the people whom Jim Quinn calls the "pop grammarians": John Simon, William Safire, Edwin Newman, and others who write for newspapers and magazines[2].

One of these new rules is the prohibition against using *hopefully* in an indefinite sense. This prohibition dates from only about 1960. Many people object to "Hopefully, the weather will be clear" because the weather does not hope. But these same

This article is based on a paper prepared for the panel "The Role of Rules in Technical Communication" at the 31st International Technical Communication Conference, Seattle, Washington, April 29-May 2, 1984, and published in the *Proceedings* of the Conference under the title "Communication vs. Correctness" (pp. WE-15--WE-17). Both the paper and the title have been revised for this issue of *Technical Communication*.

people do not object to "Clearly, the weather will be foggy." Both *hopefully* and *clearly* remain legitimate adverbs, used in the grammatically "absolute" sense, outside the context of the sentence.

What are some of the other non-rules?

Some people also object to using nouns as verbs, as in *to host* or *to impact*. When I was young, people sometimes complained about using *contact* as a verb. We don't hear much objection about that today. We can indeed say *to contact, to lecture, to dream*. This flexibility is one of the great strengths of English, not a weakness. Shakespeare helped us realize this strength, and repeatedly used expressions like "Sermon me no longer."

We can also use nouns attributively—as adjectives—and can say *electronics engineer* without groping for a word like *electronical*. We instinctively know the difference between an electronics man and an electronic man: An electronic man goes "beep, beep, beep." We inherently know the difference simply because we know something about common sense, and the idiom of the English language—not because of a rule we were taught in sixth grade.

Incidentally, we can say *publications group* not because we put out more than one publication, but because our technology is *publications*, not *publication*—and also simply because *publication group* sounds funny.

Other popular non-rules include the occasional prohibition against the double genitive: *This old hat of John's*. Some would-be purists claim that the *'s* is redundant, because *of John* already indicates possession. Redundant it may be, but it is also good English, since to most people it sounds odd to say "This old hat of John." Or to say "This country of us."

OBSOLETE RULES

The third class of rules includes those that are not current today, even though they do indeed have historic antecedent, and even logic, on their side. Fortunately, they do not bother us greatly; the rules have been overruled by idiom. Not long ago, if we wanted to be correct, we could not legitimately say "I was very pleased." We had to say "I was very *much* pleased." We could not say "I graduated from college." We had to say "I *was* graduated," because, logically, the college, not the student, does the graduating.

One hundred years or so ago, purists argued that there was no such word as *reliable*, because, logically, it meant "capable of being relied," and *rely* required the adverb *upon*. The word *reliable* thus had to be *rely-upon-able*[3].

One time at the plant, we had a person in the Reliability Department who was a stickler for grammar. He insisted upon rigid rules. I pointed out to him that *reliability* was actually an illegitimate word. It was not logical. A suitable substitute, I suggested, was "Trustworthiness Department."

He did not laugh. Instead, he gave an answer that was significant. "I never heard of that," he said, somewhat sternly. And, of course, to him, that settled it.

Most of us are fortunate that there are indeed a lot of rules that we do not need to observe, simply because we have never heard of them. One example is the rules for the common ordinary subjunctive mood. All of us use them somewhat "incorrectly" because they are tough to master; they were probably invented after the fact anyway, to codify common usage.

THE IMPORTANCE OF IDIOM

On the other hand, among the grammatical rules we do know, any of them can be stretched to the point that they sound funny. How about "It is I," or "Men are taller than we," or even "Write as you talk." These are grammatical traps. Correct though they may be, they will not be used by a good writer. An experienced writer simply writes around them, serving his reader and himself by bowing to English idiom, and *the way* we talk.

One historical word that is changing is *data*. I have always said that it could be either way, singular or plural. But today I am beginning to emerge from the closet and am advocating firmly that the right way, the precise way, the educated way, is *data is*. I think that very soon we will consider *data are* to be a pedantic affectation, used not to communicate but to raise a deliberate barrier between the writer and most of the readers. If we say *data are*, it sets us apart from the crowd, which is exactly where we want to be set. Super-correct grammar thus sometimes has the same effect as ostentatious vocabulary.

One of the arguments against solecisms is that they distract the reader. I agree. But I also believe that, ironically, the reader is distracted by super-correct language that does not observe the conventions of standard idiom.

"WRITE LIKE YOU TALK"

I believe, along with many teachers of linguistics, that the spoken language *is* the language. When we are teaching English, we make a lot over the differences between the two, spoken and written—and oddly we sometimes preach that these differences should be preserved. You would think that our written language is being somehow polluted by our most effective means of communication: human speech.

Myself, I try—rightly or wrongly—to bridge the gap with my reader by sounding as though I am talking to him. In other words, I deliberately try to write like I talk.

"Write like I talk." That of course, is a grammatical "error" in itself. I always ask students what's wrong with it. They reply, "It should be *as*," but no one knows *why*. The answer, of course, is that an obscure rule once said that *like* cannot be used as a conjunction. Ironically, if you do find someone who knows the rule, you will also find a person who is willing to break it, because she realizes that the spoken language *is* the language, and she writes like she talks.

THE CHANGE IN LANGUAGE

But what about the degenerating language?

Well, what about it? Language does change.

We already know about *reliability*. Also, we seldom hear expressions today like "The house is *abuilding*." Since about 1870, we have preferred to say "The house is *being built*." We don't say that language is *enjeoparded*. We say that the language is *jeopardized*, ignoring the criticism against *ize* which existed then and persists today. Very few of us would ever seriously say that we *motored* to the lake in our *auto*. We would say we *drove* out there in our *car*. And what happened to that word *omnibus*? It is now applied mostly to bills in Congress, when it is used at all. This is a real, tangible decline in the language. We are losing vocabulary, but no one weeps over it. Instead, we lament the demise of *data are*.

THE FUN OF OBSERVING

But if we do not observe the rules, how then can we know what is "correct"?

I would suggest that there are two remedies. First, we can simply refuse to fret about it too much. We can sit alongside the scholars and observe how language changes. We in technical writing have a ringside seat. We are right there where the new words come from. We are sitting where the languages of science and technology and computers and psychology and education and business and the military all meet each other in grinding new locutions—each of these disciplines blaming television for the decline of the language.

Second, instead of complaining, it is more fun to observe. As students of language, we ought to have pleasure in watching it grow and change. And we ought to realize what we all knew, even before we became as educated as we are now: English is not consistent.

By the way, as students of language, we ought to realize, too, that the common verb *ought*—unlike other verbs in English—has no infinitive form and no past tense [4]. Where is our consistency?

WRITING TO YOUR AUDIENCE

The second remedy is this. We can still tailor our speech or our writing to the audience. This is a basic rule and a good one. If we are editing for a research laboratory, we may want to continue to use *data are*. If we are teaching high school, or college, we may want to be strict about *he don't*. In short, we can continue to observe the distinctions we like, such as the difference between *infer* and *imply*, but we do not need to use unnatural idiom, or, worse, invent distinctions which set us apart from the people we are trying to reach.

Words like *rise* and *raise* can co-exist, along with *got* and *gotten*, *proved* and *proven*, both used in their natural places, and we can let other words disappear entirely, like *auto* and *abuilding*, without too much regret.

Just as we know that some words get favorable reaction and others do not, depending on our audience, we can tailor our grammatical usage by using our natural sensitivity to people's feelings and be proud of the way we use our language to talk to people who will listen to what we have to say. Ω

REFERENCES

1. Hugh Blair, *Lectures on Rhetoric and Belles Lettres* (London, 1783), Vol. I, p. 239.
2. Jim Quinn, *American Tongue and Cheek* (Pantheon, 1980).
3. Richard Grant White, *Words and Their Uses* (Houghton Mifflin, 1899; Thirty-seventh impression), pp. 202 ff.
4. A.G. Kennedy, *Current English* (Ginn and Company, 1935), p. 463.

DON BUSH *has been a proposal analyst at McDonnell Douglas Corporation, St. Louis, since 1962.*

The Economics of Writing

YVONNE LEWIS DAY

Abstract—Differences exist between oral and written communication. For example, as a spoken message passes from one person to another, the accuracy of the message declines. Furthermore, we soon lose 50 percent of what we hear and within two days we'll lose another 25 percent. This article argues that written communication, thoughtfully and thoroughly revised, is a better way. The author asserts that the average length of a letter or memo can be cut in half with no change in meaning by purging it of three forms of clutter: redundancy, deadwood, and roundabout phrases.

We can lick gravity, but sometimes the paperwork is overwhelming.
—Wernher von Braun

MOST executives in business and government, if cornered, will tell you that writing is one of the most important skills that an ambitious young person can have.

In her column, "Your Money's Worth," financial analyst Sylvia Porter said recently that companies today single out communication skill ahead of production, financial, or marketing abilities as the executive talent they value most.

In other words, the person who is able to write simply and clearly is more likely to get the job of his or her choice and to succeed quickly in that position.

People who hate to write, or are afraid to try, often attempt to get by with talk instead. Sometimes this works, but more often than not, oral communication causes problems and wastes time, money, and energy.

Why?

There are two reasons. One of these concerns is what is called "communication efficiency" by those who have a compulsion to make things more complicated than need be. Simply put, the idea is that the accuracy of a spoken message declines as the message is passed from one person to the next.

Any child who has played the game of gossip knows that this statement is true. Learning the truth at such an early age may explain why the child, when grown, demands that everyone put everything "in black and white."

The accuracy of oral communication declines because the average person loses 50 percent of what he hears as soon as he hears it (assuming that a person is listening to begin

with). If that were not true, you could repeat verbatim everything that was said to you yesterday.

If you're like the average person, within 48 hours you will forget an additional 25 percent of what you hear. This means that even when you pay close attention, you will probably remember only one-fourth of what was said to you two days earlier. This fact is at the basis of the exasperated cry, "But I *told* him ... in plain English!"

The written word, then, provides an accurate account of what is said and done. Not only that, but it allows us also to share knowledge, ideas, and feelings with posterity, to say to readers a century from now, "This is the way things are."

The second reason is that people who write poorly often fail in their spoken efforts. Good writing involves logic and discipline. Without these attributes, communication of any sort comes out half-baked—ill-conceived and ill-expressed. Writing helps you think. A few courageous souls say bluntly that poor writing is a sign of poor thinking.

INFLATION FIGHTER

There are many other good reasons for learning to write better but you've heard them all. Right?

Wrong.

There is one reason that a few dare mention at a time when the annual rate of inflation is higher than Lady Astor's eyebrows. That reason is simple economics.

Do you have any idea what it costs the average organization or business today to turn out a single one-page letter or memorandum? Take a guess.

If you guessed any amount under $7, you are an incurable optimist and there's no hope for you.

According to the National Office Management Association, the cost of producing one typewritten page ranges from $7 to $12, depending on the subject matter and the number of managers who have to approve the correspondence. In large businesses and in government, the typical letter is written by an underling for someone at a higher level to sign. The signed copy is then reviewed by someone in the next echelon, and perhaps by someone else a notch higher, and so on.

Although many offices are now equipped with word processors that do everything but let out the cat, the time saved in preparing correspondence is not nearly so much as you

Reprinted with permission from the August 1982 issue of *The Toastmaster*, published by Toastmasters International in Mission Viejo, Calif. (714) 858-8255. A nonprofit organization with 7,400 clubs worldwide, Toastmasters International teaches its members skills in public speaking and leadership.

The author is a communication consultant and lecturer, 14723 Stoneberg Ave., Baton Rouge, LA 70816, (504) 272-8776.

Reprinted from *IEEE Trans. Prof. Comm.*, vol. PC-26, no. 1, pp. 4–8, March 1983.

probably imagine. And from any savings you must deduct the cost of this sophisticated equipment.

But what if I tell you that the average length of letters and memoranda that you write can be reduced *by at least 50 percent* without any change in meaning? (I heard that, but I'll tell you anyway.) The length can be reduced by at least half by simply eliminating unnecessary words, roundabout and evasive constructions, and meaningless jargon. Words like these, which contribute nothing to meaning, are called *clutter*.

Peer over any editor's shoulder and you'll find that we spend 90 percent of our time striking out clutter and rearranging the ideas that are left. We find that the average person who takes pen in hand is like the man who prompted Lincoln to remark, "That fellow can compress the most words into the smallest idea of any man I ever met."

But the problem started long before Lincoln's time. Pliny the Younger, Roman statesman and writer, ended a letter to a friend with this statement: "I apologize for the long letter; I didn't have time to shorten it." Nineteen centuries later, Sir William Osler, Canadian physician and medical writer, echoed the thought. "It is harder," he said, "to boil down than to write."

The problem is still with us. In a survey of business executives and government officials, conducted by Dr. Fred H. MacIntosh of the University of North Carolina, 179 of the 182 respondents listed "wordiness" as one of the problems with writing today.

It doesn't take a super sleuth to find evidence to support their statement. The opening lines of business letters are choked with unnecessary words: *At the present time*, we are *in the process of* accepting applications *from interested individuals* for *the position of* senior accountant.

Fourteen of the 21 words in that sentence are unnecessary. If you are accepting applications, it goes without saying that this action is occurring "at the present time." And how many *un*interested individuals do you think would express interest in a job they don't want? For that matter, why mention individuals at all? Who else but a person would submit an application? The implication is that the great apes and other critters need not bother.

Besides increasing length, clutter robs the statement of the warmth and simplicity that characterize good conversation. If a friend calls and asks what you're doing, would you say "At the present time, I am in the process of viewing television news on the electronic medium known as TV"?

The 21-word sentence, therefore, can be reduced *66 percent* without changing the meaning: We are accepting applications for senior accountant.

BACK TO BASICS

The secret to writing concisely is to strip every sentence to its essential parts. Sydney Smith, English essayist, offered this advice: "When you've finished writing, go through your manuscript and strike out every other word; you have no idea how much vigor it will give to your composition."

Begin by throwing out (1) redundancies, (2) deadwood, and (3) roundabout phrases.

Department of Redundancy Department

Much of the clutter in writing consists of words that mean essentially the same thing and phrases that make a point that is implicit in what has already been said. The repetition of meaning in this way is called redundancy.

Fighting redundancy is a little like fighting weeds or the Internal Revenue Service—you're always slightly behind. A cabinet member speaks of "a hypothetical situation that does not exist," and soon all of America is echoing the phrase.

At his first press conference as secretary of state, Alexander Haig spoke of "careful caution" (which is preferable, I suppose, to careless caution). Within the week I heard a manager at the corner drugstore reprimand a clerk for lack of "careful caution" in handling a prescription request.

Haig's statements at the press conference were so laden with clutter that *Time Magazine*, in its February 23, 1982 issue, accused him of "conducting a terrorist campaign of his own—against the English language."

Secretary Haig is certainly not alone in his "campaign." Our speech and writing are strewn with redundancies. Many of us acquire the nasty habit in childhood when, to achieve a greater emphasis in our sparse vocabulary, we repeat the obvious: "Bang! I killed you dead." The habit is reinforced by listening to adults who never learned to let the obvious speak for itself.

Comedian Dave Gardner points out how insidious the habit is in his satire of Little Bo Peep who "lost her sheep and couldn't tell where to find them." Says Gardner, "If she lost them, it's only natural that she wouldn't know where to find them." And if she doesn't know where to find them, she'll certainly have to leave them alone. As for wagging their tails behind them, sheep seldom wag their tails anywhere else.

"Dumb," you say? Then obviously you've never said, "I've lost my keys," only to hear someone say, "Where did you lose them?" And you probably don't snicker when you pass the pet store that advertises "live pets" (as though there is a store somewhere that sells dead ones).

Stating the obvious is one of the main causes of redundancy. Unless you're from another planet or have sadly lost contact with this one, you know that nothing can be green in size or visible to the ear or large in smell. You know, too, that no man is bald-legged, that no one has a big smile on his foot, and that nothing ever falls to the ground above. Aside from adding clutter, therefore, nothing is gained by saying green in color, visible to the eye, large in size, or bald-headed.

Kill Them Dead?

Redundancies also stem from misuse of words that are absolute—that is, words that cannot be compared or modified. "Dead" is an example. One person cannot be deader than another. Similarly, a woman cannot be slightly pregnant. A thing that is endless is without end, period. It cannot be made more endless.

Of the many words that are absolute, one deserves special mention: "unique." The word means "one of a kind." Precious few things in this world are unique. These few things are known as the Seven Wonders of the Ancient World. A business lunch is not one of the seven and is not likely to become the eighth wonder unless it consists of Martian rock soup and Venusian fly roast.

Yet in the course of a day you will hear the word applied to anything that is even slightly different. The word is also absurdly modified. To say that one thing is more unique than another is to refute the laws of science. If a thing is one, it cannot be more so. Those who say that this thing is more unique than that compound the absurdity by adding that something else is the most unique of all and that some things are uniquely unique, or (heaven help us) most uniquely unique.

"'As a matter of fact' precedes many a statement that isn't." Mark Twain said that a century ago, yet there are people who persist in using that phrase and dozens of other platitudes that add nothing to their business writing. To begin a statement with "to be honest" implies that you haven't been. The phrase is often used when what is meant is "to be blunt." That, too, is unnecessary. If you write "Get lost, turkey," the reader isn't likely to need an explanation.

Other examples of deadwood include *I might add, it is interesting to note, it should be pointed out, it should be remembered, it is significant, it is worthy to say, for your information, may I say, may I call to your attention, may I take the liberty, permit me to say,* and *with your kind permission.*

Simply state the information and let the reader decide whether it is interesting or important. Telling the reader in advance that something is interesting is like telling someone that a joke is really funny—it rarely is.

Deadwood also includes these vague and meaningless modifiers: *appreciably, approximately, comparatively, considerably, definitely, evidently, excessive, fairly, nearly, negligible, rather, reasonably, relatively, significant, somewhat, substantial, sufficient, suitable, undue* and *various.*

The trouble with these words is that they take up space without telling the reader anything. What is a "significant" number? Ten? Fifty? A million? At best the word leaves the reader wondering whether that "significant" increase in insurance premiums will wipe out his mad money or leave him with a second mortgage on his first-born. Give the reader specific information or admit frankly that you don't know. At worst the reader will assume the latter, anyway.

Pretentious Phrases

The use of roundabout, evasive expressions is called "circumlocution." A circumlocution is clutter at its worst: a laborious phrase that has pushed out the short words that mean the same thing. These windy phrases are a drag on energy and momentum. Even before John Dean gave us "at this point in time," the average American had stopped saying "now." In its place came "at the present time," "currently" or "presently" (which doesn't mean "now," it means "soon"). None of these is as effective as "now," meaning the immediate moment; or "today," meaning the present; or simply forms of "to be," as in "It is raining." Only a windbag would say, "At the present time we are experiencing precipitation."

The urge to inflate simple sentences has become an obsession throughout business and the professions. My dentist used to ask, "Does it hurt?" Now he wants to know whether I am experiencing any pain. If I weren't so busy trying to experience respiration (breathe), I could tell him

Dog Puppies

This is a list of redundancies compiled from business and technical writing that have fallen victim to my red ink. The word or words in parentheses should be deleted.

—Y. L. D.

(a bolt of) lightning
(a distance of) ten yards
(a) myriad (of) sources
(absolute) guarantee
(absolutely) essential
(absolutely) sure
(actual) experience
add (an additional)
(advance) planning
(advance) reservations
(advance) warning
all meet (together)
alongside (of)
(already) existing
aluminum (metal)
(and) moreover
(as) for example
ask (a question)
(as to) whether
(as) yet
(at a) later (date)
(at) (about)
at (the) present (time)
at (12) noon
at (12) midnight
at some time (to come)
(awkward) predicament
(baby) boy was born
bald(-headed)
(basic) fundamentals
blend (together)
bouquet (of flowers)
(brief) moment
burn (down)
burn (up)
(but) (however)
(but) nevertheless
came (at a time) when
cancel (out)
(chief) protagonist
climb (up)
(close) proximity
(close) scrutiny
(cold) facts
collaborate (together)
combine (together)
commute (back and forth)
(complete) monopoly
(completely) destroyed
(completely) filled
consensus (of opinion)
continue (on)
(continue to) remain
(current) fad
(current) trend

(currently) being
dates (back)
(definite) decision
descend (down)
(different) kinds
(difficult) dilemma
(direct) confrontation
do (over) (again)
drop (down)
during (the course of)
dwindled (down)
each (and every)
earlier (in time)
either (and/or both)
(empty) space
(end) result
enter (in)
equal (to one another)
eradicate (completely)
(established) fact
estimated at (about)
estimated (roughly) at
(every) now and then
(exact) opposites
face (up to)
(false) pretenses
(fellow) classmates
few (in number)
filled (to capacity)
(finally) ended
(first) began
first (of all)
follow (after)
for (a period of) 10 days
(foreign) imports
forever (and ever)
(free) gift
(free) pass
(future) plans
gather (together)
(general) conclusion
(general) custom
(general) public
(glowing) ember
golden (wedding) anniversary
(grand) total
(guest) speaker
had done (previously)
(hard) facts
heat (up)
(hostile) antagonist
(hot) water heater
I (myself personally)
indicted (on a charge)
(integral) part

introduced (a new)
introduced (for the first time)
(invited) guests
(ir)regardless
is (now) pending
join (together)
(just) exactly
(just) recently
kneel (down)
last (of all)
lift (up)
(local) residents
look back (in retrospect)
lose (out)
(major) breakthrough
(mass) media
may (possibly)
mean it (sincerely)
(mental) telepathy
merged (together)
meshed (together)
(midway) between
might (possibly)
mix (together)
(mutual) cooperation
my (personal) opinion
(native) habitat
(natural) instinct
never (at any time)
never (before)
(new) beginning
(new) bride
(new) construction
(new) record
(new) recruit
no trespassing (allowed)
none (at all)
(null and) void
off (of)
(official) business
officiated (at the ceremony)
(old) adage
(old) cliche
(old) pioneer
(old) proverb
(one and the) same
(originally) created
over (and done with)
(over) exaggerate
over (with)
(pair of) twins
(partially) damaged
(partially) destroyed
(passing) fad
(past) experience
(past) history
(past) memories
(past) records
permeate (throughout)
penetrate (into)
(perfect) ideal
period (of time)
(personal) charm

(personal) friendship
(personal) opinion
(pitch) black
pizza (pie)
plan (ahead)
(possibly) might
postponed (until later)
(pre-)plan
(pre-)recorded
(present) incumbent
(private) industry
probed (into)
proceed (ahead)
protest (against)
protrude (out)
(rate of) speed
recur (again)
refer (back)
reflect (back)
repeat (again)
reply (back)
reported (to the effect) that
revert (back)
rose (to his feet)
(rough) rule of thumb
(rustic) (country)
(same) (identical)
(separate) entities
share (together)
since (the time when)
skipped (over)
soaked (to the skin)
(specific) example
spell out (in detail)
stacked (together)
start (out)
started (off) with
(still) persists
(still) remains
(suddenly) collapsed
(suddenly) exploded
sufficient (enough)
(sum) (total)
summer (season)
swoop (down)
(sworn) affidavits
talking (out loud)
(temporary) reprieve
(therapeutic) treatment
(thorough) investigation
together (at the same time)
(true) facts
2 a.m. (in the morning)
undergraduate (student)
(underground) subway
(unexpected) surprise
(unintentional) mistake
(usual) custom
(when and) if
whether (or not)
written (down)
(young) foal
(young) lad

that I "definitely have the sensation of being in an unhealthy condition" (feel sick). Though he does good work, I cannot force myself to say, "The type work performed by him is excellent in nature."

After years of editing and rewriting materials written by executives in business and government (an average of 12,000 pages a year), I still cringe at the sight of an oxygen-deprived sentence like this: "During the period of time in which certain acute emergencies arose so that a crisis situation could be said to obtain, individuals rose to the occasion with singular examples of exemplary behavior toward co-workers of close proximity."

Those 37 words can be reduced to eight: "During emergencies, employees behaved admirably toward their co-workers." That is a 78 percent reduction in copy!

CONCLUSION

Writing concisely not only saves money (in production costs) but also time. Imagine how much time your readers will save if they have to read only half a page to receive your message instead of two, three, or even six pages. The time saved is even greater if you write a memorandum or letter that will be read by dozens or even hundreds of employees. The following memorandum, written when time *and* money were at stake, got straight to the point:

THE WHITE HOUSE

TO: John Dean

FROM: Charles Colson

Now what the hell do I do?

Equally direct are these instructions written by an anonymous (but delightfully original) worker: "In case of fire, stand in the hall and shout 'Fire!'"

Besides saving time and money, this kind of concise writing helps conserve our dwindling natural resources. Prune your prose this year and save a tree. Get others to help you and save a forest.

An Engineer's Guide to Clear Language

RICHARD L. NAILEN

Abstract—Obscure language and verbiage can destroy the usefulness of a technical report. A balance is needed between the technical words that tell the facts and the words that explain what the facts mean. Some of the faults to be avoided are using the wrong words, too many words, passive words, and needless words. Examples from technical literature are used.

BRANDISHING his slide rule, with diploma and drafting board as shield and buckler, the engineer sallies forth unafraid to slay a technical dragon. Yet there could be no more painful contrast to his clear, direct attack on technical problems than his attempt to explain in writing what he did.

Why should this be so? Students in other disciplines, according to a recent survey of college graduates, are anxiety-ridden, whereas engineering students are a comparatively carefree group. Supposedly the technical expert is an unemotional person, but in writing about his work the engineer seems bound up, so tangled in obscure language as to be unintelligible. As a result he may bury the usefulness of his work in verbiage. Engineers in particular seem to have two destructive habits: using the wrong words, and using too many.

THE WRONG WORDS

There are times, of course, when you must use an obscure technical word. But always ask yourself what contributes most to the meaning. By constantly applying this test to your writing, you will ensure a reasonable balance between words that tell us the facts and those that explain what the facts mean.

"I never write 'metropolis' for seven cents," Mark Twain, who was often paid by the word, once said, "when I can get the same price for 'city'." And neither should you. Extra syllables aren't worth what they cost in reader interest. Why, for example, write *recapitulate* instead of *sum up*, *initiate* for *begin*, or *optimum* for *best*?

Also be sure the word really means what you think it does, regardless of how simple or complex it is. A few common errors can change a reader's impression of you from merely incompetent to downright stupid. For example, *infer* and *imply* are frequently misused. *Imply* is an active word; the speaker or writer implies, or makes implications, in what he says or writes. But *infer* is passive; the listener or reader infers, or draws inferences, from what is heard or read. If, after you've finished speaking, I accuse you of inferring that I'm an idiot, you can answer by saying that only an idiot would use *infer* incorrectly.

How about this one, from the technical literature: "...not in a position to undertake such a project during the intermediate future." *Intermediate* means *in the middle*, or *between*. The

correct word is *immediate*, meaning *nearest* or *close by*. Bulging syllables do not necessarily make for clearer meaning. Similar to that problem is the use of *comprised of* instead of the correct phrase *composed of*, meaning *made up of* or *put together from*. The word *comprised* means *included* or *contained*, and it also means *made up of*. If you say *comprised of*, you've really said *made up of of*, which is silly no matter how you look at it. Thus, X comprises a, b, and c—or X is composed of a, b, and c.

Then there are sound-alike words. Consider *eminent* and *imminent*. The first means *outstanding* or *prominent*. *Imminent*, not even close in meaning, means *threatening* or *about to happen*. Yet we find this in an enginering publication: "This is imminently unfair and we feel it invites possible government intervention." Such wording does beg for intervention—not by the government but by a dictionary.

And how about *its* and *it's*. Schooled to remember that the apostrophe denotes possession, the hasty writer often forgets that this is true only for the noun, not the pronoun. Remember *his*, *hers*, and *theirs*. The word *it's* is not possessive, but a contraction for *it is*. Yet in a learned technical paper we see "It's evolution has traversed an epoch of time." Make sense? Not to mention that *epoch* means *period of time*. So we get this gibberish: "It is evolution has traversed a period of time of time."

Reprinted from *New Engineer*, March 1974, p. 29; copyright 1974 by MBA Communications, Inc.

The author is a professional engineer with Louis Allis Div. of Litton Industrial Products, Inc., P. O. Box 2020, Milwaukee, WI 53201, (414) 481-6000.

Reprinted from *IEEE Trans. Prof. Comm.*, vol. PC-24, no. 3, pp. 117–119, September 1981.

Both sound-alike and look-alike words lead us to a third category, which might be called nonwords. These gems crop up repeatedly in engineering documents. *Circumvent*, for example, is a legitimate word, meaning *to go around* or *avoid*. The similar word *circumference* also has to do with going around; it means the distance around a periphery. The careless writer concludes that they must somehow be the same word, so he writes *circumvential* when he means *circumferential*. Thus he creates a nonword, which if it existed could be translated as *avoidable*. And so would this writer's problems be were he to consult the dictionary again. Other nonwords are created by misusing word endings: *accidently* for the correct form *accidentally*, or *coincidently* for *coincidentally*. The correct word is derived from the adjective ending *al*, not from the noun form.

Concern for word meanings leads us to the problem of jargon, the overuse of obscure words having specific but little known meanings. Though tiresome to the outsider, such language may be essential to clear communication in a technical field. But if your audience is outside the field, beware of too many fancy terms.

A more troublesome form of jargon is language that seems to mean something but actually doesn't, or words that cloak a kernel of meaning in a husk of imprecision. Three of the most offensive such words are *optimize*, *interface*, and *time frame*. Or as H. L. Mencken once said about the use of *implement* as anything other than a noun meaning *tool*, "The next man to use *implementation* will be shot." While *implement* as a verb has come into common usage, there are better ways to say it.

TOO MANY WORDS

As one horrible example of how to bury a simple idea under a suffocating mass of jargon, consider this passage, which could be matched in many an engineering report: "While bryographic plants are typically encountered in substrata of earthy or mineral matter in concreted state, discrete substrata elements occasionally display a roughly spherical configuration which, in presence of suitable gravitational and other effects, lends itself to combined translatory and rotational motion. One notices in such cases an absence of the otherwise typical accretion of bryophyta."

Notice the length of that first sentence. There's a good reason for that contest requirement for answers in 25 words or less. Engineers often turn out 50- to 100-word sentences. If ever understood at all, they're easily forgotten. Big words, used to convey an exaggerated sense of precision, obscure more than they enlighten. Attention wanders, and the elusive thought escapes while the reader consults his dictionary. The anonymous author of the above paragraph meant to say "a rolling stone gathers no moss."

Maybe that wasn't fair. A trick, you say; nobody would really go that far to confuse a simple idea. All right then try this, from a technical paper on computer compilation of a dictionary: "Let m_j denote the jth member of a string of lexemes; let C_n denote the substantive subcategory of the total lexeme inventory; let $op(P,x)$ be an operation acting on a unitary lexeme to transpose it into a nonunitary lexeme; then we may state with a probability in excess of 0.95 that if $E(m_j)$; $m_j \supset C_n$; then $op(P,m_j)=(m_j)+s$." What does it mean? "Most English nouns form their plurals by adding 's'."

As a British engineering editor commented a few years ago, "Surely professional people who cannot, or dare not, present one another with clearly expressed thoughts should question their calling.... A scientist, or anyone else, who clothes his thoughts in jargon and gibberish is like a man in a loud-check sports coat at a funeral; his heart may be in the right place, but he will inevitably be misunderstood."

Redundancy is an especially familiar sort of jargon. Engineers recognize the concept; it simply means use of several parts, systems, or words where one will do. Use two or three adjectives or nouns instead of one, the writer thinks. Even if one of them is misunderstood the point is bound to get across.

For instance, from a paper on electrical system short circuit protection: "Total and complete protection of all control components is impractical under all conditions." "Total," "complete," and "under all conditions" are three different ways of saying the same thing in this sentence. One way would suffice.

Or this, from a past edition of the *National Electric Code*, a document singled out by several authorities as a prime example of bad technical writing: "Locations...in which combustible dust is or may be in suspension in the air continuously, intermittently, or periodically...." What the authors meant is a location where the dust might be present at any time. Yet they threw in three polysyllabic words to do the job.

Even short words may appear in far too great a quantity. Take the phrase "at this point in time." This pomposity sends up before every reader a clear warning: windbag at work. A simple word like "now" or "today" does a far better job. Another good example is "the fact that," a phrase which according to one advocate of clear writing "should be revised out of every sentence in which it occurs."

PASSIVE WORDS

Technically trained writers often alienate luckless readers by pompously donning the mask of anonymity. It's easy enough to agree or disagree with what you read. But it's hard to make any contact at all with a writer who simply leaves himself out of his material, so you can't get at him. Some examples:

• "It is believed that...." By whom? Do you, as author, "believe" what you've written? If so, say it without hedging.

- "In the writer's opinion..." or "In my judgment...." As readers, we can assume that what you write is what you think. We don't need to be told so.
- "It was determined that...." Why not just say "I (or we) decided" or "I discovered"?

Writing comes to life when the active voice is used rather than the passive. Don't describe "what was done." Instead, describe "what you did." If an engineer cannot report his conclusions in a forthright manner, one is entitled to wonder if his work was forthright.

NEEDLESS WORDS

The late William Strunk's advice should be easily understood by engineers: "A sentence should have no unnecessary words," he said, "just as a drawing should have no unnecessary lines, and a machine no unnecessary parts." Wasted words aren't always obvious until you think about them. The needless words in some of the following examples are set in italics.

- "The preliminary studies required *a total of* 345 man hours."
- "*Located* at the top of the transformer are the line terminals."
- "It was decided *previously* that analog measurements would be used for on-line control of process parameters."

- "The philosophy of terminal operation has *historically* been to keep operating personnel informed through extensive use of graphic panels." Also, why write "operating personnel" when "operators" would do quite well? And why not "inform" them instead of "keeping them informed"?
- "Steps should be exercised to avoid large lower-order harmonics." How do you "exercise a step"? Is it like walking a dog? How much clearer and simpler to write, "Designers should avoid large lower-order harmonics," or merely "Avoid large lower-order harmonics."
- "The use of large air gaps for excitation damping is recommended." Passive, not active, voice again. Why not "Use large air gaps for excitation damping," or at least "Large air gaps are advisable for excitation damping"?

Words do cost money. Anyone familiar with drafting room operations knows the importance of reducing hand lettering on drawings. Business executives wring their hands over the mounting cost of correspondence. And at least one technical society, the Institute of Electrical and Electronics Engineers, has placed limits on the number of technical papers it will publish, as well as on the length of individual papers.

If you don't learn to express yourself well, not only will nobody understand you should your work get published, but you may not even get the opportunity. A brief, clear message is preferable to no message at all.

Put Clarity in Your Writing

DOUGLAS MUELLER

Abstract—Understandable messages may take the author longer to produce but they save time and effort and often money for all the readers. Ten Principles of Clear Statement can prevent written fog. For example, tie in with the reader's experience; write to *ex*press, not to *im*press; and avoid unneeded words.

A rod cutter examines a written order calling for valuable radioactive rods to be cut in "ten foot long pieces." The hyphen is missing. Did the order call for "ten-foot long pieces" or "ten foot-long pieces"? The cutter takes the order to mean ten pieces each a foot long. Result: a serious misunderstanding that cost the U.S. government a bundle of money.

Some chemists tell in a thick report how they synthesized a new group of compounds that show promise as insecticides. But this important message is buried on page 24, under a bushel of extraneous words and confusing sentences. After the report is channeled to many people, the company repeats this costly research unnecessarily as it begins a search for new insecticides.

The Employee Relations Department of a company sends out a confusing overtime notice to 20 employees. Later, the office gets 18 phone calls asking for an interpretation. It takes Employee Relations nearly a week to straighten matters out.

DAMAGING FOG

Wordy (foggy) writing accounts for millions of dollars of waste every year. It wastes the time and energy of a company's executives, customers, and employes. But far more serious is the money loss resulting from the mistakes of those who fail to get the meaning of foggy reports, letters, and memos.

Most business writing is watered down with words that don't count: words that create tons of paper and ink and account for months of stenographic time that could be put to better use. Foggy writing is far removed from the spoken word that makes communication easy. Would you use words in conversation that a hotel used when it posted this sign:

In order to substantiate our desire to accommodate our guests, we will appreciate your cooperation in anticipating your credit requirements prior to your departure date. (Translation: Please let us know in advance if you want to cash a check when you leave town.)

Or words like this:

The company's Employee Retirement Plan has been

Reprinted with permission from *Hydrocarbon Processing*, vol. 60, no. 6, p. 143, June 1980; copyright 1980 by Gulf Publishing Co.

The author is President of Gunning-Mueller Clear Writing Institute, Inc., 736 El Rodeo Road, Santa Barbara, CA 93110, (805) 964-3386. The Institute owns the Fog Index and the Principles of Clear Statement given in answers 13 and 14 of the quiz section; both are used here with permission.

amended to permit each participating employe to share his annuities with certain contingent annuitants.

An employee of a midwestern utility wrote the sentence about the retirement plan. When he presented a memorandum at a writing seminar held for his company, we asked if he could sum it up in his own words. He said: "You see, we have had a retirement plan around here for a long time. People like it. But some have felt that a wife or child should be able to share a pension if an employee died. We have fixed it now so they can do this." The employee said he had used the term "contingent annuitants" in his memo because "that's what the insurance people call them."

In 1977 a representative of a leading executive recruitment firm asked chief executives of the top 500 corporations to define the most important requirements for future business leaders. The executives put "communications skills" at the top of their list, ahead of technical, financial, and marketing abilities.

Improving your writing begins with understanding certain principles of clear statement such as keeping sentences short, preferring the simple to the complex, using familiar words, and avoiding unnecessary words.

HOW TO IMPROVE

Enliven Style

Put action into verbs, make full use of variety, use terms the reader can picture, and write with a conversational tone while avoiding business jargon.

Tie your writing in with the reader's experience. Write to *ex*press ideas, rather than *im*press the boss with your vocabulary by using needlessly complex words. (Chances are good he won't be impressed; few people admire stuffy, long-winded prose.)

Wasted Words

The "mock lawyer" is responsible for many surplus words in business writing. Put yourself in the shoes of a busy executive who complained recently: "I have to wade through 50 letters, memos, and reports every day. Most of them don't get read thoroughly. They are needlessly long. Many I don't fully understand. They have too much specialized detail. Still others fail to come to the point. I don't have time to dig for meaning and main ideas."

It is a typically hectic business day at this person's company, and the executive's in-basket is filling up fast. He sees a bulletin board notice marked "for your approval":

BULLETIN BOARD NOTICE

The purpose of this notice is to outline a procedure to assure that the oil in the crankcases of the factory motor scooters is changed at the properly designated intervals. The scooters are

Reprinted from *IEEE Trans. Prof. Comm.*, vol. PC-23, no. 4, pp. 173–178, December 1980.

TABLE I
CUTTING VERBIAGE

Before—99 words

This is to inform you that we have your order dated April 23, 1979, for four (4) Acme 78 X 15 fiberglass radial tires in blackwall, for which we thank you.

We regret to advise you that we are out of this size in the Acme brand, and also must advise you that we are now producing this tire only in the steel radial type.

However, we do have the size you want in our Samson brand tire, at approximately a 5 percent higher price than the Atlas. Please advise whether you wish us to ship your order in the Samson brand.

After—45 words

Thanks for your April 23 order for four Acme 78 X 15 blackwall fiberglass radial tires. We now produce this tire only in the steel radial type and have no more Acmes in your size.

May we substitute our Samson brand tires for about 5 percent more?

TABLE II
TIE IN WITH YOUR READER'S EXPERIENCE

Before

As the separation of personnel is bound to be necessary in any business, the company has established certain policies and practices to assure that such separations are made on an equitable basis and only after full and complete consideration of the individual case, as follows:

When an individual leaves the employ of the company it may be due to one of a variety of causes such as voluntary resignation, misconduct, unsatisfactory service, the curtailment or discontinuance of operations, etc. Some of these causes are within the control of the employee; others are not. It therefore becomes desirable to establish separation classifications based on the causes of the separation so that the employee's equities may be adequately considered . . . etc.

After

We want you to like us. We want you to fit in. We want you to belong. But if the day does come when we must part company, we want you to understand your rights and privileges.

Before you consider resigning for any reason, talk with your department head. There are many opportunities with this company. You may miss a good one if you leave us. If you do resign, we'd appreciate your giving us at least two weeks' notice.

It's an unpleasant task, but sometimes for the good of an employee, his fellow workers, and the company we must discharge or dismiss one of the men or women who work for us . . . etc.

equipped with four-cycle gasoline engines having crankcases that contain oil in the amount of one quart. It is recommended by the manufacturer that the oil in these scooters be changed every 400 miles, but the scooters are not equipped with speedometers and, therefore, it will be necessary to establish a time interval for oil change. Inasmuch as it is estimated that the maximum probable mileage is approximately 400 miles per month, it has been decided to have the oil changed at that interval.

It will be the responsibility of the operator of each vehicle to see that such vehicle is driven or otherwise transported to the auto garage at 30-day intervals for the purpose of obtaining a change of lubricant.

At a training session, we suggested this revision:

"TO OPERATORS OF THREE-WHEEL SCOOTERS: Please bring your scooters to the auto garage every 30 days to have the oil changed."

Quizzes

To illustrate the principles of clear statement, the Clear Writing Institute gives quizzes based on a discussion of a particular principle. Then we talk about the revisions everybody came up with. We show what *we* would have done with a particular piece of copy. For example, to illustrate the principle that needless words should be avoided, we asked one group to rewrite a letter to a business customer. The original version, 99 words, was reduced to 45 words, as you can see in Table I.

In working with companies, we have pinpointed such needless expressions as "in connection with," "with regard to," "with reference to," "prior to the start of," and "in the sum of."

Remove such worn-out phrases from letters:

Original: We wish to bring to your attention the fact that our invoice . . . .

Better: Our invoice . . . .

Original: We take this opportunity to advise you that we now produce . . . .

Better: We now produce . . . .

Original: Reference is made to your letter of May 10 in which you asked . . . .

Better: On May 10 you asked . . . .

Wooly Words

A polysyllable malady infects most business writing. Long wooly words do to business prose what rolls of fat do to a girl in a bikini. I have developed a list of 41 polysyllable words, for which simpler ones can be substituted. For example, *gather* instead of *accumulate*; *tell* instead of *acquaint*; *added* instead of *additional*. From only these three, you can get the general idea.

Complexity Level

The hallmark of good writing is to prefer the simple to the complex. But you'll also want to *measure* the complexity of your writing. Apply a readability yardstick we developed (Fog Index) to check progress in simplifying prose. It works this way: *First*, select a sample of at least 100 words. Count the average number of words per sentence. *Second*, count the number of words with three or more syllables in passages totaling 100 words. *Third*, add the average number of words per sentence to the percentage of polysyllables. Multiply the sum by 0.4. The result is a number roughly equivalent to a school-reading-level test score.

If your copy tests 13 or more, you are writing on the college level of complexity. Your reader is likely to find it heavy going. But you really need to deal with more than the complexity issue. Consider also the attitudes and feelings of those who read business copy. Judge for yourself how an employee would react to the messages in Table II. The first one was written before we worked with the company's writers. The second incorporates the writing principle: *Tie in with your readers' experience.*

QUICK IMPROVEMENT

Can bad writers learn how to write clear, readable copy in one day? They *can*. But long-term success calls for practice and persistence. Ethyl Corp. researcher John Lane put it this way: "This is an ongoing thing with us. We make sure that our new employees understand the principles of clear writing. The success of this training approach has been measured in progress reports at our company. We noticed a marked difference during the 6-month period following a 1-day training seminar. All office writing was more clear, concise and readable."

Good writing is hard work. But the problem with most businessmen is lack of confidence. Hardly anyone judges his schooling in practical English to be proper preparation for the type of writing he is required to do in a company. Most people are uneasy about their writing when they enter the business world. So they use jargon, even though many prefer to call it "professional language."

Now, you're invited to take the following quiz . . .

Can You Write Like the Pros?

This test measures your knowledge of writing techniques. Answers are on pp. 177–178.

In the case of the writing exercises, there may be more than one correct answer. But you can tell if you're on the right track by comparing your writing with the revisions suggested.

1. Fill the blanks with short words that can often take the place of the longer ones.

anticipate _____

cognizance _____

compensation _____

demonstrate _____

endeavor _____

initiate _____

locality _____

modification _____

objective _____

optimum _____

subsequent _____

voluminous _____

2. Reduce this 24-word sentence to eight or fewer words without changing the meaning:

Should the supply of manuals sent you not be sufficient to meet your requirements, application should be made to this office for additional copies.

3. The most universal principle of good organization can be described in four words. Those words are _____ _____ _____ _____.

4. The following bulletin board item violates the universal principle of good organization. Rewrite this safety notice with the principle in mind.

During the past year there were in our industry 162 lost-time leg injuries that resulted from falls. Of these more than half were caused by the injured persons slipping on floors.

In our own plant last month Jason Jones broke his right ankle as the result of slipping in B corridor. An overturned pop bottle, carelessly left at the side of the hall, was the direct cause of the accident. It was overturned and wet the floor.

Racks are furnished beside all softdrink dispensing machines in this plant. Use them.

5. Reduce this 40-word paragraph to two sentences totaling 20 words. The bold-faced words should disappear in the revision.

It has been customary in most foreign markets to lend **various sums** of money to dealers **for the reconstruction of existing** retail outlets. **Up to the present the total** amount loaned **for this purpose** is in the millions of dollars.

PRE-PLAN REPORTS

There's a *right* way and a *wrong* way to write, edit, and approve business reports in a large company. Unfortunately, many companies follow wasteful, outdated procedures that cause some reports to become lost for long periods in the rewriting-and-editing maze.

Douglas Mueller and his partner, the late Robert P. Gunning, found over the years that the glaring deficiencies of the outdated system have been most apparent in research work. Here's what happens:

An employee and a supervisor agree to put certain findings into a report. They have a brief discussion, and then the employee writes the report. When the supervisor reads it, he notices that many important points were not covered. Often he tells the employee, "We'll have to start from scratch again. It's all wrong. Even the emphasis is wrong." The employee, discouraged by this time, rewrites his report. Then the supervisor makes further changes and approves the document.

The report goes to different division heads for further approval. The tone, emphasis and facts are changed, again and again. This cycling and recycling process continues for months. Often the complaint is heard that someone changed sentences back to what they were in the first place. The final result: a poorly organized, badly written piece of "lawyered" copy.

There is a much better system of writing, editing, and approving copy. It is called "pre-planned report writing," and was developed by Gunning. With some variations, it works this way:

First, the employee prepares a brief outline of the points he intends to cover. He discusses this outline with his supervisor and they agree on the material for an expanded outline. This discussion alone saves hours of writing and rewriting later.

The employee prepares the expanded outline. But the next step is the greatest time saver of all. There is a "fact-editing" conference in which the department heads all review the expanded outline and edit it for facts, policy and focus. This meeting is attended by everybody who will approve the final draft. The writer and supervisor present their views, and the department heads have ample opportunity to suggest and criticize.

Finally, the employee writes his report. He may turn it over to a professional writer employed by the company or edit the final draft himself. The employee or professional corrects grammatical mistakes and spelling. Also he may revise paragraphs to improve clarity. The final result: a clear, readable document.

Supervisors may complain that they have no time to attend "fact-editing" meetings. They discover, however, that these meetings usually last only an hour. Even when the meeting lasts a day and a half, it saves time formerly spent on the rewriting and recycling of reports.

---

6. In a good piece of writing all sentences are short and run about the same length. True _____ False _____.

7. Reduce this sentence to 10 words that convey the same idea in a more conversational style:

In industrial communities, the principal motivation for the purchase of curtains is practicality.

8. Underline the three needless words in the following sentence and substitute the appropriate word.

We would like to talk to you with reference to our new line of products.

9. Improve this long-winded prose by substituting 33 short words in its place:

In our endeavor to ascertain whether or not the proposals we have formulated are fundamentally sound, we anticipate engaging a survey organization to determine whether our con- *ception of the market can be substantiated by information accumulated in the field.*

10. Good writers try to _____ press rather than _____ press when they write.

11. Poor writers often make three mistakes when they write business letters. The use words that are _____, _____, and _____.

12. One *fault* of much business writing is that important facts are left out. Describe the *cure* in six words. _____ _____ _____ _____ _____.

13. List as many principles of clear statement as you can and define each briefly.

14. Apply the Fog Index to measure the reading complexity of the following passage by W. Somerset Maugham. (Note

that the third sentence consists of three complete thoughts linked by a comma and a semicolon. These should be counted as separate sentences.)

I have never had much patience with the writers who claim from the reader an effort to understand their meaning. You have only to go to the great philosophers to see that it is possible to express with lucidity the most subtle reflections. You may find it difficult to understand the thought of Hume, and if you have no philosophical training its implications will doubtless escape you; but no one with any education at all can fail to understand exactly what the meaning of each sentence is. Few people have written English with more grace than Berkeley. There are two sorts of obscurity you will find in writers. One is due to negligence and the other to willfulness.

15. A steel company salesman sent this letter to a man who wanted to buy a prefabricated swimming pool:

"Dear Sir: We don't make swimming pools, and don't intend to." The writer obeyed the principle of conciseness, but broke a higher principle of friendly courtesy. He could have fulfilled both by writing . . .

_____

_____

_____

Now, see answers.

Answers to Writing Quiz

1.	anticipate	foresee
	cognizance	knowledge
	compensation	pay
	demonstrate	show
	endeavor	try
	initiate	begin
	locality	place
	modification	change
	objective	aim
	optimum	best
	subsequent	next
	voluminous	bulky

2. If you want more manuals, ask for them .

3. Come to the point.

4. For the tenth time: Put soft drink empties back in the racks. This is essential for safety and neatness.
Jason Jones broke his right ankle last month because someone failed to do this. He lost three weeks of work because a fellow worker carelessly left a half-filled bottle by the corridor wall. It was overturned. Jason slipped on the wet spot.
Last year 162 lost-time leg injuries occurred in our industry. Most of them were caused by slippery floors. Don't be a bottle culprit.

5. In most foreign markets we have lent money to dealers to rebuild retail stores. These loans total millions of dollars.

6. False. Such writing violates a principle of good writing: Make full use of variety in sentence length.

7. In factory towns, housewives buy curtains because they wash well.

8. with reference to, about.

9. In our effort to find out if our plans are sound, we expect to hire a survey group to check our view of the market against facts they will gather in the field.

10. ex, im.

11. long, unfamiliar, abstract.

12. Put yourself in the reader's place.

13. Ten Principles of Clear Statement

• **Keep sentences short.** On the *average,* most sentences should be shorter than 25 words. But sentences should vary in length and structure.

• **Prefer the simple to the complex.** Avoid complex sentences and phrases. Write "try to find out" rather than "endeavor to ascertain."

• **Prefer the familiar word** but build your vocabulary. If a reader doesn't understand your words, he can miss your meaning. But you may want to use long words in some cases—to clarify your point.

• **Avoid words you don't need.** Extra words weaken writing. Make every word carry its own weight.

• **Put action into your verbs.** Passive verbs tire the reader. Write "We intend to write clearly" not "Clarity in composition is our intention."

• **Use terms your reader can picture.** Choose short, concrete words your reader can visualize, not abstract terms. Don't say "industrial community" when you're describing a "factory town."

• **Tie in with your reader's experience.** The reader probably won't get your *new* idea unless you link it with an old idea he already understands. If you're describing how a new pump works, compare its operation with that of an old, standard pump.

• **Write the way you talk,** or at least try for a conversational tone. People rarely use business jargon when they talk.

• **Make full use of variety.** Vary the length of words and sentences and arrange them in different ways. Avoid monotonous patterns of writing.

• **Write to EXpress, not to IMpress.** Don't show off your vocabulary by using needlessly complex words.

14. This passage has a Fog Index of 10.9. (Words of three syllables or more are italicized.)

I have never had much patience with the writers who claim from the reader an effort to *understand* their meaning. You have only to go to the great *philosophers* to see that it is *possible* to express with *lucidity* the most subtle *reflections.* You may find it *difficult* to *understand* the thought of Hume, and if you have no *philosophical* training its *implications* will doubtless escape you; but no one with any *education* at all can fail to *understand exactly* what the meaning of each sentence is. Few people have written English with more grace than Berkeley. There are two sorts of *obscurity* you will find in writers. One is due to *negligence* and the other to *willfulness*.

You will find the Fog Index by taking the following steps: First, compute the average sentence length. To do this, count the words in each sentence (20 - 23 - 11 - 13 - 20 - 10 - 11 - 10). (Because the third sentence consists of three complete thoughts separated by punctuation, it is counted as three separate sentences.) The total number of words is 118. Divide this figure by 8 (the number of sentences) and you get the average sentence length: 14.5 words.

There are 15 words of three syllables or more. Thus, the percentage of polysyllables in the passage is 12.7.

Add the average sentence length and the percentage of polysyllables. The total is 27.2. Multiply this figure by 0.4 and you get a Fog Index of 10.9, about the level of TIME.

15. "Dear Sir: I am sorry, but we don't make swimming pools and do not plan to enter this market. A sheet-metal firm in your community might be able to help you."

Add Style to Your Technical Writing

DANIEL L. PLUNG

Abstract—Technical writing need not be dry either by design or default. Without neglecting precision and conciseness, language and phrasing can be used to assist rather than impede understanding. Putting style in technical writing can be aided by awareness of these devices: alliteration, anaphora, antemetaboly, antithesis, climax, colon, epistrophe, metonymy, and simile. Examples of each are included.

TECHNICAL writing is typically recognized for its devotion to precision, conciseness, and concreteness. Yet, the delivery and success of technical communication also require that each author develop a *style*. Style is an ability to use language and phrasing to assist rather than impede the reader's efforts to understand the information being communicated. Such a style for technical writing is accommodated by only one readability formula, the Orwell Writing Success Number (OWSN; see the appendix) [1, 2]. In addition to emphasizing the precision and conciseness not accounted for by other popular formulas, the OWSN is also the only readability formula that encourages the development of a true style.

WHAT IS STYLE?

There is certainly no shortage of quotations by great writers about style. For instance, consider Henry James' astute remark that "It is by style that we are saved." Or the comment by E. B. White that style is "the sound words make on paper." However, enlisting those authors as spokesmen for the cause of improving technical writing might seem too strained for the skeptical author who has been repeatedly advised to stress objectivity and impartiality and who senses intuitively that he should not embellish technical wisdom with fanciful words and phrases. Rather, it seems obvious when one considers the paucity of information on style provided by most texts on technical writing that such discourse has no use for style.

Yet, that assumption is definitely incorrect. It is that mistaken inference that originally led to technical writing's unwarranted reliance on first-generation readability formulas such as the Fog Index. And the Fog Index has been the willing godparent of a subsequent generation of readability formulas—computerized editors that input good writing and output lifeless, tedious prose.

Reprinted with permission from *Hydrocarbon Processing*, May 1983, vol. 62(5), pp. 123–135; copyright 1983 by Gulf Publishing Company, Houston, Texas.

The author is technical editor for Exxon Nuclear Idaho Co., Inc., P. O. Box 2800, Idaho Falls, ID 83401; (208) 526–3084.

Consider the following revision of Lincoln's Gettysburg Address produced by Bell Laboratories' computerized system [3]:

Eighty-seven years ago, our grandfathers created a free nation here. They based it on the idea that everybody is created equal. We are now fighting a civil war to see if this nation or any similar nation can survive. On this battlefield we are dedicating a cemetery to those who died for their country. It is only right. But in another sense, the task is impossible, because brave men, living and dead, dedicated the place better than we can. Hardly anyone will notice or remember what we say here, but nobody can forget what those men did. We should continue the work they began, and make sure they did not die in vain. With God's help, we will have freedom again, so that the people's government will endure.

TOWARD BETTER STYLE

Nonetheless, this view that technical writing must be boring to be good is not held by all authors of technical literature. Rather, one of the most eloquent statements about the need for true style in technical communication was offered more than 30 years ago by Dr. J. Robert Oppenheimer, the renowned American physicist [4]:

"Maybe you should explain to him what you meant by developing style."

Reprinted from *IEEE Trans. Prof. Comm.*, vol. PC-27, no. 1, pp. 20–24, March 1984.

The problem of doing justice to the implicit, the imponderable, and the unknown is, of course, not unique to politics. It is always with us in science, it is with us in the most trivial of personal affairs, and it is one of the great problems of writing and of all forms of art. The means by which it is solved is sometimes called style. It is style which complements affirmation with limitation and with humility; it is style which makes it possible to act effectively, but not absolutely; it is style which, in the domain of foreign policy, enables us to find a harmony between the pursuit of ends essential to us and the regard for the views, the sensibilities, the aspirations of those to whom the problem may appear in another light; it is style which is the deference that action pays to uncertainty; it is above all style through which power defers to reason.

This same conclusion about the need for a broader appreciation and application of style in technical writing was also noted by other scientists. For example, in a recent book about language, V. V. Nalimov, a Russian mathematician and information scientist, had these thoughts about style and technical writing [5]:

> The transmission of thought is carried out on a logical level, but its perception is greatly influenced by some psychological factors which are not entirely understood. An idea is perceived more readily if it is shocking and requires an intellectual effort. A good scientific paper ought to (require this intellectual effort) .... Use of the metaphorical structure of language is only one of the techniques used to create intellectual strain.

We can infer from the representative selections more precisely what technical writing style should be:

1. It is writing that displays a literary quality.
2. It is writing that uses the best words, expressions, and phrases to communicate.
3. It is writing that takes advantage of all devices or techniques that can enliven the prose without interfering with the communication of a precise, concise message.

And we should be able to recognize that this is the definition of writing Orwell is promoting when he advises us to choose the best terms and voice, and when he advises us to avoid the "barbarous." When trying to define more particularly what we are to "avoid" according to Orwell's injunction, we might consider what one philosopher noted: "Subject without style is barbarism" [6].

CREATING STYLE

Therefore, we need to know what devices or techniques we can use to add this element of style to our writing. Accordingly, I offer the following primer of ten stylistic devices.

For each device I supply only a definition and an example or two of its use. Extensive detail is not warranted since the examples clearly suggest the strength and potential application of each device. Further, I have principally selected examples from political speeches; my reason for selecting political rather than scientific examples is simple: They will be more readily recognized and remembered than scientific quotations. Yet, to demonstrate the devices' use in a scientific text, I also include a summary of a recent article from *Scientific American* [7] in which all ten devices are incorporated (see box).

Only when discussing metaphors do I give a more detailed analysis; this device warrants more coverage because the

---

Summary of "Highly Excited Atoms"

The test tube has provided many significant breakthroughs in the chemical sciences (*metonymy*).* In atomic physics, advances have been due in great part to Niels Bohr's hydrogen theory and to quantum physics. Now a new field of study has emerged: Rydberg atoms. "It is the physics of atoms in which an electron is excited to an exceptionally high energy level" (*alliteration*).

"Rydberg atoms have been detected whose diameter approaches a hundredth of a millimeter ... . The Rydberg atoms are so large that they can engulf other atoms. Rydberg atoms are also remarkably long-lived" (*anaphora*). A particular difference between Rydberg and other atoms is their responses to electric and magnetic fields. "Ordinary atoms are scarcely affected by an applied electric field or magnetic field; Rydberg atoms can be squeezed into unexpected shapes by a magnetic field" (*epistrophe*).

"Rydberg atoms are like hydrogen in their essential properties" (*simile*). And "according to the Bohr theory, the hydrogen atom is a solar system in microcosm" (*metaphor*). However, quantum physics needed to refine this model; "Bohr's simple model accounted for the most conspicuous features of the spectrum of hydrogen, but this model incorporated such a jumble of traditional concepts and radical ideas that it could not be generalized or refined" (*antithesis*).

Using this refined model, Rydberg atoms, particularly those of alkali metals, were investigated. "These elements are commonly chosen because they are easily turned into a gas, because their spectral absorption lines are at wavelengths conveniently generated by laser light and because they absorb light efficiently" (*colon*). Once produced, these atoms were studied in electric and magnetic fields.

In electric fields, the atoms' hydrogen-like qualities were evaluated, their degrees of degeneracy examined, and ultimately their resultant Stark effects analyzed (*climax*). This allowed successful definition of the relationship between experimental conditions and the specific shape assumed by the Rydberg atom in an electric field. Yet, attempts to define the electron's motion in a magnetic field were not as successful. Though significant progress was made, a complete explanation of our data is lacking. But, efforts continue. For, as we continue to excite these electrons, so these electrons continue to excite us (*antemetaboly*).

---

* Devices used are identified in parentheses following their use; quotation marks identify examples of devices taken verbatim from the original article [7].

use of metaphors is as common in scientific writing as in any other form of expository prose—although its use may not be understood as thoroughly by authors of technical literature. The discussion of metaphors is therefore a reasonable place to begin my review of these stylistic devices: It illustrates the literary heritage technical writing shares with all other types of writing.

Metaphor

A metaphor is an implied comparison of dissimilar things; words of explicit comparison, such as "like" and "as," are omitted. Metaphors are quite common in technical writing. For instance, consider the following metaphors, based on the shapes of letters of the alphabet [8]:

A frame
C clamp
D ring
F head engine
H beam (structural engineering)
I beam (structural engineering)
J stroke (canoe paddling)
L head engine, pipe el (overlapping elbow)
N strut (interwing strut on biplanes)
O ring
P trap
S curve
T bolt
U bolt
V block, groove, brace
X brace
Y pipe
Z section, cut.

In the book quoted previously [5], the use of metaphor in scientific discourse is lucidly explained by V. V. Nalimov:

> If reading a scientific text, we stop for a moment and ponder the character of terms in our field of vision, we shall find that they are metaphorical. We have become so used to metaphors in our scientific language that we do not even notice it. We keep coming across such word combinations as "course of time," "the field of force," "temperature field," "the logic of experiment," "the memory of a computer," which allow us to express new notions with the help of rather unusual combinations of old, well-known, and familiar expressions. Recognizing the right of metaphors to existence in scientific language, scientists have permitted rather different senses for old terms with the emergence of these new theoretical conceptions. In science, theories are continuously changing, but the change does not cause a waterfall of new words. The new phenomena are interpreted through the old, familiar ones, through the old words for which the prior distribution function of meaning is slightly, but continuously, changed. Something remains unchanged but becomes of less importance, something new appears, entirely different from, and to a certain extent contradictory to, the former meaning of the word.

Here, in addition to the scientific metaphors already noted, are two examples of how politicians used metaphors:

1. Adlai Stevenson: "The anatomy of patriotism is complex,"

2. Lyndon Johnson: "This is what America is all about. It is the uncrossed desert and the unclaimed ridge. It is the star that is not reached and the harvest sleeping in the unplowed ground."

Simile

A simile is an explicit comparison of dissimilar things in which such words as "like" and "as" *are* used:

1. Thomas Paine: "Until an independence is declared, the Continuent will feel itself like a man who continues putting off some unpleasant business from day to day, yet knows it must be done."

2. Abraham Lincoln: "Sending men to McClellan's army is like shoveling fleas across a barnyard—they don't get there."

Anaphora

Anaphora is the repetition of the same word, or group of words, at the beginning of successive clauses:

1. Winston Churchill: "We shall fight on the beaches, we shall fight on the landing grounds, we shall fight in the fields and in the street, we shall fight in the hills."

2. Martin Luther King: "Let freedom ring from the mighty mountains of New York. Let freedom ring from the heightening Alleghenies of Pennsylvania. Let freedom ring from the snow-capped Rockies of Colorado."

Epistrophe

Epistrophe is the repetition of the same word, or group of words, at the end of successive clauses:

1. Abraham Lincoln: " ... and that government of the people, by the people, for the people, shall not perish from the earth."

2. Martin Luther King: "With this faith we will be able to work together, to pray together, to struggle together, to go to jail together, to stand up for freedom together, knowing that we will be free one day."

Alliteration

Alliteration is the repetition of the initial sounds of words:

1. Winston Churchill: "The Battle of France is over. I expect the Battle of Britain is about to begin."

2. Spiro T. Agnew: "Hysterical hypochondriacs of history," "nattering nabobs of negativism," "troubadours of trouble."

Colons

Colons are the division of an idea into equal grammatical parts, done in successive clauses, to impart a rhythm to the sequence:

1. Charles Dickens: "It was the best of times, it was the worst of times, it was the age of wisdom, it was the age of foolishness, it was the epoch of belief, ..."

2. Abraham Lincoln: " ... that from these honored dead we take increased devotion to that cause for which they gave the last full measure of devotion; that we here highly resolve that these dead shall not have died in vain; that this nation, under God, shall have a new birth of freedom; and that government of the people, by the people, for the people, shall not perish from the earth." (In this quotation, there is a quadracolon—clauses beginning with the word "that"—and a tricolon—"of the people, by the people, for the people.")

Antithesis

Antithesis is the contrast of ideas by means of parallel arrangements of words or groups of words:

1. John F. Kennedy: "We observe today not a victory of party, but a celebration of freedom—symbolizing an end, as well as a beginning—signifying renewal, as well as change."

2. Abraham Lincoln: "The brave men, living and dead, who struggled here, have consecrated it far above our poor power to add or detract."

Climax

Climax is the arrangement of words, or groups of words, according to their increasing value or strength:

1. Julius Caesar: "I came, I saw, I conquered."

2. John F. Kennedy: "All this will not be finished in the first 100 days. Nor will it be finished in the first 1,000 days, nor in the life of this administration, nor even perhaps in our lifetime on this planet. But let us begin."

Antemetaboly

Antemetaboly is the repetition of words in successive clauses, but in reverse order:

1. Thomas Paine: "For as in absolute governments the King is law, so in free countries the law ought to be King."

2. John F. Kennedy: "Mankind must put an end to war—or war will put an end to mankind." "Let us never negotiate out of fear. But let us never fear to negotiate." "And so, my fellow Americans, ask not what your country can do for you: Ask what you can do for your country."

Metonymy

Metonymy is when a word or image associated with a larger idea or concept is made to serve for the expression of that idea:

1. Press Release: "The White House (used here to serve as substitute for the President or the Executive Branch of government) said today that a formal statement would be issued later this week."

2. Richard Nixon: "In Europe, we gave the cold shoulder to DeGaulle, and now he gives the warm hand to Mao Tse-tung."

3. John F. Kennedy: "To those peoples in the huts and villages across the globe struggling to break the bonds of mass misery, we pledge our best efforts to help them help themselves, for whatever period is required."

USING STYLE

The first step in animating writing is developing an understanding and knowledge of these techniques. The second step, which is of equal importance, is using them judiciously. This step can be divided into two seemingly contradictory components, caution and practice. Caution is necessary to ensure their use is judicious, not whimsical or improper. Improper use produces a turgid style and detracts from any professional communication.

Caution

A few commonsense guidelines are in order:

1. Make certain the device and its use are appropriate to the subject, purpose, and audience.
2. Don't use greatly exaggerated devices that call attention to themselves; subtly interweave them into the fabric of your material. If in doubt about whether something is too exaggerated, leave it out.
3. Don't overuse the devices. Effectiveness is not measured by how many devices you use or how many times you use them. It is better to have only a few than to clutter your communication.
4. Although the poetic requirement of freshness is not essential, avoid using devices that are overworked, commonplace, or cliche.

These simple cautions should allow you to make effective use of the devices while also focusing on your foremost responsibility: the presentation of an accurate, precise, and concise message. You must remember that the literary merit must function as a support to—not a substitute for—the overall clarity, reasoning, organization, and straightforward presentation of your information.

Practice

Now, with these cautions in mind, let us briefly examine the other facet of this second step: practice. As all the great writers and speakers have known, literary ability must be developed; the art must be practiced. This is something the writers and speakers quoted in this essay knew well. Thomas Paine said, "Fit the powers of thinking and the turn of language to fit the subject, so as to bring out a clear conclusion that shall hit the point in question and nothing else." Similarly, Sir Winston Churchill stated, "There is no more important element in the technique of rhetoric than the continual employment of the best possible word."

This concern for language and the diligence with which these men practiced their art paid off. Thomas Paine's pamphlet "Common Sense" became the most discussed material in America prior to the Revolution. Winston Churchill was awarded the 1953 Nobel Prize in Literature for his "historical and biographical presentations and for his scintillating oratory."

The Kennedy Speech

Yet, surely the best-known quotations in this essay are those taken from John F. Kennedy's inaugural address. Kennedy is a prime example of the tangible results of diligent practice and study, as his inaugural address attests. Prior to drafting this speech, Kennedy reviewed all the previous Presidential inaugural addresses to uncover their stylistic strengths; he also requested his special counsel, Theodore Sorensen, "to study the secret of Lincoln's Gettysburg Address." That Kennedy learned the "secret" is evident from the many stylistic points the two speeches have in common. Kennedy also assiduously reworked his manuscript. For example, his famous sentences "Ask not ... " were revised and tested numerous times; they were reworked from speeches delivered on September 5 and September 20 and were again revised the morning of the inauguration.

As a result of Kennedy's commitment to this writing, his inaugural address (whose length, coincidentally, was approximately that of the average conference paper, 14 minutes) was praised both for its substance and its language: "distinguished for its style and brevity as well as for its meaty content."

This compliment from *The New York Times* is the type to which all communicators should aspire. It recognizes that the author has dutifully observed the primary mission—to impart information; but it also credits the author's ability with language and desire to communicate in a manner that elevates as well as enlightens.

APPENDIX: THE ORWELL WRITING SUCCESS NUMBER (OWSN)

The Orwell Writing Success Number system [1] is predicated on the six writing principles discussed by George Orwell in his essay "Politics and the English Language" [2]. He neatly summarized the questions the "scrupulous" writer should ask:

1. What am I trying to say?
2. What words will express it?
3. What image or idiom will make it clearer?
4. Is the image fresh enough to have an effect?
5. Could I put it more succinctly?
6. Have I said anything that is avoidably ugly?

These questions he then developed into six rules—rules that serve as the cornerstone of the OWSN system:

1. Never use a metaphor, simile, or other figure of speech that you are used to seeing in print.
2. Never use a long word where a short word will do.
3. If it is possible to cut a word out, always cut it out.
4. Never use the passive where you can use the active.
5. Never use a foreign phrase, a scientific word, or a jargon word if you can think of an everyday English equivalent.
6. Break any of these rules sooner than say anything outright barbarous.

In the OWSN, these rules are weighted in terms of their contribution to sound technical writing and then coupled with a component called the "Superfluity Ratio": the number of words used divided by the number of words necessary to communicate the thought. With these principles in mind, the reader can evaluate and thereby improve his technical writing.

This system differs markedly from the popular readability formulas such as Robert Gunning's Fog Index and Rudolf Flesch's Reading Ease Score: (1) It is designed specifically for evaluating technical writing. (2) It accounts for the value of conciseness and precision. (3) It recognizes the need for technical terminology. (4) It allows each author to develop a "style" of writing.

REFERENCES

1. Plung, D. L. "Evaluate Your Technical Writing." *Hydrocarbon Processing.* July 1981; 60(7):195–212.
2. Orwell, G. "Politics and the English Language." In *Shooting an Elephant and Other Stories.* London: Secker and Warburg; 1950.
3. Angier, N. "Bell's Lettres." *Discover.* July 1981; 2(7):78–79.
4. Oppenheimer, J. R. "The Open Mind." *Bulletin of the Atomic Scientists.* January 1949; 5(1):3–5.
5. Nalimov, V. V. *In the Labyrinths of Language: A Mathematician's Journey.* R. G. Colodny, Ed. Philadelphia: ISI Press; 1981.
6. Quoted in Kasher, Asa. "Style! Why Bother?" In *Scientific Information Transfer: The Editor's Role.* Miriam Balaban, Ed. Boston: D. Reidel; 1978:299–301.
7. Kleppner, D.; Littman, M. G.; Zimmerman, M. L. "Highly Excited Atoms." *Scientific American.* May 1981; 244(5):130–149.
8. Harris, J. S. "Metaphor in Technical Writing." *The Technical Writing Teacher.* Winter 1975; 2(2):9–13.

Create time to write reports

These 11 guidelines can free you from a task that many technical professionals view as drudgery

J. K. Borchardt, Shell Development Co., Houston

HUNDREDS OF BOOKS and articles have been written on better business and professional writing. But few of these address a critical problem. There probably isn't a single engineer or manager in the hydrocarbon processing industry who feels he has enough time to write reports. Some spend evenings, weekends and even vacation time writing in a never-ending effort to catch up.

Such drastic measures aren't necessary if one takes control of time and makes a concerted effort to change work style. The key is a productive use of available time since opportunities to create more time are limited.

1. Create time by getting up earlier. Most people can easily get by on 30 to 60 minutes less sleep per night. Use this time to beat the rush hour traffic. By getting to the plant or office early, you will have been able to sit down and write with few interruptions. If you live close to work, you could put in a stint of writing at home before the rest of the family wakes up. In either situation, the absence of other people and interruptions will improve your productivity.

2. Start writing before or immediately after dinner when you bring work home. If you wait too long, your commitment will weaken and physical fatigue will reduce your productivity. Besides, it's harder for a tired mind to create brilliant thoughts and develop a logical chain of argument.

3. Take control and organize your time. Determine the time of day when you are most productive. Reserve some of this period for writing assignments. Establish a daily undisturbed writing period. This can actually become more difficult as you progress in your career and become involved in more projects and corporate activities.

So let your co-workers know this reserved writing time is off-limits except for emergencies. Have your secretary screen your telephone calls or get an answering machine. If possible, avoid scheduling meetings during this time.

4. Shut off distractions. Some engineers have gone so far in their efforts to shut off distractions that they have found hideways such as a secluded library carrel and retire there to write uninterruptedly. However, being completely out of contact can cause problems. It should be a last resort.

5. Enlist help. Make sure co-workers know when your regular writing period is. Put a brief and humorous sign on your office door to let casual visitors know you are busy and do not wish to be disturbed. If you bring work home, set up rules so that the family does not disturb your work period.

6. Organize your writing. Start by outlining your report. This will help you divide a writing assignment into sections. Working on each section separately can make even a massive report seem less intimidating. Don't be afraid to write the different sections out of sequence. The feeling of accomplishment you will get when finishing one section helps provide the motivation to tackle the next.

Keep in mind your upcoming need for reference materials and gradually assemble these while you are working on the project. It is very convenient to have these collected when the time arrives to begin writing the report.

Write around missing information. Don't stop writing.

7. Schedule your writing tasks to fit your time. Short periods can be used in writing short memoranda, assembling data or working on a short section of a report. By having your resource materials well organized and readily available, even "free time" can be effectively used.

8. Use travel time effectively. Whether or not you have a laptop computer, time in airports, on planes and in hotel rooms can be used for writing. By dividing your report into sections, you can carry only the materials needed to work on specific sections. Also, you can pick a "lighter" section such as the introduction to work on in the airport where your concentration may be less than complete.

9. Delegate tasks if you can. If possible, have a librarian or a secretary locate and assemble references for you. Have someone else make photocopies of references. Your support staff will find these requests much less aggravating if they come at one time in an organized fashion than if you send them a stream of individual requests as they occur.

Your operators and technicians can assemble and organize data, ideally while the work is still in progress. A microcomputer and spreadsheet program can be a great organizational tool for recording large volumes of data and performing calculations using these data.

10. Establish a deadline for your report even if your supervisor does not. Allow time for report writing when setting up project schedules. Robert Moskowitz, author of *How to Organize Your Work and Your Life,* recommends that you "estimate how long you expect a project to take and then schedule backwards from your deadline. That will tell you when you have to get started."

Don't use this estimated starting time as an excuse to delay getting started. Also, do not establish too tight a schedule. Try to allow for slippage which can arise due to the need to perform additional tests to verify data or fill in gaps in your results, order references from other libraries, and unforeseen circumstances.

11. Reward yourself and express appreciation to others. Do something to make completing the report a special occasion. This will make it easier to tackle the next big writing project. When you finish that big report, take a break from writing for a couple of days. Celebrate by going out to lunch. Take your operators or technicians along to express appreciation. Take the family out to dinner.

You'll be surprised how good these activities will make you feel. They will also build cooperation and toleration for the inconveniences that occur during your next writing project. ∎

Part 2
Text and Graphics
Presenting Information Visually

A S AN EFFECTIVE writer, you will sometimes need to consider devices that can make your information even more accessible to your readers, such as page organization, font choice, graphs, tables, lists, and symbolic material. This section deals with some of these visual tools and how you can use them to your advantage.

APPEARANCE MAKES THE DIFFERENCE

Most of what Philippa Benson states in the opening article applies to writing of any length, from short memos to lengthy published reports. Supporting her observations by a great deal of documented research, she focuses on how visual factors such as typeface, line length, margins, white space, headings, and illustrations can work for you as a writer. Her final topics, graphs and tables, are then more fully developed in the next two articles by Kathryn Szoka and Eva Dukes.

CHARTS AND TABLES TELL THE TALE

First, Szoka considers some choices available if you use charts in oral or written reports. She discusses the main kinds of charts, indicating where each is best used and warning how they can be misused. Inadequate planning can result in a poor choice of chart, but a focus sentence can guide you to the right chart for a given situation. Dukes's article on table construction then illustrates how well-arranged tables can efficiently present information. Since tables are so useful as communication tools, Dukes's six pointers for table construction, her cautions on pitfalls to avoid, and her illustrations make this a valuable article.

LISTING YOUR DATA

In the next article, Saul Carliner investigates and illustrates why lists can be considered "the ultimate organizer for engineering writing." How do you construct lists? How do lists improve on prose versions of the same information? How long can a list be before subdivision becomes necessary? How

should a list be punctuated? Carliner not only answers these questions but also illustrates by the format of his material just how accessible listed information can be.

WARNING YOUR READER

Another kind of visual that stands apart from written text, particularly in operator manuals, procedures, and instructions, is the safety alert. If you write such documents you will need to design warning labels to reveal hazards, prevent injury, and avoid litigation. Some of Christopher Velotta's advice on word choice, symbols, colors, and readability can be applied not only to safety labels but also to other kinds of notations and labels you might want to visually separate from your text for emphasis, such as directions, headings, pointers, or captions.

ABBREVIATING WITH CARE

If you glance over a page of technical writing containing no illustrations, you might still find in the text such visual units as acronyms or symbols. The popular press regularly expresses amusement or frustration over technology's addiction to acronyms, but in science and engineering they are indispensable. Thus Joseph Mancuso's article compiles a list of rules and practices enabling you to use acronyms to clarify and expedite your writing rather than obscure it.

SENSIBLE MATH SYMBOLS

In the final article, Barry Burton provides guidelines on what to watch for when your writing includes mathematical symbols. Although the guidelines are written for editors, many technical writers can benefit from Burton's observations on eliminating unneeded symbols, defining or redefining symbols, and following standard grammar when using them. As always, you should use symbols (and the same must be said for all graphics, lists, labels, and acronyms) only after you have first spent some time evaluating the technical level of your readers.

Writing Visually: Design Considerations in Technical Publications

Philippa J. Benson

RESEARCH IN DOCUMENT DESIGN SUGGESTS GUIDELINES *that both writers and designers can apply to make their publications easier for readers to understand and use. The author discusses the merging roles of writers and graphic designers, and provides some guidelines that document designers can use to reveal and reinforce the structure of a text. The article serves as a refresher for experienced technical writers and as a primer for those writers who are new to the field of document design.*

Writers and graphic designers are working more closely together as each learns more about how the other approaches the task of planning and structuring a document. The field of document design has developed from this synthesis of the processes of writing and design: writers and designers have come to agree that the most effective documents are those that use both words and design to reveal and reinforce the structure of information in a text.

The synthesis of writing and design is supported not only by the developing bond between writers and designers, but also by the conclusions of applied research. Research in cognitive psychology, instructional design, reading, and graphic design indicates that documents are most usable when the information in them is apparent both visually and syntactically. In particular, the results of usability tests of documents are providing concrete evidence that readers find documents most usable when the visual aspects of a document support the hierarchy of the information in a text.[1-5]

The synthesis of writing and design has also been evolving because computers increasingly put control over typography and page format in the writer's domain. Although many text-formatting programs limit the design decisions that we can make, they do allow us to simultaneously write and design information.

The increasing use of computers has also led to the development of a new concern for document designers: online information. Designers of online documentation should be concerned with many of the same issues as page designers—that is, the legibility and accessibility of information. However, because the constraints on screen design are different from those on page design, and because readers use online documentation differently than they use a printed document, the discussion of screen design is in a separate article of *Technical Communication*. (See "Application of Research and Document Design to Online Displays," *Tech. Comm.* vol. 32, no. 4, p. 29, 1985.)

UNDERSTANDING CONSTRAINTS ON DESIGNING A DOCUMENT

How effectively writers and designers can work together depends not only on how well they can communicate but also on the practical constraints of their working environment. Before you begin planning a document, find out about the limits of your schedule and budget, and the flexibility of the method you'll be using to produce the end product (printer, typesetter, printing process). Your schedule, budget, and method of printing may limit what you can do with the design of your document. Constraints may affect these aspects of your book:

- Page count, page size
- Packaging (binding, cover, separators, etc.)
- Type face, type size
- Range of available fonts
- Use of color and graphics
- Number of heading levels available
- Kinds of graphics you can use (tables, flow charts, illustrations)

If you find that the constraints depart from the principles of good document design, you should consider trying to change them. For example, you may be told you must use a type size of smaller than 8 or 9 points or set a sizable portion of text in upper case. The research cited in this article provides evidence you can use to convince your managers to change the constraints that would make your document less usable. Once you know your constraints, you can choose design options that are compatible with them.

DESIGNING FOR YOUR AUDIENCE(S) AND THEIR TASKS

A primary step in planning a document is understanding the audience for the document and their purpose in reading it. For example, if you are writing a book about how to service a software product on a computer system, you may have several audiences. Some readers may simply want to know the procedures they need to follow to service the product, while other readers may want to know how these procedures affect the computer system as a whole.

Research indicates that document designers should consider how readers will use the document. In a document with multiple audiences, each with dif-

Reprinted with permission from *Technical Communication,* published by the Society for Technical Communication. November 1985, vol. 32, no. 4, pp. 35–39.

ferent purposes and tasks in mind, research suggests that you should design the document so that members of each audience can quickly find the specific information they need.[6-8]

Depending on the reading situation, readers come to a technical document with assumptions and expectations about its structure, in terms of both its language and its design. For example, readers of a tutorial manual may expect to see step-by-step procedural instructions, while readers of a reference manual may expect to see lots of definition lists. The tutorial or reference manual you design, however, may be very different from what the reader initially expects. The design options you use can affect the assumptions and expectations readers have when they begin using a book and, in doing so, can affect the way the readers understand and use it. To design a document well, you need to imagine what linguistic and visual organizers will help readers understand how the text is structured. [9-10]

Here are some guidelines you can use to help readers find, understand, use, and remember the information in a document:

Guideline: Provide readers with road maps through the document.

Think about how you can design the tables of contents, glossaries, indexes, appendixes, and other materials that support the body of text so that readers can quickly find the information they need. One type of road map is a separator between each section of a document, marked with icons followed by text explaining the information that will be covered in that section.

 add automatic reminders

One type of roadmap is a section heading with an icon.

Another type of road map is a table that directs readers to specific information. Here is an example of a table road map:

If you want to	Go to
Install product x	Section A
Use product x	Section B
Service product x	Section C

Guideline: Make your document esthetically pleasing.

Many readers are motivated to use a document that is esthetically pleasing. Documents should be attractively bound and packaged, have well-balanced page layouts, contain some graphics and illustrations to break up the text, and be legible.

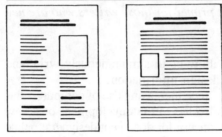

A good and a poor page layout.

Guideline: Be consistent in the visual formats and prose styles you use.

Readers can use a document more effectively when it contains consistent visual patterns, such as use of boldface or italic type, and a consistent linguistic style, such as use of the active voice or direct address. Readers use these patterns and styles to help them structure, comprehend, and remember the relationships between different levels and types of information in a text.

DESIGNING THE VISUAL STRUCTURE OF YOUR DOCUMENT

You can think of the structure of a document on several levels: as a whole, as a set of interrelated parts (chapters/sections), as a series of pages, and as text. Because the audience, purpose, and information for each document are unique, the relationships between these levels are different for each document you produce. Your challenge as a document designer is to choose design options that will support the document on all levels.

Research on design options suggests guidelines on how to graphically reinforce the structure of a text and therefore, how to improve its usability. These guidelines cover

- **Serif vs Sans-serif Typeface**
- **Lowercase vs Uppercase Type**
- **Type Weight and Font**
- **Type Size, Line Length, and Leading**
- **Justification of Lines, Size of Margins, Use of White Space**
- **Headings**
- **Graphs, Tables, Flowcharts, and Diagrams**

Serif vs Sans-serif Typeface

Guideline: Choose between a serif and sans-serif typeface according to the visual tone of the document you want. Note: Use a type size 10 points or larger, 2 or more points of leading between lines, and a moderate line length.

Serif typefaces have small strokes added to the edges of each letter. Serifs reinforce the horizontal flow of letters and make each letter distinct. Letters in sans-serif typefaces are less distinct from each other not only because they lack the extra, ornamental strokes of serif type, but also because the widths of the lines that constitute each sans-serif letter are often uniform. Sans-serif typefaces are generally thought to have a cleaner, more modern look than serif type.

Some studies indicate that readers can read text set in serif typefaces more quickly than text in sans-serif type. Readers also seem to prefer text set in serif type.

However, research also shows that the legibility of serif and sans-serif typefaces is approximately the same if the text follows general standards for type size and leading (10-point or larger type size with 2-4 points of leading). Because the legibility of serif and sans-serif print is so similar when properly designed, many researchers and typographers suggest that document designers choose a serif or sans-serif typeface according to what visual texture they want the text to have.[11-13]

Choose a serif or sans-serif typeface according to the visual texture you want your document to have.

Choose a serif or sans-serif typeface according to the visual texture you want your document to have.

Serif and sans-serif typefaces.

Lowercase vs Uppercase Type

Guideline: Avoid using use all uppercase type. When you want to emphasize a word or portion of text, use boldface type or italics.

Research supports the conclusion that in printed text words set in lower case are faster to read than words set in upper case.[14-17] Text set in lower case is easier to read than text in upper case because

1. Lowercase letters take up less space and, therefore, allow readers to take in more words as they scan a line of text.
2. Each lowercase word has a distinct outline that aids recognition and recall of the word.

However, some research does indicate that in some situations, such as when readers are scanning text or reading short informational statements, words set in upper case are easier for readers to locate and read.[18-19] Therefore, you can put words in uppercase, like ''WARNING!,'' when it is appropriate.

LOWERCASE WORDS TAKE UP LESS SPACE THAN UPPERCASE WORDS AND ALSO HAVE A DISTINCT OUTLINE.

Lowercase words take up less space than uppercase words and also have a distinct outline.

Text set in all uppercase and mixed upper- and lowercase.

Type Weight and Font

Guideline: Use a medium type weight.

Guideline: Use boldface type to emphasize words or short portions of text.

Research suggests that when a text has a logical structure, typographic distinctions such as changes in type weight or typeface may help readers understand the structure. Research specifically indicates that readers notice changes in type weight (heavy, medium, light) more readily than they notice changes in typeface and that readers find very light or very heavy type tiring and difficult to read.[20]

Readers may find large blocks of text set in very heavy or very light type difficult and tiring to read.

Readers may find large blocks of text set in very heavy or very light type difficult and tiring to read.

Text set in a very heavy type and a light type.

Type Size, Line Length, and Leading

Guideline: Use 9- to 11-point type for text.

Guideline: Use between 2 and 4 points of leading between lines of text.

Guideline: Use a moderate line length, approximately twice the length of the alphabet of the typeface you use.

Type size, line length, and leading all affect the legibility of print. The actual point size of a typeface does not necessarily reflect the perceived size of a letter. Words set in the same type size may appear larger or smaller because of variations in the height of the lowercase letters.

And then there was light
And then there was light
And then there was light

Same typesize with different typestyles.

However, research supports a conclusion that type sizes larger than 9 points are easier to read than smaller type sizes—when they are set with an appropriate amount of leading, usually 2 to 4 points.[21-22]

Although researchers have developed some rules of thumb for considering combinations of line length and leading, these rules are not hard and fast.[14, 16, 21, 23] For example, if a text is set in a typeface that appears small, reducing leading between lines may well have no effect on the legibility of the text.

When you consider options of type size, line length, and leading, remember:

- Sans-serif typefaces may be easier to read with slightly more leading.
- Long lines may be easier to read with more leading between them.
- As type size increases, the effect of leading on legibility decreases.

Justification of Lines, Size of Margins, Use of White Space

Guideline: Use ragged-right margins (unjustified text).

Guideline: Use white space in margins and between sections.

Many researchers and typographers advocate the use of ragged-right margins both because research indicates that justified text is more difficult for poor readers to read and comprehend, and because many readers prefer text with ragged-right margins.[24-26] Unjustified text is also less costly to produce and easier to correct than justified text.

Unjustified text is less costly to produce and easier to correct than justified text. Also, many readers prefer unjustified text.

41

However, many publications, still set text with both left and right justified margins.

Unjustified text is less costly to produce and easier to correct than justified text. Also, many readers prefer unjustified text. However, many publications still set text with both left and right justified margins.

Unjustified and justified text.

Headings

Guideline: Use informative headings.

Guideline: Make all headings consistent and parallel in structure.

Research suggests that readers understand and remember text preceded by titles and/or headings better than text with no headings or titles. Poor readers are significantly aided by headings that are full statements or questions.[27]

Clear, active, and specific headings can also alter how readers comprehend a text. Headings can affect both what sections of text readers choose to read and how readers remember the organization of the text.[28-32]

For example, which of the following headings do you find easier to understand and remember?

    Subpart B: Applications and Licensee
        83.20 Authorization required.
        83.22 General citizenship requirements.
        83.24 Eligibility for license.

                    or

    2 - How to get a license
        2.1 Do I need a license?
        2.2 How do I apply for a license?
        2.3 May I operate my car while my
            applications are being processed?

Graphs, Tables, Flowcharts, and Diagrams

Guideline: When you are considering using graphics to convey quantitive or procedural information, you should consider
- **the reader**
- **the amount and kinds of information the graphics must display**

- **how the reader will use the graphics**
- **what graphic forms you can use (tables, graphs, charts)**

If you decide it is appropriate to use graphics, keep the following guidelines in mind:

Guideline: Reinforce all graphics with supporting text.

Guideline: Use typographic cues (changes in type weight and style) to distinguish between different types of information, such as headings or different kinds of data.

Guideline: Use 8- to 12-point type and adequate spacing between printed items.

Guideline: Include informative labels with all types of graphic devices.

Guideline: Use only those lines, grid patterns, and other ink markings necessary to make the information clear.

Guideline: Whenever possible arrange items vertically so they can be easily scanned.

Because readers' skills in interpreting graphics vary so widely, and because graphic displays can be designed and used in so many different ways, a tremendous amount of research has been done to explore the most effective ways to present quantitative and procedural information.[33-40]

This research strongly suggests that many readers can understand and use complex information more quickly and accurately when it is presented in well-designed tables, graphs, flowcharts, or diagrams rather than solely in text. However, these graphic devices are even more effective when they are reinforced with explanatory text.[41-44]

Some research suggests that bar charts are easier to understand than line, pie, or surface charts; or text.[36, 45]

(A)

Range (in miles)		Payload (in lbs)
878	100	710
630	200	598
459	300	471
388	400	326

(B)

Range (in miles)	Payload (in lbs)	Payload (in lbs)	Range (in miles)
100	878	100	710
200	630	200	598
300	459	300	471
400	388	400	326

Table B is easier to use than Table A because the columns are explicitly labeled.

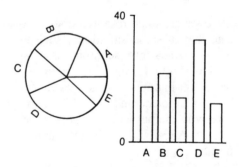

This bar chart reveals the difference in proportions more clearly than the pie chart.

WHAT DOES THE RESEARCH SUGGEST TO THE PRACTITIONER?

Although the relationship between writers and designers is often a difficult one to develop, research in document design (and the practical experience of designers and writers) supports the conclusion that to produce good technical documentation, designers and writers must work together from the beginning of the document design process. As you begin to plan a document, you should not only consider who your audience is and how they will use the document, but also how you can use the visual presentation of the text to reinforce the structure of the information.

To develop a sound working relationship, writers and designers should make conscious efforts to learn about how the other approaches planning and preparing a document so they can communicate effectively. Ω

REFERENCES

1. J. Hartley, *Designing Instructional Text* (New York: Nichols Publishing Company, 1978).

2. D.H. Jonassen, ed., *The Technology Of Text* (Englewood Cliffs, NJ: Educational Technology Publications, 1982).

3. P. Wright, "Usability: The Criterion For Designing Written Information," *Processing Visible Language, Vol 2*, ed. Koler, Wrolstad, and Bouman (New York: Plenum Press, 1979), pp. 183-205.

4. L.T. Frase, "Reading Performances And Document Design," Bell Laboratories, Paper given at Society for Applied Learning Technologies, Washington DC, June, 1976.

5. R. Waller, "Typographic Access Structures For Educational Texts," *Processing Visible Language, Vol 1*, ed. Koler, Wrolstad, and Bouman (New York: Plenum Press, 1979), pp. 175-187.

6. E.E. Miller, *Designing Printed Instructional Materials: Content and Format* (Alexandria, VA: Human Resources Research Organization, 1975).

7. W. Diehl and L. Mikulecky, "Making Written Information Fit Workers' Purposes," *IEEE Transactions on Professional Communication* PC-24 (1981).

8. J. Redish, R. Battison, and E. Gold, "Making Information Accessible To Readers," *Research In Non-Academic Settings*, ed. Odell and Goswami (Guilfor Press, in press).

9. P. Carpenter and M. Just, "Reading Comprehension As Eyes See It," *Cognitive Processes in Comprehension*, ed. Just and Carpenter (Hillsdale, NJ: Lawrence Erlbaum Assoc., 1977), pp. 109-139.

10. A. Marcus, "Icon Design Requires Clarity, Consistency," *Computer Graphics Today*, November 1984.

11. R. McLean, *The Thames and Hudson Manual of Typography* (London: Thames and Hudson, Ltd., 1980), pp. 42-48.

12. D. Robinson, M. Abbamonte, and S. Evans, "Why Serifs Are Important," *Visible Language* 4 (1971).

13. R.F. Rehe, *Typography: How To Make It Most Legible* (Carmel, IN: Design Research International, 1981), p. 32.

14. A. Seigel, *Designing Readable Documents: State Of The Art* (New York: Seigel and Gale, 1978).

15. M.A. Tinker and D.G. Paterson, "Influence of Type Form on Speed of Reading," *Journal of Applied Psychology* 7 (1928).

16. J. Hartley, "Space and Structure In Instructional Text," paper given at the NATO Conference of the Visual Presentation of Information, Het Vennenbos, Netherlands, 1978.

17. E.C. Poulton and C.H. Brown, "Rate of Comprehension Of Existing Teleprinter Output And Possible Alternatives," *Journal of Applied Psychology* 52 (1968).

18. E.C. Poulton, "Searching for Newspaper Headlines Printed in Capitals or Lower-case Letters," *Journal of Applied Psychology* 51 (1967).

19. A.G. Vartabedian, "The Effects of Letter Size, Case, and Generation Method in CRT Display Search Time," *Human Factors* 13 (1971).

20. H. Spencer, L. Reynolds, and B. Coe, *A Comparison of the Effectiveness of Selected Typographic Variations* (Readability of Print Research Unit, Royal College of Art, 1973).

21. M.A. Tinker, *Legibility of Print* (Ames, Iowa: Iowa State University Press, 1963), pp. 88-107.

22. D. Felker, *Document Design: A Review Of The Relevant Research* (Washington, DC: American Institutes for Research, 1980), pp. 104-106.

23. M. Gray, "Questionnaire Typography and Production," *Applied Ergonomics* 6 (1975).

24. M. Gregory and E.C. Poulton, "Even Versus Uneven Right-hand Margins and the Rate of Comprehension in Reading," *Ergonomics* 13 (1970).

25. J. Hartley and P. Burnhill, "Experiments With Unjustified Text," *Visible Language* 3 (1971).

26. T.E. Pinelli, et al, "Preferences On Technical Report Format: Results Of Survey," *Proceedings*, 31st International Technical Communication Conference (Society for Technical Communication, 1984).

27. J. Hartley, P. Morris, and M. Trueman, "Headings in Text," *Remedial Education* 16 (1981).

28. E. Kozminsky, "Altering Comprehension: The Effect Of Biasing Titles On Text Comprehension," *Memory and Cognition* 5, (1977).

29. J. Hartley, J. Kenely, G. Owen, and M. Trueman, "The Effect Of Headings On Children's Recall From Prose Text," *British Journal of Educational Psychology* 50 (1980).

30. H. Swartz, L. Flower, and J. Hayes, "How Headings In Documents Can Mislead Readers," (Pittsburgh, PA: Document Design Project, Carnegie-Mellon University, 1980).

31. P. Wright, "Presenting Technical Information: A Survey Of Research Findings," *Instructional Science* 6 (1977); pp. 96-100.

32. L.T. Frase and F. Silbiger, "Some Adaptive Consequences Of Searching For Information In A Text," *American Educational Research Journal* 7 (1970).

33. W.S. Cleveland and R. McGill, "Graphical Perception: Theory, Experimentation, And Application To The Development Of Graphical Methods," *Journal of the American Statistician* 79 (1984).

34. W.S. Cleveland, "Graphical Methods For Data Presentation: Full Scale Breaks, Dot Charts, And Multibased Logging," *Journal of the American Statistician* 38 (1984).

35. A.S.C. Ehrenberg, "What We Can And Can't Get From Graphs, And Why," paper presented at American Statistical Association Meeting, Detroit, 1981.

36. M. MacDonald-Ross, "Graphics In Texts," in *Review Of Research In Education*, Volume 5, ed. Shulman (Itasca, IL: F.E. Peacock Publications, 1978).

37. P. Wright and K. Fox, "Presenting Information In Tables," *Applied Ergonomics*, September 1970.

38. P. Wright and K. Fox, "Explicit And Implicit Tabulation Formats," *Ergonomics* 15 (1972).

39. P. Wright, "Tables In Text: The Subskills Needed For Reading Formatted Information," (Cambridge, UK: MRC Applied Psychology Unit).

40. E.R. Tufte, *The Visual Display Of Quantitative Information* (Chesire, CT: Graphics Press, 1983).

41. P. Wright and F. Reid, "Written Information: Some Alternatives To Prose For Expressing The Outcomes Of Complex Contingencies," *Journal of Applied Psychology* 57 (1973).

42. P. Wright, "Writing To Be Understood: Why Use Sentences?" *Applied Ergonomics* 2 (1971).

43. R. Kammann, "The Comprehensibility Of Printed Instructions And The Flowchart Alternative," *Human Factors* 17 (1975).

44. D. Felker, *Guidelines For Document Designers* (Washington, DC: American Institutes for Research, 1981).

45. D.D. Feliciano, R.D. Powers, and B.E. Keare, "The Presentation Of Statistical Information," *Audio-Visual Communication Review* 11 (1963).

PHILIPPA J. BENSON *is a Senior Research Associate in the Document Design Center of the American Institutes for Research (AIR) in Washington, D.C.*

A Guide to Choosing the Right Chart Type

KATHRYN SZOKA

Abstract—Data displayed in a chart rather than a table are easier to understand and trends or patterns are easier to identify. Four basic data types that are easily charted are changes over time, percentages of a whole, comparisons, and relations between variables. Displays discussed are simple and double-scale curves; scatter and line plots; column, bar, and step charts; pie charts; patch maps; and combinations of these. Before making a chart develop a focus sentence, one that summarizes what is to be shown, and select a type of chart which illustrates that focus. Present only key points; extras detract from the message.

THE success or failure of a presentation usually hinges on how the speaker is perceived by the audience. Graphics can significantly help the speaker if they portray him or her as a careful, accurate person who is interested in effective communication. They are an important tool for clearly illustrating key points in presentations.

If poorly designed charts overloaded with data are used, they may serve to retard rather than speed the audience's acceptance of key points. Good charts cannot make a good presentation out of a poorly organized one, but bad charts can ruin an otherwise effectively designed briefing.

More and more, people are relying on computer-generated graphics because they can be produced in a timely, economical way with the quality and variety an art department would offer. The principles discussed in this article, however, should be helpful to those who design and prepare charts either manually or by computer.

The speaker who wants to support a presentation with graphics should always use well designed charts and avoid the common excuse that "these charts are just for analysis." Too often analytical charts are pressed into service for presentations or attached to technical reports. Every time someone sees those charts, the producer and his work are being judged. The benefits of good chart design far outweigh the time and attention necessary to produce them.

Selecting the right chart involves two key decisions:

1. Choose or create a focus sentence—a sentence that summarizes what is to be shown.
2. Select a type of chart that illustrates that focus sentence.

There are four basic types of relationship or pattern in chart design from which to choose. The focus sentence should lead the graphics producer to the type of chart best suited to the particular requirements. Beginning without a focus sentence is likely to result in selection of an inappropriate chart.

Reprinted with permission from *Industrial Engineering*, vol. 13, no. 10, pp. 74–79, October 1981 (part 2 of "Graphics as Aids to IE Communication"); © copyright 1981 by the American Institute of Industrial Engineers, Inc., Norcross, GA 30092.

This work was done while the author was advertising manager at Integrated Software Systems Corp., San Diego, CA; her current address is Megatek Corp., 3985 Sorrento Valley Blvd. San Diego, CA 92121, (714) 455-5590.

TIME SERIES

The first type of pattern typically shown via graphics is a time series. Monitoring change over time is an essential function of management, and graphics often focus on time series information. These comparisons show trends, increases, decreases, or fluctuations in some quantity.

Some sample focus sentences calling for time series charts are

> Productivity per employee has increased more than 30 percent during the past year.
>
> Minicomputers are taking an increasing share of the data processing market.
>
> Employment declines follow the onset of recessions.
>
> Fluctuations in cost of living are due mainly to housing prices.
>
> Projected improvement of environmental quality is out of line with recent experience.
>
> The northern district has been the star performer for the last three years.

Once a time series chart or graph has been selected, the designer is faced with four primary methods of displaying time series data.

The Curve Chart

A curve connecting values for each period is the most common method of showing change, but not the only method. Curves are usually used when trends over many periods are sought.

One common use of the curve chart is to show planned sales versus actual sales. In this particular case, two line patterns are used. When comparing one curve against several others, it is best to single out that curve for emphasis, or else the chart may be difficult to read. On the other hand, several charts should be used together to compare one curve with each of the others individually, with the same scale being used in each of the plots. Multiple charts per page are also useful when larger volumes of data are being reviewed or presented (see Fig. 1).

Use of double-scale curve charts should be limited because people tend to misread them. Also, a scale change on either axis can alter interpretation. For analytical use, however, they can be effective for comparing relative changes. When scales coincide, a single set of grid lines should be used; otherwise avoid grid lines (Fig. 2).

There are a number of rules to keep in mind when working with curve charts. Always make the curves thicker than the grids. Not only will the chart be easier to read but also there will be no mistaking what is part of the curve. It is better to label the curves right on the chart than to use a legend.

Different line thicknesses or different line patterns with long dashes can be used for comparing yearly data. Dashed curves are often used for plans or budgets.

Reprinted from *IEEE Trans. Prof. Comm.*, vol. PC-25, no. 2, pp. 98–101, June 1982.

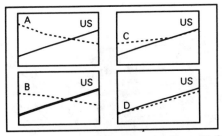

Fig. 1.　Curve charts used to compare one data line with four others.

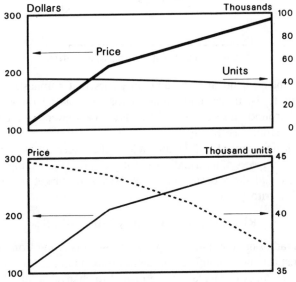

Fig. 2.　Double-scale curve charts.

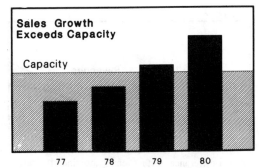

Fig. 3.　A four-year column chart with reference line.

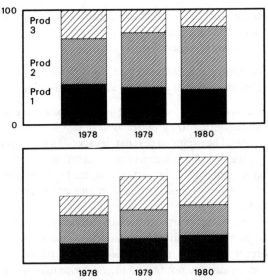

Fig. 4.　A 100-percent column chart showing relative shares and a stacked-column chart representing actual values.

The Column Chart

A column chart may be more effective than a curve chart when there are few time periods. Column charts are especially effective in showing large changes from one period to the next.

If there is one period that is more important than the others, changing the shade pattern draws attention to the column with the different pattern. The use of a reference line is effective when the area under it is shaded because the columns extending out of the shaded area appear to be breaking through the barrier (see Fig. 3).

To compare movement of items over time, groups or clustered columns should be used. This type of chart also illustrates differences between the two items in each period. Generally, clustered columns should not be used when more than two items are being compared; curves usually improve comprehension of trends for comparison of three or more items over time.

A combination column-and-curve chart is especially useful for looking at data that span several years. It uses a curve to show current data on a monthly basis and columns to show annual data for prior years.

Changes in shares of a total over time are well represented in a 100-percent column chart. However, when showing values rather than percentages, use a stacked-column chart (see Fig. 4).

Net-difference columns show the difference between two series of data. By changing shading and overlaying a range of acceptable limits, net-difference charts can be used effectively as management tools for monitoring performance versus budget.

Always make the bars and columns wider than the space between them and order the shade patterns from darkest to lightest, avoiding garish patterns.

The Surface Chart

Surface charts are curve charts with the area between or below the curves shaded. They emphasize volumes and generally should be used for showing quantities, not for showing indexes and similar non-volume indicators. Order the data so that irregular lower layers do not intrude on upper layers.

Net-difference surface charts show the difference between two series of data. Changes in shading pattern can emphasize the shift from loss to gain. It is always better to put labels in the shaded areas of surface charts when the bands are wide enough (see Fig. 5).

The Step Chart

A step chart is a column chart with no space (or lines) between columns. It is useful when a column chart is needed to reflect large changes from period to period but there are so many periods that separated columns would resemble a picket fence.

PARTS OF THE WHOLE

The second pattern shown by graphics is known as parts of the whole. When the words "market share" or "portion" or

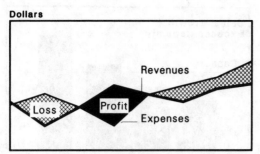

Fig. 5. A net-difference surface chart showing alternation between loss and profit.

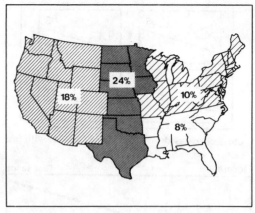

Fig. 6. A patch map showing regional data.

"percentage" are used in focus sentences, a comparison of parts of a whole is taking place.

Sample focus sentences are

> Administrative costs are less than 15 percent of total cost.
> Our market share is larger in the small cities.

Pie charts are useful for this type of comparison because they automatically bring to mind the concept of a whole separated into its parts. The exploded pie chart is especially useful because it focuses the eye on one segment of the chart, as does changing the shade pattern in important segments. One thing to remember is to avoid using more than four or five segments in pie charts.

Column and bar charts are useful when showing segments of several totals. To extend the comparison over multiple areas, a 100-percent column is useful. If there are too many columns, however, labels may not fit and the audience may incorrectly perceive a time series. The 100-percent bar chart is useful when labels are large and when there are too many labels to put effectively on a column chart.

COMPARING SEVERAL PLACES OR THINGS

The third type of chart is a comparison chart used to illustrate differences between dimensions other than time, such as areas, products, people, companies, and the like.

Sample focus sentences calling for comparisons among several items are

> Sales are greater for Product A than for others.
> All four companies have approximately the same profit margin.
> The central region contributes more energy than other regions.

Bar charts are often used for item comparisons but column charts can also be used. Pie charts and pictograms can be used effectively for item comparisons, especially when the comparison has a geographic dimension.

On a bar chart, one bar can be made to stand out regardless of the order of the bars by using a different shade pattern for it. A sorted bar chart is an effective tool to use when showing where one item stands relative to others.

Another type of bar chart, the floating bar chart, can be used with a calendar to show project schedules and to isolate projects that need attention.

The shaded, or patch, map presents one piece of information for each area. The keys to good shaded maps are boundaries that correspond to areas of interest (sales territories, for example); legend and shading categories that emphasize the point to be made; and numbers that are placed on white backgrounds inside the areas to give the viewer additional information (see Fig. 6).

Placing symbols on the map allows the presentation of two or more pieces of information about each area. For example, a map could show both total market size and market share in each territory.

RELATION BETWEEN VARIABLES

Traditional computer graphics—those emanating from the statistical and scientific communities—are often used to show how changes in one factor are related to changes in another. This is the fourth type of relationship from which to choose in chart design.

Sample focus sentences calling for comparison of relationships between two variables are

> Viscosity decreases with temperature.
> The relationship between salary and performance is hard to identify.

Curves can be used to show such relationships, as can scatter plots. In addition, multiple bar charts may prove effective. Two variables can also be compared using time series and item comparison charts.

The dual bar chart or pyramid chart offers an effective means of comparing two sets of data about the same places or items. The scatter-and-line plot offers both a symbol at every data point and a curve to show patterns.

Scatter plots are sometimes used in an attempt to show the lack of a relationship. This attempt is rarely successful. Most viewers find a pattern whether there is one or not. It is usually better to use correlation coefficients instead of graphics when trying to illustrate a lack of relationship.

Figure 7 summarizes the four types of relationship or pattern that charts are used to illustrate and the chart types best suited to each. Axis and other labeling has been excluded to make the individual character of each type stand out better. Effective and complete labeling is, of course, an important characteristic of a good chart.

POINTERS AND COMMON ERRORS

• *Chart selection* The first common error arises when the analyst selects the wrong chart from among too many choices.

46

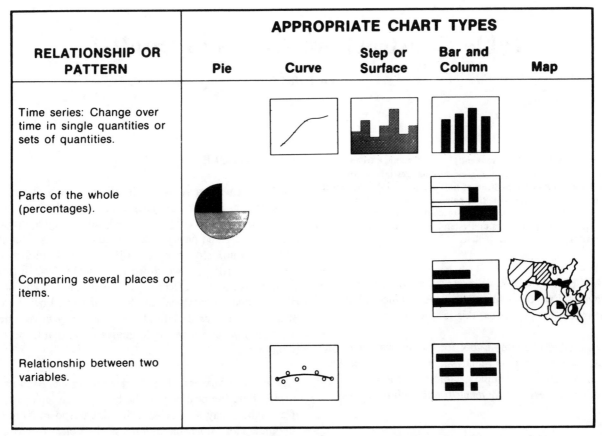

RELATIONSHIP OR PATTERN	APPROPRIATE CHART TYPES				
	Pie	Curve	Step or Surface	Bar and Column	Map
Time series: Change over time in single quantities or sets of quantities.		✓	✓	✓	
Parts of the whole (percentages).	✓			✓	
Comparing several places or items.				✓	✓
Relationship between two variables.		✓		✓	

Fig. 7.   The four basic uses of data-display graphisc with chart types for each.

When an analyst tries to make a chart from a table, the analyst has two choices: Put all the data on the chart or be selective. Putting all the data on the chart nearly always results in a busy, unreadable chart. On the other hand, selecting a subset of data requires the analyst to choose one aspect of the table to illustrate.

Chart selection should depend on what you want to say. The relationships or points you are trying to make or show should be determined before choosing a chart.

The fact that the outcome of a chart is uncertain should not stop you from creating one. Often a general idea of the relationship is all that is needed to select the right chart. ("Change over time" or "relative performance" is specific enough.) Then produce charts from the tables to find or verify the key patterns in the data.

• *Axis*   It is usually better to write axis labels horizontally and make the axis scale large enough to read easily.

• *Scale*   When comparing levels or totals, include the zero point. When zero is omitted, let the reader know. Use the same scale when using multiple charts to compare trends, and use scales that make interpolation easy (add tick marks, if necessary).

• *Shade pattern and type*   Order shade patterns logically and progressively on map scales to facilitate remembering which ones represent the end values, and use a single family of typeface on a chart. Simple sans serif fonts are preferable. Try to maintain consistency of style from chart to chart in a presentation.

CONCLUSION

In the entire computer field, few tools offer more potential for improving executive productivity than computer graphics. If used correctly, graphics can legitimize the investment in data gathering and data processing by making the data more visible and more understandable and therefore more useful to management.

Effective chart design is the most important prerequisite for ensuring that executives use these new graphics tools and that the potential productivity gains are actually realized. Don't make a chart until a focus sentence summarizing what you want the chart to say, or the pattern you wish to display, is created; make charts look professionally designed; and limit the charts you present to those that illustrate key points. Extras only detract from your message.

ACKNOWLEDGMENT

The illustrations in this article were produced using Tell-A-Graf®, a proprietary software product of ISSCO Graphics, San Diego, CA.

FOR FURTHER READING

A. Paller, K. Szoka, and M. Nelson, *Choosing the Right Graph*.   ISSCO Graphics, 4186 Sorrento Valley Blvd., Suite G. San Diego, CA 92121.

Table Construction: Do's and Don'ts*

EVA DUKES

Abstract—A table is a tool frequently used to arrange technical data in a meaningful way. A good table summarizes information by avoiding the unnecessary repetition of unit symbols, test conditions, or other data compressible into stub, column, or spanner heads. Whereas poor table construction tends to confuse the reader, sound table construction helps readers understand the significance of technical information in a minimum of time.

BECAUSE FAMILIARITY breeds contempt, most of us believe that we can construct a proper table without giving it much thought. Alas, this is seldom so.

Basically, a table is a simple tool for listing exact data for purposes of comparison and analysis. However, when the data are numerous and complex, table construction becomes an art: it needs refinement to attain clarity.

THE BASIC TABLE

As Voltaire once said [1], "If you would dispute with me, define your terms." In the same spirit, let us first agree on naming the parts of a table. Figure 1 [2] shows a rather simple table, comparing it with a graph. Mary Fran Buehler has compared the two by saying, "A graph may not give so much data, or such exact data, but it gives an instant picture of relationships and trends in a way that a table cannot." (Of course, a table may also show trends [3]—as when all, or most, of the figures in a column can be grouped in ascending or descending order.) Figure 2 [2] shows a more complex structure, identifying all the essential parts of a table.

The following points are important rules for constructing and reading a table:

- Information always reads *down* from the boxhead, all the way to the end of the table.
- Information should also read *down* from the stub head.
- But information controlled by the stub head reads *across*.

* A shorter version of this article was published in 1983 in *Guide for RCA Engineer Authors*, and is used courtesy of General Electric Company.

Eva Dukes is a technical writer and editor with experience in training and supervising personnel in communication skills.

THE INVISIBLE BOX

The easiest way to visualize the structure of a table is to think of it as a box. For typeset tables, or those set on the more sophisticated computer-aided publishing equipment, the box is created easily, perhaps automatically. But every table has a box structure—even if the box is not drawn, as in the typewritten or word-processed table. Drawing vertical lines is not cost effective, and even the top and bottom rules (above the boxhead and below the last line, or row, of data) are not essential. That is why many government specifications and other style guides insist that these rules be omitted.

Figure 3 is a boxless table. Underscoring has been used to differentiate the heads from the body of the table; italics, if available, may also be used for this purpose. Note that the underscore for each column head extends from the beginning of the longest word in the head (in this case, *Column*) to the end of that word; lines above or below the longest word are centered in regard to it or, in some cases, flush left to it.

The line underscoring the spanner head extends from the beginning of the longest line in the first column head it spans to the end of the last column head it spans. The spanner head itself is centered over that bar.

Styling a table with a box differs significantly in one respect from styling one without a box. In the former, column heads should be centered vertically as well as horizontally within their own boxes [3]. This vertical centering is difficult to achieve on an ordinary typewriter or word processor. In this case (the boxless table), column heads should be aligned horizontally with the bottom of the lowest column heads under the spanner head (as in figure 3). That is, the heads should be "flush bottom" in the invisible box for the header. In this way, each head is close to the data it controls.

Even in the most crowded table, there should be a space (at least one pica) between the underscores of each pair of column heads and between the underscores of the stub head and the first column head. The latter space may be a little larger than the between-column-heads spaces, which should be equalized (unless good design dictates otherwise).

Reprinted from *IEEE Trans. Prof. Comm.*, vol. 32, no. 1, pp. 36–40, March 1989.

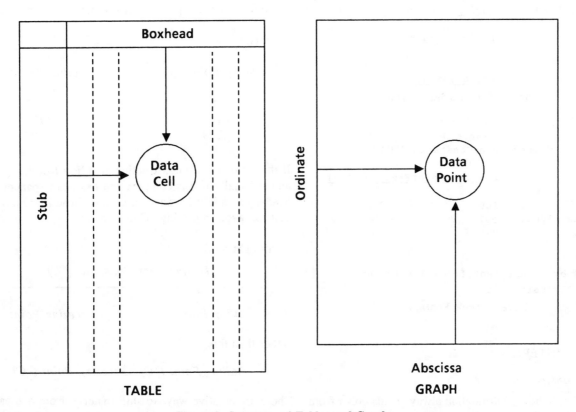

Figure 1. Structure of Tables and Graphs

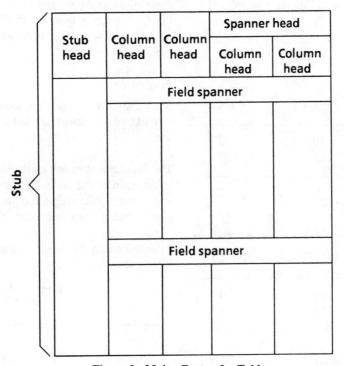

Figure 2. Major Parts of a Table

		Spanner Head		
Stub Head	Column Head	Column Head	Column Head	Column Head
		Field Spanner		
		Field Spanner		

Figure 3. The "Boxless" Table

Table 0
Temperature Constants (°C)

Liquid	Boiling Point	Freezing Point
Water	100	0
Ethylene glycol	197	−20[a]
Glycerol	290	0[b]

[a] In 44% by volume (antifreeze) solution
[b] Gradually solidifies

Figure 4. Table Footnotes

USEFUL POINTERS

Rules of Construction

As pointed out above, information always reads down from the boxhead, all the way to the end of the table; information from the stub head also reads down; information reads across from the stub column. The following examples illustrate these points:

Do this:

Liquid	BP (°C)	FP (°C)
Water	100	0
Ethylene glycol	197	−20

Don't do this:

Characteristics:	BP (°C)	FP (°C)
Water	100	0
Ethylene glycol	197	−20

Independence From Text Material

A table, like a figure, should be able to stand alone. The caption, heads, and (if necessary) footnotes should tell the whole story, in case the reader skips the surrounding text.

Avoiding Clutter

Applying the law of paucity does wonders for clarity. Therefore, never repeat a unit symbol if you don't have to.

Do this:

Boiling Point (°C)

100
197
290

Don't do this:

Boiling Point

100°C
−20°C
0°C

If the table you are constructing also lists freezing points, and especially if it is complex and has other spanner heads, you might introduce a spanner head here, too. Then you can avoid repeating °C even once.

You might do this:

Temperature Constants (°C)

Boiling Point	Freezing Point

rather than this:

Boiling Point (°C)   Freezing Point (°C)

There are creative ways to shift material from one part of the table to another to achieve maximum paucity and clarity. It helps to ask yourself: "How much data does this description apply to, and what is the clearest way to show that?" A lengthy description that applies to a small amount of data—a few entries, perhaps, or one column—could be placed in a footnote.

Footnotes

Since the data in tables are usually numerical, many editors like to use lowercase letters to refer to footnotes (figure 4).

The footnotes themselves should be placed flush with the left margin of the table, not the left margin of the page, and directly following the table, not at the bottom of the page. That is, they fall within the box.

A footnote may be used to indicate a shift in units of measurement or in conditions:

Degree of Hardness
(at 20°C)[a]

9
5 (at 10°C)
2

[a] Unless otherwise indicated.

	Amsterdam	Brussels	Frankfurt	Madrid	Paris	Rome	Vienna	Zurich
Amsterdam		126	285	1092	303	1033	752	563
Brussels	126		239	966	177	931	729	518
Frankfurt	285	239		1156	361	813	466	304
Madrid	1092	966	1156		789	1300	1494	1169
Paris	303	177	361	789		915	774	397
Rome	1033	931	813	1300	915		750	806
Vienna	752	729	466	1494	774	750		436
Zurich	563	518	304	1169	397	806	436	

Figure 5. Matrix of Distance Between Cities

Consistency

Small inconsistencies hurt the appearance of a document and may interfere with comprehension. For instance, when identical or comparable column heads appear in different form in the same table or in tables within the same document, the reader is left to wonder if the information contained in the columns is the same or different.

It is important to use a consistent convention for handling missing, inadequate, or negligible data. Many editors use the word *none* or *N/D* (no data) or leave an empty space for missing data and insert a dash (or double hyphens) for inadequate or negligible data. *N/A* may be used for *not applicable*.

Alignment

Usually, numerical data are aligned on the decimal point, or wherever the decimal point would be if one were given. In the case of numbers less than 1, insert a zero before the decimal point (0.95, not .95). It is not advisable to add zeros after the decimal point unless you known that this is warranted by the degree of accuracy of the measurement.

Legibility

Tinker analyzed the results of 238 legibility studies [4]; the following of his observations are particularly applicable to tables:

- Arabic numerals are easier to read than Roman numerals.
- Lowercase type is easier to read than all capitals type.
- Leaving a space every five or ten lines down a column aids legibility; five is better.
- The smallest type recommended is 10 points. Although 8-point type may be readable, it does not fulfill the legibility requirements for micropublishing.

If at all possible, it is best to set a table vertically rather than horizontally, so that it will fit on a right-reading page. Similarly, avoid foldouts if possible.

COMMON PITFALLS

Lack of a Stub Head

Authors often omit a head for the stub column because the entries in this column are so disparate that a common denominator is not easily found. This in itself points to a weakness in the table. If a common denominator cannot be applied—even a ''portmanteau'' word like *parameter*—you should consider revising the table.

Misuse of the Field Spanner

It makes no sense to use only one field spanner; two is the minimum. If you have only one, you are using the field spanner for information that should go into the caption or into the boxhead of the table.

The field spanner spans all column heads but should not extend into the stub column.

THE INFORMAL TABLE

Informal tables are essentially lists. Data may be arranged in tabular form to give the reader a better overview, even though the material does not lend itself to full tabular treatment. Such informal tables may have no stub column; they may merely present data for comparison under adjacent column heads, such as the following [5]:

Advantages *Disadvantages*

Much of the advice given above applies to informal tables as well as to formal tables—for instance, the principles of alignment, legibility, and so forth.

THE MATRIX

The matrix is a special type of table that makes each item in the stub column interact with each item in the column heads.

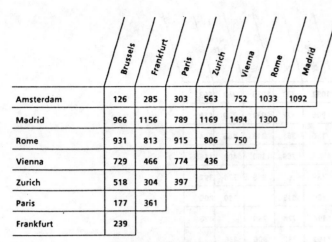

	Brussels	Frankfurt	Paris	Zurich	Vienna	Rome	Madrid
Amsterdam	126	285	303	563	752	1033	1092
Madrid	966	1156	789	1169	1494	1300	
Rome	931	813	915	806	750		
Vienna	729	466	774	436			
Zurich	518	304	397				
Paris	177	361					
Frankfurt	239						

Figure 6. Simplified Matrix

	Dollars	Yen	Francs	Pounds
Dollars		125	6.08	0.563
Yen	0.00803		0.0490	0.000454
Francs	0.164	20.4		0.0926
Pounds	1.78	220	10.8	

Figure 7. Nonsymmetrical Matrix

In figure 5, a matrix of distances between cities, the same entries may be found in the stub column and in the column heads. The diagonal block of entries is blank because, obviously, the distance between Amsterdam and Amsterdam, or Paris and Paris, is zero. Each entry occurs twice—for

instance, under both Brussels—Frankfurt and Frankfurt—Brussels; therefore, this matrix can be read across and down *or* down and across.

A simpler presentation of the same information is shown in figure 6. This matrix has no blank spaces and no repetitions (at the cost of having a peculiar stepped appearance). The economy has been effected by a rearrangement of the stub and column heads.

Similar matrixes can be drawn for, say, currency conversion factors (figure 7). In a case like this, however, there is no repetition of data, because dollars expressed in yen is not the same as yen expressed in dollars.

Thus the matrix fits the definition of a table: an arrangement of data that makes the data easier to read and to understand.

REFERENCES

1. Voltaire, F. M. A., private communication (via an instructor who knew Voltaire's writings well).
2. Buehler, M. F., "Table Design—When the Writer/Editor Communicates Graphically," *Proceedings* 27th ITCC, 1980, pp. G-69 to G-73.
3. Arnold, C. K., "The Construction of Tables," *IRE Trans. Engineering Writing and Speech EWS-5*, 1 (1962), pp. 9–14.
4. Tinker, M. A., *Legibility of Print*, Ames, IA: Iowa State University Press, 1963.
5. Buehler, M. F., "Report Construction: Tables," *IEEE PC 20*, 1 (1977), pp. 29–32.

Lists: The Ultimate Organizer for Engineering Writing

SAUL CARLINER

Abstract—Lists are a useful device for engineering writing in that they help readers to see what is important, to see how important items relate to one another, and to remember what they see. This article presents guidelines for the use and construction of lists in technical material.

LISTS are well-known tools of the compulsively organized. These people make grocery lists, to-do lists, guest lists, wardrobe lists, and even lists of their lists.

Lists are equally valuable organizational tools for engineering writers. Lists may be used to present information, from very small amounts to very large amounts, in a manner which makes the information easier to understand. For example, lists may be used for needed materials, supplies, tools, personnel, and so forth; general notes, items for consideration, precautionary items, and the like; and chronological facts, step-by-step procedures, etc. This article discusses the nature of lists, reasons for their use, and methods for making them.

WHAT ARE LISTS?

A list is a device used in writing, to separate information into logical parts for easier comprehension by the reader. Although even a simple series of words such as 'The IEEE symbol includes a rising arrow surrounded by a rotating arrow' constitutes a list in a technical sense, for the purposes of this paper, a list is defined as a series of two or more items which are set apart in text or in a columnar format, and highlighted by numbers or symbols.

Lists often begin with a lead-in sentence, such as, 'The five steps in preparing a printed circuit board are as follows:' What follows is a group of items, listed underneath the lead-in sentence.

Lists offer these benefits:

- They show relationships among pieces of information, such as whether the list items follow a sequence [1].

- They condense information and make it more accessible to the reader by presenting it in highly readable ''chunks'' rather than buried within a paragraph.

Atlanta-based writer Saul Carliner is Associate Editor of *Technical Communication*, Program Manager of the 1989 International Technical Communication Conference, and author of numerous articles on technical communication.

- They eliminate redundancies, such as repeated words used to create sentences of parallel structure.

Readers especially like lists. Typical readers of technical information read not for pleasure, but to learn important information [2]. Lists help readers find that information quickly and easily because readers can scan lists more quickly than they can scan paragraphs. Readers can also extract information more easily from lists than from paragraphs.

Consider, for example, the difference between these two versions of the same information:

Paragraph Version

When cleaning relay contacts, first ensure that all power is disconnected. Then, remove the cover from the relay cabinet. Release the pressure on the relay contacts. Wipe the relay contacts with a cloth moistened with carbon tetrachloride.

Afterwards, reset the pressure on the relay contacts. Replace the relay cabinet cover. Power may now be restored.

List Version

When cleaning relay contacts, do the following:
1. Ensure that all power is disconnected.
2. Remove the cover from the relay cabinet.
3. Release the pressure on the relay contacts.
4. Wipe the relay contacts with a cloth moistened with carbon tetrachloride.
5. Reset the pressure on the relay contacts.
6. Replace the relay cabinet cover.
7. Power may now be restored.

USING LISTS

The ultimate goal of lists is to help readers better understand your writing. Notice how the list format above serves that purpose.

The List version states the procedure more succinctly than the Paragraph version. The numbers immediately tell us we're looking at a step-by-step process. In the paragraph version, readers have to wade through extra words (such as *then* and *afterwards*) to find that out. The numbers also act as a bookmark that readers can return to easily when they leave the book to clean the relay contacts.

The list format helps readers do the following:

Reprinted from *IEEE Trans. Prof. Comm.*, vol. PC-30, no. 4, pp. 218–221, December 1987.

- Organize the instructions into a sequence [steps (1)–(7)].

- Identify relationships between items of information ['Step (3) is dependent upon successful completion of step (2).']

To understand how lists help readers organize information and identify relationships within it, we need to understand how people process information. In general terms, processing happens like this [3]:

Mind reads information.

↓

Mind decides it must remember the information.

↓

Mind temporarily puts information into short-term memory.

↓

Mind permanently stores information in long-term memory for future use.

To save information in their long-term memories, readers must first understand it. In the process of understanding information, readers relate new information to information already saved in long-term memory. New information is then stored in relation to the old.

We can help readers understand information by giving them clues as to how it relates to information they already know—that is, putting it into context [4]. One device that helps readers put information in context is the lead-in sentence to a list. For example, this lead-in sentence...

You need to pack the following items into the shipping container:

...tells readers that the information that follows relates to products to be shipped.

During the time that readers check long-term memory to see how new information relates to existing information, new information is kept in short-term memory. Information cannot be saved in long-term memory unless it is first saved in short-term memory. When information moves to long-term memory, short-term memory is "erased."

Short-term memory must be erased because it cannot hold much information—generally, only five to nine pieces. A telephone number, for instance, fits into short-term memory. However, short-term memory can hold more information, such as a telephone number and its area code (10 digits), if the information is suitably "chunked," or divided into pieces [5]:

(area code) exchange—extension
1 2 3

A 10-digit number can be held in short-term memory if it is divided into three chunks like this, because memory treats each chunk of information as one piece and stores them together. A list is a system for chunking information as necessary to reduce the number of pieces to be held in short-term memory.

WRITING LISTS

Engineering writers can apply these ideas in writing by using lists which help readers understand, find, and save information.

Use the type of list best suited to the situation.

Does the information being presented follow in a sequence? Or is each item of equal importance? The type of list used is an immediate signal of these relationships to readers. There are three main types of lists:

- Numbered lists, in which each list item is preceded by a number. (Letters, either upper- or lowercase, may be used instead of numbers; however, numbers are more likely to stand out—to herald the appearance of a list—in text that is made up of letters.) Use numbered lists when the items on the list follow a certain order [6]. For example, use a numbered list when writing a procedure for using a multimeter:

1. Turn ON an ABC model XYZ multimeter.
2. Connect the ground lead (black) to the equipment ground.
3. Set the multimeter scale to an appropriate voltage range.
and so forth.

The numbers may placed in parentheses or followed by periods. (Using both periods and parentheses is overkill!)

- Bulleted lists, in which each list item is preceded by a black circle or other printer's device, called a bullet. Use bulleted lists when the items on the list are equally important and should not be considered to be in any order. This list is itself an example of a bulleted list.

When bullets are unavailable, the lower case *o* may be used, or the asterisk.

- Checklists, in which each list item is preceded by a little box that readers check off. Use checklists when readers must attend to all the items on the list—for example, at the beginning of a test procedure directing the readers to ensure that they have all required equipment before beginning the procedure:

☐ Multimeter, with 1.5, 15, and 150 vdc and vac scales
☐ Audio signal generator which will produce 60 Hz to 20 kHz

☐ Oscilloscope which will display 60 Hz to 20 kHz with amplitudes of 0.1 to 15 vac

It is important to choose the type of list carefully. It is confusing to the reader to use a numbered list when the sequence of the items has no significance.

Keep bulleted lists to five items, if possible.

The longer your lists, the more difficult it is for readers to process and use the information, even allowing for the advantages of chunking. As noted above, readers can keep only five to nine items in short-term memory.

As a rule of thumb, when lists grow beyond five items, it is time to further chunk the information into main points and subordinate points. Put subordinate points into subordinate lists, called sublists. For example:

If you have this list:

Equipment list

- Audio signal generator
- Capacitance meter
- Decade altimeter
- Junction box
- Multimeter
- Oscilloscope
- Power supply
- Q-meter
- Volt-ohmmeter

Rewrite it like this:

Equipment list

- Test setup equipment
  —Audio signal generator
  —Decade altimeter
  —Junction box
- Test measurement equipment
  —Capacitance meter
  —Multimeter
  —Oscilloscope
  —Power supply
  —Q-meter
  —Volt-ohmmeter

Notice the punctuation introducing the sublist items. Here, an em dash is shown; if the em dash is not available, use two hyphens instead.

Dividing information into sublists offers these benefits:

- It helps the writer to express ideas more succinctly, because the more important points are separated mechanically (rather than verbally) from the less important points.

- It helps readers to connect information more readily, because he has fewer main points to remember and can see which items relate to main points in the text and which are subordinate. For example, rather than remembering that they have to obtain and arrange nine pieces of equipment, readers remember that they have to provide two groups of equipment, one to be measured and another to do the measuring.

Keep numbered lists to ten items, if possible.

As a rule of thumb, it is a good idea to limit procedures, which are what numbered lists describe, to no more than ten steps each. If your procedures are longer, perhaps you need to reorganize the information. For example:

If you have this list:

1. Remove the four screws holding the cover.
2. Remove the cover slowly and place it and the screws in a safe place.

Rewrite it like this:

1. Remove the cover as follows:
   a. Remove the four screws holding the cover.
   b. Remove the cover slowly and place it and the screws in a safe place.

Punctuate lists correctly for maximum comprehension.

There is a great deal of disagreement over the correct punctuation of lead-in sentences and list items. Capitalization is another point of debate. Although it is possible to make a valid case for several different styles, it is best to pick one and use it consistently. Many authoritative style guides prescribe answers to list punctuation and capitalization problems; the following set of guidelines has been synthesized from these answers.

The most common punctuation mark found at the end of a lead-in sentence is the colon (:). A colon is used before material which extends or amplifies that which precedes it, according to one definition; according to another, at the end of a sentence which is grammatically, but not rhetorically, complete, and to which the logical (rhetorical) completion is then added. Because the material which preceded the colon should be 'grammatically complete,' it is best not to use the colon after a lead-in like this:

The conclusions reached are

but instead after a complete sentence such as

- The following conclusions were reached:
- The conclusions of the study are listed below:

After a lead-in which is not grammatically complete, it is better to use a dash (—), an ellipsis (...), or no punctuation at all.

List items should begin with capital letters, if for no reason other than that readers are accustomed to seeing capital letters at the beginning of chunks of information. When the first word isn't capitalized, it can interfere with processing of the information. There is no need to capitalize other words in the items unless they would normally be capitalized (outside of a list).

List items should close with a period (or other sentence-end punctuation) only if they are complete sentences. If they are not, no punctuation should be used—again, because unnecessary punctuation is more likely to interfere with than to aid comprehension.

SUMMARY

The points in this article may be summarized by the following list:

- Lists are an effective tool for nonfiction prose. They help readers in many ways:
  —Organizing information
  —Showing relationships between points
  —Condensing information
  —Eliminating excess verbiage

- Lists help readers remember information more easily by helping them to chunk information.

- Types of lists include the following:
  —Numbered lists, used to show a sequence
  —Bulleted lists, used to show equally weighted information
  —Checklists, used to list things a reader should do

- There are a few guidelines for writing understandable, usable lists:
  —Choose the appropriate type of list.
  —Keep list items to a manageable number.
  —Use sublists when list items exceed the recommended number.
  —Punctuate and capitalize consistently, for maximum comprehension.

REFERENCES

1. Frank, D., *Silicon English: Business Writing Tools for the Computer Age*, San Rafael, CA: Royall Press, 1985, p. 65.
2. Flower, L., *Problem-Solving Strategies for Writing*, Pittsburgh: Carnegie Mellon University, 1978, pp. 1.1–1.2.
3. Turnbull, A. T., and Baird, R. N., *The Graphics of Communication* (4th ed.), New York: Holt, Rinehart, and Winston, 1980, pp. 19–33.
4. Reed, S. N., *Cognition: Theory and Applications*, Monterey, CA: Brooks/Cole Publishing Company, 1982, p. 37.
5. ibid, p. 70.
6. *Chicago Manual of Style* (13th ed.), Chicago: University of Chicago Press, 1982, p. 245.

Safety Labels: What to Put in Them, How to Write Them, and Where to Place Them

CHRISTOPHER VELOTTA

Abstract—Current standards developed by organizations like the American National Standards Institute and the Westinghouse Electric Corporation can help technical writers design effective safety labels. According to such standards, safety labels should contain a signal word, a hazard alert symbol, a specific color, a symbol or pictograph, a hazard identification, a description of the result of ignoring the warning, and a description of how to avoid the hazard. In addition, the safety label should be clear, concise, forceful, descriptive, and well-organized. Finally, safety labels usually should be placed in the operator manual and on the product, and they should appear before the operator encounters the hazard. Considerations involved with this placement include reading distance, viewing angle, and available space on the product.

WITH legal precedents that often seem to contradict themselves, the consumer products industry is facing a dilemma concerning how to construct effective safety labels and how to avoid product liability suits. For example, Joseph G. Manta cites many cases where the courts have found that a failure to warn was not the *proximate* cause of injury; therefore, the companies in question were not liable for the injuries: "plaintiff has the affirmative burden of establishing causation between the failure to warn and the injury." According to Manta [1], the focus can now be diverted "from the defendant and its product to the plaintiff and his conduct and knowledge."

However, Harry M. Philo quotes The Metal Forming Subcommittee of the Machine Tool Builders Association's Accident Prevention and Safety Committee as reporting, "The proper use of warning signs can also reduce possible exposure to product liability suits." Philo himself asserts that [2]

> If there is a failure to warn, then in every instance that failure is a cause of the injury or death complained of unless the injured person was attempting suicide, was blind or blindly intoxicated, or was intentionally undertaking an unreasonable risk.

Although he does not cite specific cases, Philo argues that the law should be moving toward holding the defendant liable in instances of failure to warn, but he believes that "neither lawyers, judges, nor the law are up to date when it comes to the law and warnings."

Chris Velotta is a technical writer at NCR Corporation, E&M—Wichita.

This unstable legal climate does little to guide a company in the process of constructing effective safety labels and in avoiding product liability suits. Current research in this area seems to offer the most hope for a standard set of guidelines to follow. For example, Michael Ursic's study of 91 undergraduates showed that a product with a safety label projects a significantly safer image than a product without a safety label: "Therefore, it seems that a safety warning, instead of causing a person to perceive a product as more dangerous, may cause an individual to believe that the product is safer." This finding should help reassure marketing specialists who are worried that putting safety labels on a product will cause consumers to perceive that product as hazardous [3].

Although this development is encouraging, because now safety labels are thought to enhance a product's image, the rest of the problem is yet to be solved. What techniques can we use to ensure the effectiveness of safety labels? In the previously mentioned study, Ursic manipulated three variables—pictograms, signal words, and capital letters—to find out if they affect a consumer's perceptions of a product and also to see if they increased a consumer's recall of a safety label. The results of these manipulations showed that the subjects did not demonstrate any increased perceptions of product safety and effectiveness or any increased recall of the safety label. Unfortunately, these results do not offer any immediate guidelines to follow, but Ursic is quick to point out that a great deal more research needs to be done in this area, such as in the effects of color or in the way a safety label affects the way a consumer uses the product.

The potential contribution that such research could offer to those who design safety labels seems boundless. For example, the American National Standards Institute (ANSI) has already developed a set of guidelines for constructing effective safety labels. In addition, many corporations have developed their own suggestions for designing safety labels. One such organization is Westinghouse Electric Corporation (WEC).

To assist those who must design effective safety labels, this paper discusses "Specifications for Product Safety Signs and Labels," ANSI Draft Z535.4, 1984 [4] because ANSI is widely respected in the consumer products industry and because many organizations follow its standards.

Reprinted from *IEEE Trans. Prof. Comm.,* vol. PC-30, no. 3, pp. 121–126, September 1987.

This discussion is supplemented with material from the WEC *Product Safety Label Handbook: DANGER, WARNING, CAUTION* [5], which corroborates the ANSI standards in many instances. The WEC handbook contains a disclaimer that bears repeating, and it applies to both the quoted material from the handbook and to any other suggestions contained in this article:

> Westinghouse makes no representations or warranties, express or implied, including warranties of merchantability or fitness for purpose as to the accuracy, completeness, or legal sufficiency of the recommendations or information contained herein. Westinghouse assumes no liability arising out of any use of this book.

Also presented are sample safety labels from The Charles Machine Works, Inc. (CMW), because they were developed after much research into current practices and because they exemplify the proper use of ANSI guidelines.

Because constructing effective safety labels is a complex process, I must limit my discussion to a few main stages of the process. The areas covered include what to put in safety labels, how to write them, and where to put them. Other stages that are worth further investigation include the identification of hazards, the specific layout and design of safety labels, and the differences between safety labels in the manual and safety labels on the product itself.

WHAT TO PUT IN SAFETY LABELS

Before actually assembling the ingredients of an effective safety label, the technical writer should first determine what levels of warnings to use (DANGER, WARNING, CAUTION, or NOTICE) and what situations to use them in. These decisions should be easier to make if the technical writer uses the following definitions from the September 1986 revisions [6] to the ANSI Z535.4 Draft, "Specifications for Product Safety Signs and Labels":

- DANGER: indicates an imminently hazardous situation which, if not avoided, will result in death or serious injury. This signal word is to be limited to the most extreme situations.

- WARNING: indicates a potentially hazardous situation which, if not avoided, could result in death or serious injury.

- CAUTION: indicates a hazardous situation which, if not avoided, may result in minor or moderate injury. It may also be used to alert against unsafe practices.*

Based on these definitions, the difference between DAN-

---

* DANGER or WARNING should not be considered for property damage accidents unless personal injury risk appropriate to these levels is also involved. CAUTION (and NOTICE) are permitted for property-damage-only accident hazards and for unsafe practices.

GER and WARNING is in the possibility for occurrence and in the extremity of the hazard. For example, the hazard is *imminent* in the definition of DANGER but is only *potential* in the definition of WARNING. In addition, death or serious injury *will* result in the definition of DANGER but they only *could* result in the definition of WARNING. DANGER's definition also indicates that it should only be used in "the most extreme situations"; however, the definition of WARNING does not carry this restriction. To see this difference, compare the following safety labels developed by CMW:

DANGER: TURNING SAW can kill or cut off arm or leg. STAY AWAY.
WARNING: MOVING PARTS can cut off hand or fingers. DO NOT TOUCH.

Whereas both safety labels warn against serious hazards, the danger from a turning saw is more emphatically threatening and more extreme than the danger from moving parts in general. Therefore, DANGER is the appropriate choice for the turning saw, and WARNING is used correctly for moving parts.

Compared to DANGER and WARNING, CAUTION discusses minor injuries and warns "against unsafe practices." For example, the following safety label from CMW accomplishes both objectives:

CAUTION: FALL POSSIBLE. People may slip or trip and fall from operator's area, causing broken bones. KEEP AREA CLEAN.

The injury being described here is minor compared to a lost life or hand as described in the previous DANGER and WARNING examples. In addition, this CAUTION warns against the unsafe practice of having a messy work area.

Finally, NOTICE is used only with alerts that warn against property damage. NOTICE is inappropriate when any type of potential injury is involved. The following example from CMW shows an appropriate use of a NOTICE:

NOTICE: If engine does not start in three tries, find out what is wrong. Correct the problem. Overheating the starter can damage it.

Using these definitions as a guide, the technical writer can now decide what to put into individual safety labels.

Seven basic elements of an effective safety label have been described by WEC in the *Product Safety Label Handbook: DANGER, WARNING, CAUTION:*

- The signal word
- The hazard alert symbol
- The color
- The symbols and pictographs

- The hazard identification
- The result of ignoring the warning
- The description of avoiding the hazard

Signal Word

The four signal words listed in the handbook are DAN-GER, WARNING, CAUTION, and NOTICE. In their respective safety labels, these words are larger than the rest and are the first words in the label. This word catches the operator's attention and orients him or her to the possibility and extremity of the hazard. By defining this set of words and advising the operator to learn them, the technical writer will be helping the operator to easily recognize the severity of individual hazards. Based on the nature of the hazard, the technical writer should use the signal words consistently. For example, after reading a few safety labels and the definitions of the signal words, operators will quickly become aware that DANGER signals imminent and extreme hazards while CAUTION signals minor injuries and unsafe practices.

Hazard Alert Symbol

Immediately preceding the signal word is the hazard alert symbol, which consists of an exclamation point inside a triangle, ⚠. This symbol should be used with the signal words DANGER, WARNING, and CAUTION but not with NOTICE. The symbol is recognized internationally and is a reminder to be alert. It combines with the signal word to attract the operator's attention to the safety label and the hazard being warned against.

Color

Encompassing both the hazard alert symbol and the signal word is the color field. (In a NOTICE, the color field includes only the signal word because the hazard alert symbol is not used.) Each level of safety label has its own distinct color. According to both the WEC handbook [5] and the ANSI Z535.4 Draft [4], DANGER should have a red background with white lettering, WARNING should have an orange background with black lettering, and CAUTION should have a yellow background with black lettering. The handbook adds that a NOTICE should have a blue background with white lettering. This consistent color-coding increases the operator's ease in distinguishing among the different levels of hazard being warned against in an operator manual.

Symbols and Pictographs

The technical writer should also add an appropriate symbol or pictograph to give the operator an immediate idea of the hazard. The symbol or pictograph should consist of a universally recognizable graphic that displays the potential injury involved with the hazard. Designing an effective set of symbols or pictographs is a complicated process that is best left to experts in symbol recognition. Experts have developed symbols for many common hazards, and technical writers can incorporate these symbols in their own safety labels. One such set of symbols is in the WEC

handbook. ANSI Z535.3, "Specifications for Safety Symbols," also presents guidelines for symbol use.

Hazard Identification

The next task in constructing a safety label is to describe the hazard. This description should be "the first verbal message following the signal word." Bold type and all capital letters can emphasize the type of hazard being warned against [5]. Some examples from CMW safety labels include ROLLOVER POSSIBLE, DEADLY GASES, and HOT PARTS. This segment can be its own sentence (as in ROLLOVER POSSIBLE), or can be part of a sentence (as in "HOT PARTS can cause burns"); but it should always come first in this section of the safety label. Therefore, the technical writer should not say, "Burns can be caused by HOT PARTS," but rather "HOT PARTS can cause burns."

Results of Ignoring the Warning

Bold type and upper- and lower-case letters can make this section easy to read, and it should not be separated from the first segment of the verbal message. It tells the operator what harm a rollover, for example, will cause or what harm deadly gases or hot parts will cause. For example, in a CMW safety label, "DEADLY GASES" is followed by "Breathing exhaust gases can cause sickness or death," which describes the result of ignoring the warning.

Description of How to Avoid the Hazard

A final and important element in a safety label, this segment should be separate from the rest of the verbal message; it can either start a new sentence or come after a skipped line. In addition, this segment can contain either bold or regular type. At CMW, for example, bold type and all capital letters were used:

> DANGER: DEADLY GASES. Breathing exhaust gases can cause sickness or death. BREATHE FRESH AIR.

In this example, "BREATHE FRESH AIR" describes how to avoid the hazard of deadly gases.

HOW TO WRITE SAFETY LABELS

After deciding what ingredients to include in a safety label, the technical writer should decide how to write the verbal message. The writing style should allow the technical writer

- To say as much as possible in as few words as possible
- To say it forcefully and directly
- To say it in a way that is as well-organized as possible

Concise Messages

To say as much as possible in a safety label that has limited available space, the technical writer should first be sure not to waste any words. Only then will there be enough room to effectively communicate the warning. For example, the WEC handbook offers a helpful guideline in this area: avoid using adjectives unless they provide crucial information in the safety label. Most writers are already familiar with this principle; an example is *very,* which is commonly known to carry little meaning and is, therefore, not often used. Another guideline from the handbook describes the proper use of prepositions: do not use strings of words where one preposition will convey the same meaning. For example, *in the event of* does not convey any more meaning than *if*; therefore, *if* should be used instead.

Another way of pruning unnecessary words is to write in a telegraphic style. The WEC handbook suggests usually eliminating the articles *a, an,* and *the* in addition to eliminating pronouns such as *that, this,* and *they* and all forms of the verb *to be.* This technique is illustrated in the following safety label from CMW:

> CAUTION: HOT PARTS can cause burns. DO NOT TOUCH UNTIL COOL.

Four words could be added to this safety label by including the missing article, the missing form of the verb *to be,* and the missing pronouns. However, the following rewrite shows that these insertions do not add any meaning to the safety label.

> CAUTION: *THE* HOT PARTS can cause burns. DO NOT TOUCH *THEM* UNTIL *THEY ARE* COOL.

When the operator needs quick access to the information, as in safety labels, every word should carry meaning. Therefore, the telegraphic style is appropriate for writing safety labels.

Forceful Messages

Not only should every word carry meaning, but the entire message should be forceful. Writing the message in active voice helps ensure that the message comes across forcefully; there is a clear connection between the cause of the hazard and the effect of the hazard. The following example from CMW shows an effective use of active voice:

> CAUTION: BATTERY ACID can cause burns. DO NOT SPILL.

Writing the first sentence of this safety label in passive voice would add two unnecessary words to the message and would reduce the force with which the message comes across to the operator:

> DANGER: Burns can *be* caused *by* BATTERY ACID. DO NOT SPILL.

Using strong auxiliary verbs will supplement the active voice and will help ensure that the operator knows the true nature of the hazard. For example, if a hazard "will" happen as the result of a certain action, the auxiliary *will* should be used as opposed to *may,* and *can* should be used

as opposed to *could* to increase the forcefulness of the message. Finally, technical writers should avoid using contractions because, for example, *don't* is much less forceful than *do not.*

Direct Messages

In addition to being forceful, safety labels should be direct. A direct message starts with appropriate word choice. Nouns should carry precise and easy-to-understand meanings. For example, in "LEARN HOW TO USE ALL CONTROLS" from a CMW manual, *CONTROLS* is the correct choice because it encompasses the entire range of levers, pedals, and switches that an operator should know about. Using jargon can also cloud the issue and diminish the directness of the communication; *machine* is much more direct and easy to understand than *modularmatic trenching unit.*

When telling the operator what action to perform, the technical writer should choose verbs with precisely the intended meaning. For example, "FEEL FOR LEAKS WITH CARDBOARD" could mislead operators into using their hands to *feel* for leaks and could leave operators wondering what to do with the cardboard. The CMW version, "CHECK FOR LEAKS WITH CARDBOARD," eliminates this ambiguity because of correct choice of the verb. In addition, the technical writer should choose the simplest form of the verb: *utilize* and *recognize* do not say anything more than *use* and *know,* and the operator will more readily understand the latter, more direct verbs.

Choosing the most precise modifiers can also increase the directness of a safety label. In an example from CMW, the true nature of the hazard is clear and direct: "Striking *electric* lines may cause death." Rewriting this warning with a less precise modifier shows the ineffectiveness of such ambiguous constructions: "Striking utility lines may cause death." The operator may have no idea that electricity flows through the lines and may think instead that water flows through the lines, which would be a much less severe hazard.

The technical writer should also strive for detailed descriptions of the results of the hazards. Operators are more likely to avoid moving parts if they read the direct warning from CMW that such parts "can cut off hand or fingers" than if they read the indirect rewrite that "MOVING PARTS can cause serious injury."

Organized Messages

One final stage for the technical writer is to organize the safety label. The previously discussed order of the seven elements of a safety label shows a progression from general (signal word) to specific (how to avoid the hazard). The WEC handbook goes on to suggest that using a chronological order will help the operator understand how to avoid the hazard. A safety label from a CMW manual illustrates this concept: "LEARN HOW TO USE ALL

CONTROLS and OPERATE ONLY FROM OPERATOR'S POSITION.'' After reading this line, operators should be aware that they ought to learn how to use the controls before operating the machine.

Another way to ensure effective organization is to pay attention to line breaks and to keep the style consistent from one sentence to the next. A line like "TURNING SAW can cut off arm or leg" should appear all on one line to invest the communication with the proper impact; and "STAY AWAY" should not be broken after "STAY" so that operators do not misunderstand and think that they should stay near the hazard.

Finally, the technical writer should use parallel grammatical constructions throughout the safety label to set up a recognizable pattern for the operator to follow. For example, the following CMW safety label exhibits a consistent use of active voice, imperative mood:

> CAUTION: HOT PARTS can cause burns. DO NOT TOUCH UNTIL COOL.

Changing the second sentence to passive voice, indicative mood, decreases the label's readability and shows inconsistent organization, which is harder for the operator to follow:

> CAUTION: HOT PARTS can cause burns. THEY SHOULD NOT BE TOUCHED UNTIL COOL.

WHERE TO PUT SAFETY LABELS

Typically safety labels appear in two places:

- In the operator manual
- On the product

When placing a safety label in an operator manual, the technical writer should determine when and where a hazard could occur and then include a safety label immediately before the section of the instructions that deals with the hazardous situation. Placing a safety label on the product involves that consideration plus several others, such as where the operator will be located in relation to the safety label and the available space on the product.

Placement of safety labels depends a great deal on the sequence of steps needed to operate the product. To ensure that the operator will see a safety label before encountering a hazard, the technical writer should think chronologically through the operation procedure. If technical writers rely on what they think they know about using the product or on memory to identify potential hazards, "shop blindness" can cause them to gloss over some potentially hazardous situations. Such omissions can be avoided if the technical writer actually performs the operation and takes notes on every step.

After defining the hazards, the technical writer usually should locate a spot close to each hazard and then place a safety label in those places. In this way, the operator will be able to observe the label while using or maintaining the product. However, some instances require that the operator observes the safety label before coming close to the hazard. For example, a safety label warning against breathing exhaust gases should not be placed on an exhaust pipe but on a part of the machine that is visible and at a safe distance from the exhaust gases. Some instances require a pair of labels: one outside the hazardous area and another that functions as a reminder inside the hazardous area—for instance, a hazardous voltage warning on the back of a stereo amplifier and the complementary warning inside the stereo amplifier.

Technical writers should also consider the operator's position in relation to the safety label. Such factors as reading distance and viewing angle can affect whether or not the operator can easily see the safety label in time to avoid the hazard. As noted, the safety label should not require the operator to get too close to the hazard; therefore, the technical writer should determine how large to make a safety label so that the operator can read it at a safe distance. Fortunately, handbooks such as the one produced by WEC offer tables that specify reading distances for different sizes of safety labels. To help provide a proper viewing angle, the technical writer should place the safety label so that the operator can look directly at it and have a perpendicular viewing angle. The farther the viewing angle varies from perpendicular, the less readable the safety label becomes. In fact, the WEC handbook states that "A viewing angle more than 60 percent from the perpendicular will decrease readability."

In the rare instance that the product offers the exact amount of space needed for a safety label of appropriate size, the technical writer can consider this stage of the process completed. However, more often than not the product will offer either too much or too little space. Too much space is not a difficult problem to deal with; it is only important to remember not to try to make the safety label fit the available space. Instead, writers should follow the previously referred to tables, which specify safety label sizes for various reading distances. One exception is the product which is used under inadequate lighting conditions, for which the safety label may need to be larger to increase readability.

If the product offers too little space for a readable and complete safety label, the technical writer should make certain modifications to the safety label. The techniques described so far should help ensure that the safety label contains all the essential information in the minimum space. However, if space is extremely limited, the technical writer should not sacrifice readability for information. If operators cannot read the safety label, they will not receive any information at all. Therefore, when available space requires that some information be left out, the technical writer can create a readable safety label that warns of the hazard and its effects and then refers the operator to the operator manual for further information about the hazard.

CONCLUSION

Although the courts still have not conclusively defined what constitutes an effective safety label, current standards can help guide technical writers in designing safety labels. Organizations such as ANSI and WEC have developed such standards, but further research in this area can provide even more support for constructing effective safety labels. In spite of the fact that safety labels are–of necessity–short, their construction is a complex process. Stages of this process involve including the appropriate informa-

tion, writing in an appropriate style, and placing the safety label in an appropriate location.

REFERENCES

1. Manta, J. G., ''Proximate Causation in Failure to Warn Cases: The Plaintiff's Achilles Heel,'' *For the Defense,* October 1984, p. 11.
2. Philo, H. M., ''New Dimensions in the Tortious Failure to Warn,'' *Association of Trial Lawyers of America,* 1981, p. 17.
3. Ursic, M., ''The Impact of Safety Warnings on Perception and Memory,'' *Human Factors,* 26:3, 1983, p. 680.
4. ''Specifications for Product Safety Signs and Labels,'' *Draft ANSI Z535.4,* August 1984, p. 11.
5. *Product Safety Label Handbook: DANGER, WARNING, CAUTION,* Westinghouse Electric Corporation, Trafford, PA, 1981.
6. The ASNI Z535 Committee voted for these revisions in its meeting of September 16-18, 1986.

Rules for Acronyms

Joseph C. Mancuso

More and more acronyms find their way into conversations and meetings and subsequently into technical manuals and reports, and it is important that students of technical communication know how to use them. A sub-category of abbreviations, they differ from other abbreviations in that they can be pronounced (RAM for random access memory and ROM for read only memory, for example). Acronyms do not contain periods, and all their letters are capitalized [1].

Writers will find inadequate rules for acronyms even in *The Chicago Manual of Style*, the *U.S. Government Printing Office Style Manual*, *Words Into Type*, and Wrathall's *Computer Acronyms, Abbreviations, Etc.* Therefore, writers and editors should supplement these references with their own common sense and the on-the-job practices of editors.

I suggest the following rules and practices for acronyms:

1. Provide a "List of Abbreviations and Acronyms" in the front matter of technical reports and manuals [2].
   —Alphabetize the "List" [3].
   —Follow the principle that internal letters in the *definitions* of the abbreviations and acronyms have no impact on the alphabetization of the "List."

2. Define acronyms the first time they appear in each chapter of a technical manuscript.
   —This practice may be modified in a manuscript with very short chapters.
   —Do not capitalize definitions of acronyms unless those definitions are accepted proper nouns. Definitions are not capitalized simply because the letters of the acronym are capitalized.

3. Form plurals of acronyms only by adding "s" [4].

4. Do not use acronyms in the headings of technical reports.

5. Do not begin a sentence with an acronym [5].

PROVIDE A LIST

Depending on the technical awareness of the audience, a longer or shorter "List of Abbreviations and Acronyms" may be needed in the front matter of a report. Even with a highly technical audience, it is still a good idea to use a "List of Abbreviations and Acronyms" as a guide to new or shortened groups of letters used as "words." One sees too few lists of abbreviations and acronyms in technical manuals and reports, and writers and editors should provide them to create more audience-based documents.

Writers or editors generate a "List of Abbreviations and Acronyms" at the time of writing or by working through the text from the beginning and recording each new acronym. When the list is complete, it should be alphabetized. The following exemplifies the form for a "List of Abbreviations and Acronyms":

LIST OF ABBREVIATIONS AND ACRONYMS

AC	- alternating current
ACU	- acquistion control unit
AEDC	- after effective date of the contract
ASOS	- Automated Surface Observing System
BC/PS	- battery charger power supply
CDR	- critical design review
DCP	- data collection package
DOC	- Department of Commerce
DOD	- Department of Defense
DOT	- Department of Transportation
FAA	- Federal Aviation Administration
FAT	- factory acceptance test
GFE	- government-furnished equipment
I/O	- input/output
LRU	- line replacement unit
LWI	- laser weather identifier
MTBF	- mean time between failure
NDBC	- NOAA Data Buoy Center
NOAA	- National Oceanic and Atmospheric Administration
NSTL	- National Space Technology Laboratories
NWS	- National Weather Service
OID	- operator interface device
PDR	- preliminary design review

The words following each abbreviation and acronym are definitions. Note that writers and editors usually alphabetize a "List of Abbreviations and Acronyms" with an initial-letter approach; the internal letters in the definitions have no impact on alphabetization of the acronym. Hence AC (*a*lternating current) comes before ACU (a*c*quisition control unit).

DEFINE ACRONYMS THE FIRST TIME THEY APPEAR

Technical communicators should define acronyms the first time they appear in each chapter—not only in the first chapter. This practice reminds readers of the meaning of an acronym each time they first see it within a chapter. In a very short manuscript with one-to-three page chapters, writers and editors may modify this practice, supplying a definition only for the first use of the acronym. Once defined, the acronym, and not the definition, should be used throughout.

An example of the correct form for defining the initial use of an acronym in the text of a chapter follows:

> This specification describes requirements for a local data acquisition package (DAP) for use in the Automated Surface Observing System (ASOS).

Notice that the first definition in the above example is not capitalized; the second is because it is the title of a project. Editors should guard against the propensity of some authors to capitalize words in definitions which are not proper nouns or proper adjectives. Also keep in mind that definitions of acronyms are not to be cap-

Reprinted with permission from *Technical Communication,* published by the Society for Technical Communication. May 1987, vol. 34, no. 2, pp. 124–125.

italized merely because all the letters of acronyms are capitalized.

FORM PLURALS ONLY BY ADDING "S"

Writers and editors may pluralize acronyms in the text without using an apostrophe—

> The engineer configured the hardware so that two DAPs (data acquisition packages) . . .

—but the "List of Abbreviations and Acronyms" in the front matter should contain the acronym without an "s" added. Writers and editors should not use the single apostrophe ("'s") or plural apostrophe ("s'") with acronyms to avoid confusion with another symbol which might be part of the acronym.

DO NOT USE ACRONYMS IN THE HEADINGS OF TECHNICAL REPORTS

Writers and editors should not use acronyms in the headings of technical reports. Each acronym must be written out in each heading or subheading, even if the acronym has been defined in the text preceding the heading. The only exception would be, for example, an acronym containing a computer command known by its abbreviated form: "The Serial I/O Command Activates the Printer." Acronyms may be used in vertical lists. Vertical lists, although they have the appearance of headings, are still considered to be text.

DO NOT BEGIN A SENTENCE WITH AN ACRONYM

Although all technical communicators should know this rule, found in *Words Into Type*, it makes little sense to me. No confusion should result from the use of an acronym at the beginning of a sentence, unless the abbreviation is the single letter "A," or some other letter which might be mistaken for a word. Beginning a sentence "RAM means random access memory in . . ." should not confuse a reader, nor should "UNICEF today accepted a plan to . . . ." Therefore, I would adjust this rule to fit the textual situation.

Given the increasing use of acronyms in technical manuals and reports and the increasing numbers of less technical and non-technical readers of these reports, writers and editors should do their utmost to employ, fully and consistently, rules and widely accepted practices for acronyms.

REFERENCES

1. *The Chicago Manual of Style*, 13th ed. (Chicago, and London: The University of Chicago Press, 1982), p. 379 (Rule 14.15). See also rules 7.152 and 7.153 on pp. 228-229.
2. *Manual of Style*, p. 20 (Rule 1.45).
3. *Manual of Style*, p. 542 (Rule 18.97).
4. *Manual of Style*, p. 160 (Rule 6.9).
5. *Words Into Type*, 3rd ed. (Englewood Cliffs, NJ: Prentice-Hall, Inc., 1974), p. 101.

Dr. Joseph C. Mancuso is Director of Technical Writing in the Department of English at North Texas State University.

Editing Math: What to Do With the Symbols

BARRY W. BURTON

Abstract—When faced with mathematical material, editors with limited technical training often address only superficial concerns such as format and punctuation. A few simple guidelines, however, can help us do a more complete job. We first delete superfluous symbols, then make certain that the remaining symbols are defined properly, redefined where necessary, and used according to the rules of grammar.

WHAT often makes many ordinarily courageous, capable technical editors trained in the liberal arts cower in humility and submissiveness? Math. Technical editors are faced daily with reports, articles, presentations, and other material packed with mathematical symbols and expressions. And many editors have a distressingly poor background in mathematics. This deficiency often leads to a feeling of inadequacy that may prevent the editor from delving too deeply into the mathematical portions of the document for fear of making a mistake that could mislead the reader or hold the writer up to ridicule. As a result, some may concentrate only on formatting the displayed expressions to conform to the organization's style guide (if one even exists). Some may address nothing more than punctuation. In either case, the job may fall short of professional standards.

But how can we do a more professional job without going back to school and getting a degree in mathematics? Quite easily, in fact. Many of the issues raised by the use of math symbols allow themselves to be addressed by editors with minimal mathematical background. Here I will discuss what to do with mathematical symbols. Once problems with math symbols have been cleaned up, many of the other math problems become more manageable. The meaning of the math may sometimes be beyond us, but the editing can be done proficiently and professionally if we approach it systematically.

The discussion below briefly describes four simple steps in addressing the symbols in a mathematical document: eliminating superfluous symbols, defining the remaining symbols, redefining the symbols where necessary, and using the symbols in parallel constructions. Formatting of displayed expressions, punctuation, and fitting expressions grammatically into sentences are not discussed. These concerns are covered by definite and easily applied rules found in many style manuals, though those rules differ from source to source. Some very popular style guides, such as *The Chicago Manual of Style* [1] and *Mathematics into Type* [2], adequately deal with these subjects.

All the examples used here are real, taken from scientific journals. They were remarkably easy to find, evidence that these easy-to-fix problems are widespread.

DELETE UNNECESSARY SYMBOLS

In most organizations, policy dictates that the author is responsible for the accuracy and relevance of the information—math included—in his or her document. It is, however, up to the editors to enforce this policy. They must ensure that the mathematical symbols and expressions used are relevant to the discussion and help achieve the writer's goals. It makes no sense to spend time editing a document to make it readable if some of the material should never have been included to begin with.

An easy way for the editor to begin is by deleting unnecessary symbols. Many sources, including mathematicians themselves, agree that writers of mathematical material should use symbols only when necessary [3], [4]. Writers are often so wrapped up in the formal language of mathematics that they get carried away and throw in symbols they don't need. Superfluous symbols cluttering the page can make the text forbidding to nonspecialists, confuse other workers in the writer's field, and distract the reader from the document's important points.

Mathematical symbols are just names of things or abbreviations of those names. For instance, a writer may use the abbreviation $s_s(f)$ to represent the speed of sound in a fluid instead of writing it out longhand every time the term is needed. This shorthand can be a useful space saver when the term is used more than a few times. But if the term is used only once, inclusion of the symbol is gratuitous and, without a verbal definition in context, can be confusing. A symbol shouldn't be used at all unless it appears often enough to make it a worthwhile shortcut for the reader to learn it. Introducing gratuitous symbols can even be misleading: the extra emphasis placed on an idea when a symbol is assigned to it suggests that the thing being symbolized is more important that it may really be. Such misplaced emphasis can distract the reader from the writer's point.

One-time symbols are easy to find. By listing all the math symbols during the first editing pass, then checking each one off when it appears a second time, the editor can

- spot gratuitous symbols,
- keep track of the definitions of symbols, and
- find symbols that aren't defined at all.

An editor doesn't need a complete understanding of mathematical material to identify unnecessary symbols. This dictionary definition exemplifies a typical problem:

A topological space is a set X along with a collection T of

Barry W. Burton is with Burton Literary Services, Los Alamos, NM 87544.
IEEE Log Number 9035636.

Reprinted from *IEEE Trans. Prof. Comm.*, vol. 33, no. 2, pp. 62–65, June 1990.

distinguished subsets of X called *open* sets which satisfies: 1) each member of X is contained in some open set, 2) the intersection of two open sets is open, and 3) the union of a subcollection of open sets is open.

Here the first symbol, X, represents a particular set. That symbol appears twice more in the passage, distinguishing the set X from all other sets and avoiding confusion in the discussion. Just as each person has a name that sets him or her apart from everyone else, so this set is distinguished by its name, X. The reader doesn't care, however, what a particular set is called if the name isn't needed to show that it's unique. As far as the editor should be concerned, the important thing about X as a symbol is that it's used more than once. The introduction of X as a symbol in this example is fully justified because it's needed more than once and serves as a useful shorthand.

On the other hand, in the same paragraph the symbol T identifies a collection of special subsets called open sets. An editor need not even know what a distinguished subset or an open set is; he or she should note only that the symbol T is used once to define something and is never seen again. Readers shouldn't be asked to expend any more energy than necessary, so they shouldn't be required to learn the definition of T if that knowledge isn't going to help them understand the material.

One-time symbols such as T in the example above can usually be deleted without any loss of meaning and without confusion. At other times, the grammatical structure of the sentence requires that a phrase or even the whole sentence be rewritten. In either case, the correction should be fairly simple.

Just because a symbol appears only once doesn't necessarily mean it's superfluous. This expression is a good example: "Because the mesh is fine enough to approximate the derivative, $\partial u / \partial x$, both elements described above give about the same shear stress." A partial derivative is usually identified by a "curly d," ∂, as in $\partial u / \partial x$, but an editor who is unfamiliar with the notation may want to delete the $\partial u / \partial x$ in this case as superfluous. Discussion with the author, however, just might reveal that the $\partial u / \partial x$ is needed to keep the reader from confusing this particular derivative from some other one, perhaps $\partial w / \partial x$, or from the corresponding regular derivative, given by du / dx. Such misunderstandings often result from the editor's imperfect mathematical background. No rule is foolproof, but one-time symbols almost always prove to be gratuitous. If the $\partial u / \partial x$ in this example serves only to indicate that the derivative is a partial one, the sentence could be rewritten to make that point without the symbol being used.

The most common type of unnecessary symbol usually appears in a construction similar to the following:

In order to study the properties of multilayer shielding, we put a source in an iron cylinder and placed it at a distance $d = 40$ cm from the wall in a cubic tank filled with pure water.

Scientific writing often contains this kind of tautology, which might be called a "mathematical stutter." The word is first spelled out, then given as a symbol in a mathematical relation. In this example, the author assigns the symbol d to distance,

giving "... a distance $d = 40$ cm ..." An editor who is alert to the mathematical stutter can easily delete the symbol and recast the phrase as "... and placed it 40 cm from the wall ... ."

DEFINE THE SYMBOLS

Once the confusion of superfluous symbols has been eliminated, the editor must ensure that the remaining symbols are defined properly. The following excerpt illustrates how unidentified symbols can render an equation ineffective:

The frictional pressure difference ΔP_f in the return tube is

$$\Delta P_f = \rho v^2 F \frac{L}{2D} .$$

This passage tells us that the frictional pressure difference is denoted by ΔP_f and that the writer presumably knows how to calculate it. But we can't calculate it, and neither can the readers, at least not with only the information at hand. The writer's thoughts would be clearer were she to explain the meanings of the symbols in the formula, in this case, that ρ is the fluid density, v is the liquid velocity in the tube, L/D is the length-to-diameter ratio of the tube, and F is a friction factor. By defining the variables, the writer would not only demonstrate that she knows what she is doing but also make the mathematics more accessible to the readers.

Sometimes a symbol almost always represents the same quantity, such as m for mass, v for velocity, or ρ for density. But just because a symbol has a standard definition doesn't mean that the writer is using it in that sense. Even when all the symbols in a document are universally known, the author still must identify them, partly because not all readers know the standard definitions of all the symbols they might see and partly because they can never be sure whether the writer means for the symbols to represent the standard quantities or some other quantities unique to that document. For instance, most technical editors would recognize the equation $W = \int F \, ds$ as the definition of physical work, W, where F is the force and ds is the infinitesimal distance over which the force is applied. A writer might think it unnecessary to explain such a self-evident equation, but some readers need to be reminded and others need to be reassured that this is really the familiar work equation and not something else that just happens to look like it. Symbols, no matter how familiar, should always be defined.

REDEFINE THE SYMBOLS

We have all seen documents filled with such a proliferation of symbols that we can't keep them straight. It's not just us. The readers, even other mathematicians, have the same problem. Many symbols have more than one meaning, and writers often attach entirely new definitions to familiar symbols. In some long or complicated documents, a single symbol can represent different quantities in different contexts.

Sometimes a symbol introduced near the beginning of a document doesn't reappear until many pages later. Just because a symbol is defined early on doesn't guarantee that readers will remember that definition when they finally see it again. Also, many readers jump from one section to another, skipping the material in between, looking for specific information. Since they don't read the entire document in the usual front-to-back way, they miss many symbol definitions. The editor must help the readers recognize symbols wherever they find them.

A glossary of symbols can to some extent mitigate the confusion. Glossary definitions, however, are necessarily short (and therefore incomplete) and, relegated to a separate part of the document, taken out of context. Besides, if a given symbol has several meanings, a glossary can't tell the readers which meaning is used on which page.

An editor can solve some of these problems by using what might be called "strategic redundancy," that is, by redefining symbols at important locations throughout the document. A location can be "important" when

- the symbol is first used,
- the symbol is reintroduced after a long intervening discussion, or
- the symbol is redefined to represent a new quantity.

It's not easy to say just how large a gap between appearances of a symbol is tolerable. It depends partly on how well known the symbol is. Some symbols, those having standard definitions, for instance, can be defined only on their first appearance with no trouble later on; others may need to be redefined several times. Consultation with the author (often several consultations) can provide answers to many editorial questions about usage of certain symbols in a given scientific field. As a rule of thumb, though, if the editor, as advocate for the reader, needs to be reminded what a symbol means, it might be time to redefine it.

We seek not merely to redefine a symbol but to do it so subtly that the reader doesn't realize its being redefined. For example, without being too obvious one can write "the C^1 norm $\|f\|$ is small" or "both the matrix T and its Hermitian conjugate T^* are zero." In the first case, we managed to sneak in a redefinition of a previously defined $\|f\|$; in the second case, we inserted a reminder that the universal notation for the Hermitian conjugate of a matrix uses an asterisk. By applying a little thought and attention to consistency and context in which the symbols appear—just as one does with verbal phrases—an editor can redefine most symbols without attracting much attention.

USE THE SYMBOLS IN PARALLEL

The versatility of modern mathematics can be attributed to its symbolic nature. That same symbolism, however, readily leads to abuses of the underlying grammar, and without proper grammar the math can become incomprehensible. Symbolic notation requires parallelism for the same reason that verbal phrases do: the underlying grammatical structure should support the overt meaning.

Some common symbols can be one of several parts of speech depending on how they fit into a sentence. Problems with the grammar of symbols occur most often when the writer uses a symbol twice in the same sentence without regard to changes in its position (and hence its part of speech) [4]. The difference is usually the presence or absence of the verb "to be." For example, the symbol $=$ is pronounced either "equal

to" or "is equal to," the symbol $>$ is pronounced either "greater than" or "is greater than," and the symbol $\in$ is pronounced either "in" or "is in." Just which interpretation is read into a symbol depends on its position in the sentence.

This phrase exemplifies the problem: "For $x > y$, Farber showed that $u > v$." When the symbols are spelled out, this sentence reads, "For x greater than y, Farber showed that u is greater than v." Here we pronounce the symbol $>$ as both an adjective and a verb in the same sentence. The ideal version of this sentence would use the same part of speech in both places; that is, the verbal definitions of the symbol would be parallel. Probably the simplest way to make the definitions parallel is to recast the sentence in an if-then structure. The writer could just as easily have written "according to Farber, if $x > y$, then $u > v$," which is parallel with "is greater than" used in both places. The if-then construction always leads to parallel definitions of symbols.

Another way to tackle the problem is simply to avoid it by using the symbol only once. We can write out the first clause: "For x greater than y, Farber showed that $u > v$." This trick solves the problem, but it also raises the question of why we don't just write out the entire sentence, not using any symbols at all. Avoiding symbols altogether is often an acceptable tactic in writing for the general public, but most readers of scientific material need a technical explanation, one that calls for symbols. The editor must therefore balance the precision of verbal descriptions with the conciseness of symbolic notation. The if-then construction may help us out of this dilemma, but its repeated use results in very tedious reading. Occasionally replacing one symbol in a pair with its verbal counterpart can introduce a little variety and keep the text from being too repetitive.

Correcting nonparallel constructions involving symbols implies that the editor knows how to pronounce those symbols. Learning the symbols is a simple matter of consulting reference works. Mathematical and technical dictionaries [5]–[7] and even some writing style guides [1], [8] list the most common symbols and their definitions. Unfortunately, most sources assume that the reader is already familiar with the symbols' meanings and offer only the briefest of definitions. A few, on the other hand, go into more detail. The Chemical Rubber Company's *Standard Mathematical Tables* [9] offers complete definitions of almost all the symbols an editor is ever likely to run into, although it assumes some mathematical background. Two reference works that explain a few of the most common symbols are *The Technical Editor's and Secretary's Desk Guide* [10] and *The McGraw-Hill Style Manual* [11]. Schenkman's *The Typing of Mathematics* [12] discusses in considerable detail the meanings and uses of the most common mathematical symbols. A review of these works can be time well spent for a technical editor who is unsure of his or her grasp of mathematical notation.

SUMMARY

The lack of a rigorous mathematical background often leads technical editors to address only superficial issues such as formatting and punctuation when editing documents with many mathematical symbols embedded in the text. This practice can result in the publication of poorly edited documents. With some effort, however, editors can edit more effectively, tackling complex and pertinent issues without fear of making some apocalyptic mistake. Editors can attack the problems caused by poor use of symbols first by deleting unnecessary symbols, then by ensuring that the rest of the symbols are properly defined, redefined where necessary, and used according to the rules of grammar. By following these four easy steps, editors can clarify and focus the discussion of even a document that is replete with mathematical symbolism.

ACKNOWLEDGMENTS

I wish to thank M. L. DeLanoy and L. C. McFarland, Los Alamos National Laboratory, for their constructive review of an early draft of this manuscript.

REFERENCES

[1] *The Chicago Manual of Style*, thirteenth ed. Chicago, IL: The University of Chicago Press, 1982.

[2] E. Swanson, *Mathematics into Type*. Providence, Rhode Island: American Mathematical Society, 1971.

[3] E. J. Podell, "Mathematics must be effective in technical communication," *IEEE Trans. Prof. Commun.*, PC-27, no. 2, 1984, pp. 97–100.

[4] N. E. Steenrod, P. R. Halmos, M. M. Schiffer, and J. A. Dieudonné, *How to Write Mathematics*. Providence, Rhode Island: American Mathematical Society, 1973.

[5] *James & James Mathematical Dictionary*. New York: Van Nostrand Reinhold Company, 1968.

[6] W. Karush, *Webster's New World Dictionary of Mathematics*. New York: Webster's New World, 1989.

[7] *McGraw-Hill Dictionary of Scientific and Technical Terms*, S. Parker, Ed., third ed. New York: McGraw-Hill Book Company, 1984.

[8] M. E. Skillin and R. M. Gay, *Words Into Type*, third ed. Englewood Cliffs, NJ: Prentice-Hall Inc., 1974.

[9] *Standard Mathematical Tables*, S. Selby, Ed., twentieth ed. Cleveland, OH: The Chemical Rubber Company, 1972.

[10] G. Freedman and D. A. Freedman, *The Technical Editor's and Secretary's Desk Guide*. New York: McGraw-Hill Book Company, 1985.

[11] *The McGraw-Hill Style Manual: A Concise Guide for Writers and Editors*, M. Longyear, Ed. New York: McGraw-Hill Book Company, 1983.

[12] R. Schenkman, *The Typing of Mathematics*. Santa Monica, CA: Repro Handbooks, 1978.

Part 3
Construction and Organization
Putting Some Documents Together

THIS SECTION focuses on how certain kinds of writing are organized and presented. (Manuals and proposals are covered in separate sections.) The first two articles discuss how you can construct an introduction to a technical report and organize reports in general. Then trip reports, specifications, job-hunting documents, recommendations, newsletters, and persuasive papers are considered. The section ends with advice on turning a technical paper into a published article.

INTRODUCTIONS THAT WORK

Emmanuel Papadakis takes a systematic approach to writing an effective introduction, providing a five-point method for handling what may be your first obstacle in writing any report. Many writing texts suggest you should leave your introduction and conclusion until you have written the main content of your report, and there are good reasons for this. With equal justification, however, Papadakis claims that the introduction is the first step to take and shows how a carefully constructed introduction can ensure an organized and effective paper. He provides four key questions to ask when writing an introduction and gives two examples of introductions written using his approach.

ORGANIZING YOUR REPORT

Since the second article deals with the structure of technical reports in general, it will help you prepare a variety of reports. Gael Ulrich's advice is universal, although you may want to modify parts of the author's suggested outline to suit specific documents. The emphasis each part of a report should have is portrayed graphically in a table, and the comments on report length and style are to the point. A final comment on foggy writing is also worth noting, for, as Ulrich observes, your fine engineering work may go unnoticed if it is not communicated effectively.

TRIP REPORT TIPS

Marcia Petty's article on trip reports displays how she actually wrote one after attending a technical communications conference. Her method of emphasizing key ideas by using "hanging heads" in a carefully formatted document could be emulated in any trip report—and in other kinds of reports as

well. What your audience needs to know about your trip is most important of course, but Petty doesn't neglect a secondary purpose of some trip reports: persuading the boss that similar trips should follow.

SPECIFICATIONS THAT CLARIFY

Even if you never write specifications, chances are you will have to read them sometimes. Few documents demand more waterproof language and coherent organization. Tim Whalen points out that written specifications must provide the accurate technical description required of legal contracts. He categorizes specifications into four types and shows how each type has its own characteristics. Both text and graphics in this article provide you with aids to writing specifications that are free from obscure language or ambiguity.

GETTING A JOB

Ron Blicq offers a compendium of what you should know about the documents needed to get a job. The most obvious are the resume and letter of application, and these are discussed and illustrated fully. Blicq also considers other aspects of the job hunt, however, such as application forms and the dynamics of the job interview. Audience analysis is as important in writing to get a job as it is in other kinds of technical writing, a point Alan Wilcox also emphasizes in the article following Blicq's. A letter of recommendation needs a clear focus, and you should identify who your reader will be before you begin to write. The "general-purpose outline" Wilcox offers, together with his two examples of recommendation letters, will be of considerable help if you are asked to write such letters.

SPREADING THE NEWS(LETTER)

Newsletters are becoming ever more popular, and although Janet Potvin focuses on newsletters for professional societies, her advice will help you produce any kind of newsletter, including in-house letters for employees. Her lists and table make her information particularly accessible as she outlines eight steps in publishing a newsletter. A concluding comparison between engineering design and newsletter editing adds a useful insight into the process of producing this kind of document.

PERSUADING YOUR READER

Is technical writing only a matter of stating facts? J. W. Gilsdorf points out that, although technical writing's purpose is usually to convey information, you may also need to write persuasively at times. In fact, you will need an element of persuasion, ranging from slight to very strong, in a lot of technical writing you do. Gilsdorf analyzes the characteristics of persuasive writing, how it works, when it can be used, and finally considers the ethics of being persuasive in technical documents.

PUBLISHING FOR GROWTH

The last article shows why it is worth transforming some of your technical reports into published articles. Richard Manley, Judith Graham, and Ralph Baxter write specifically for engineers and scientists who may not yet have considered the advantages to their company and to themselves of publishing technical articles. The authors consider the difficulties involved, including the question of proprietary information, and then provide steps you can take to organize papers for submission. Finding out what editors want and what kinds of articles different journals look for are also important steps in the process.

Why and What for (Four): The Basis for Writing a Good Introduction

by Emmanuel P. Papadakis,*
Research Coordinator, *Materials Evaluation*

INTRODUCTION

The systematic writing of the introduction of a paper is important because the introduction outlines several important aspects of the paper and because the writing process focuses the thoughts of the author upon the logical organization of his presentation. Most scientists and engineers have received little formal training in writing and write introductions by a "seat-of-the-pants" method that involves mostly intuitive or experiential feelings as to the proper contents and sequence, causing some items to be left out or put out of order. While there is no hard-and-fast rule for writing good introductions, the author has discovered a sequence of five topics that will be helpful to most authors of technical papers. The author has attempted to describe in a systematic fashion the requirements for a good introduction, to list them in a memorable way, and to illustrate their use. The present short note describes the five-point approach and gives some examples of its applications.

BACKGROUND

"Well begun is half done," reads the old English proverb. This adage holds true in the writing of technical papers. Once the scientist or engineer has completed his technical work and sits down to face a pad of paper or a word processor, his work begins in earnest. He has plenty to write about. He has an urge to tell the whole world. Yet how to begin? He ponders, sips coffee, squirms in his chair, smokes nervously, and ponders more. Technology is his forte, not literature. It was rough in college; it is even worse now.

Writing a good introduction is the first step, the "well begun" part of the proverb. Once this is accomplished, the paper does *seem* half done. There may be reams of writing under other sections, such as theory, experiment, results, and discussion, but the introduction is the initial obstacle to overcome if the rest of the paper is to fall into place.

The introduction ought to inform the audience about the author's effort. Certainly, the work was worth doing, wasn't it? Why should the author want to tell me (the audience) about it? Why should I want to read about it? Even if he and I have a personal interest in his subject, why is it worth printing in the archival literature? What is the reader to look for as he reads onward (or decides not to)? The introduction begins to answer these questions, serves to orient the reader, and attempts to convince him to read further. It is the salesman's "foot in the door" translated into literary terms. When written effectively, it also organizes the thoughts of the author so he can write more effectively and complete his task more quickly.

THEORY

To be effective, the introduction should answer the questions "Why and What For (Four)?" Expanded, these questions are:

- *Why* is the topic of interest?
- *What* (1) is the background on the previous solutions, if any?
- *What* (2) is the background on potential solutions?
- *What* (3) was attempted in the present effort (research project)?
- *What* (4) will be presented in this paper?

This mnemonic format really works. The author formalized it by analyzing the introduction to a manuscript written by another ASNT member. That introduction was so logical and complete that it seemed worth summarizing for wider use. The resulting summary was the mnemonic, "Why and What For (Four)," cited above. After twenty years of writing introductions that were incomplete, out of order, verbose, and probably less effectual than they should have been, the author has adopted the above format and finds it very useful.

EXAMPLES

The following are very short examples of introductions written by using the "Why and What For (Four)" method for hypothetical technical papers. Any correspondence between these and the output of real people is purely coincidental. However, by using exaggerated examples, the principles are illustrated.

Ultrasonic Transducers with No Beam-Spreading, by P. Z. T. Backing

Why is the topic of interest? Ultrasonic transducers with no beam-spreading and constant amplitude and phase profiles within the beam would be useful in studies of flaw detection and materials properties. Quantitative nondestructive evaluation could be done expeditiously because the wave impinging upon a flaw would be exactly known and very simple in form. Ultrasonic attenuation and velocity measurements could be made in all sizes of specimens without recourse to beam-spreading (diffraction) corrections and without interference from sidewall boundaries.

What is the background on the previous solutions, if any? Up until the present time, available ultrasonic transducers have produced beam spreading of both the central lobe and the geometrical side lobes of energy. The

*Manufacturing Processes Laboratory, Ford Motor Company, 24500 Glendale Ave., Detroit, MI 48239

most advanced laboratory techniques to date have produced a Gaussian central lobe that spreads without side lobes. None of the present transducers can produce constant amplitude and phase fronts within the beam.

What is the background on potential solutions? With the advent of microelectronics, the possibility has arisen to build a phased array transducer on a piezoelectric macrochip using a three-dimensional, multi-layer printed circuit for a backing. A small handheld transducer with a 1000×1000 array of radiators and 10 circuit elements per array element should be capable of producing the desired phase and amplitude effects.

What was attempted in the present effort (research project)? In the present study, the theory was worked out for the proposed transducer and also for a 10×10 array pilot model. This model was built and tested experimentally. The pilot model was evaluated by several electronic and optical methods. Three critical experiments were then performed on quantitative flaw evaluation, ultrasonic attenuation, and ultrasonic velocity, respectively.

What will be presented in this paper? In this paper, the theory will be presented for both arrays. Construction details of the pilot model will be given. The most relevant results of its evaluation will be shown, and projections will be made concerning the expected improved performance of the 1000×1000 array. The results of the three experiments will also be enumerated, and comparisons will be made with results found using presently available transducers. Some fundamental limitations will be pointed out.

A New Method for Transporting Potable Water, by Aquus Transductus IV

Why is the topic of interest? Potable water is in short supply in several cities while it is abundant in the adjacent provinces, particularly in the mountains not far distant. It would be advantageous to transport the available water from the locus of supply to the locus of need without the continual burden on human and animal transport systems presently employed.

What is the background on the previous solutions, if any? Previous facilities such as wells and cisterns are proving inadequate for the urban population. Ox carts with large barrels are an interim measure, but they are of limited utility because they overburden the roads and cannot be relied upon in seasons when the oxen are needed for farming. Terra cotta pipe, used first in Crete by the Minoans, is usable only for short distances. It cannot withstand the hydrostatic pressure in a valley when the line of flow is down one hill and up another. Previous proposals for elevated aqueducts have been rejected as infeasible because of the short span bridged by a single lintel stone. See, for instance, the width of the gateway of the Temple of Agamemnon and the number of columns needed in the Parthenon.

What is the background on potential solutions? The proposed solution is an improved elevated aqueduct based on a new invention—the "arch." The "arch" is a structure of fitted stones describing a circle in the vertical plane. It can span a gap at least one order of magnitude longer than is possible with the largest practical lintel stone. Masonry structures, which can support great loads, can be built on top of the arch. The load-bearing capability of an arch is derived from the exact angles of the faces of the stones within the circle. The resolved forces from the load above and including the arch are thus transmitted perpendicularly to the faces of the stones, resulting in a stable structure because there are no shearing forces on the joints and essentially no flexural forces on any stone, the nemesis of a lintel. The arches can be built atop columns, so the gentle slope of an aqueduct can be maintained above variable terrain with only I/X as many columns as needed previously. This should result in a cost saving, making the new design feasible.

What was attempted in the present effort (research project)? In the present effort, an experimental aqueduct, XVIII stadia long, was built from the south branch of the North River to the eastern cistern of Westium. The cistern was enlarged, and control gates were constructed. The facilities were manned by the MCMXVI Engineers. This having been done, the aqueduct was operated for XIV months.

What will be presented in this paper? This paper will give the theory of the "arch," the method of calculating the angles of the faces of the stones including the "keystone," and the details of the masonry construction. Also, flow data for the rainy season and the dry season will be presented, demonstrating successful operation.

CONCLUSION

The "Why and What For (Four)" method of writing introductions to technical papers can be applied to most subjects. The reader will note that the first paragraph of this article contains five sentences that apply the five points of the method to the present subject—writing introductions. The two tongue-in-cheek introductions shown are more extensive examples. If authors choose to use the "Why and What For (Four)" method, they and their readers will reap the benefits of clarity, conciseness, simplicity, and completeness.

Write a Good Technical Report

GAEL D. ULRICH

Abstract—A *good* technical report can have an important effect on a wide range of people. Here are some techniques to help you prepare, choose a suitable structure, provide the right amount of information in the right places, and make your points with clarity. An informal style—using "I" and "we," for example—is acceptable for technical reports and publications. To improve your writing, read good writing by others and invite criticism of your own; practice is important.

IN the beginning, the story goes, God, after creating humankind, was defining the professions. "Anticipating that squabbles would ultimately develop between chemists and chemical engineers, He decided to settle the issue once and for all, dictating to His typist, 'All a chemical engineer does is *right*.' Unfortunately the typist misspelled the last word" [1]. At times, many of us might agree that all a chemical engineer does is *write*. Some feel we don't even do that very well and that we're getting worse. In a recent survey of educators and industrialists, for instance, some respondents complained that language skills among chemical engineering graduates had deteriorated severely in recent years. Others, according to the reporter, felt simply that the skills were not better than before—abominable [2].

I have heard some managers in industry claim that communication skills are more important than technical competence. I do not agree. Communication would be unnecessary if there were no technical result to report. (It does, indeed, require exceptional writing or speaking skill to camouflage an inept or incomplete engineering job.) But I do agree with a variant of the managers' statement: "Many exceptional engineering jobs go unappreciated because of poor writing or speaking." With this in mind, let us consider the elements of effective writing.

PHILOSOPHY OF TECHNICAL REPORTING

Unlike politicians, engineers should write with the hope that readers will find their errors. It is much less embarrassing and painful for an engineering mistake to be found in print before it appears in fact. Thus, a technical report should be designed with clarity as the major goal.

Basic honesty is a key ingredient of clear writing. If there is no concrete result or recommendation, say so. Perhaps your most important contribution will be to expose a question or

a mistake. Such honesty may not always pay off immediately, but a reputation for integrity is worth the wait. Reports intended to reveal rather than obscure will be better understood by others and, when deserving, will be defended by them.

MECHANICS OF REPORT WRITING

An outline does wonders to initiate the writing process. Professional or experienced writers often outline their work mentally, not formally. However, judging from the indictment in the first paragraph of this chapter, you should prepare a written outline if you are a student or an engineer. As an example, a skeleton outline of this chapter is shown in the box. (Of course, the real outline is scribbled on three sheets of paper with numerous insertions and marginal notations.) As you prepare your outline, think about the audience. Van Ness and Abbott [3] caution that readers of most technical reports

1. Are busy or at least believe so.
2. Have a background similar to yours but know much less about the project in question.

Other reader characteristics may prevail under various circumstances. In fact, the abstract and summary of a report are often designed for administrators and business people with nonengineering backgrounds.

"Joe, we need your report. Is it about ready?"

Extracted and reprinted with permission from chapter 9, "Report Preparation," of the author's new text *A Guide to Process Design and Economics for Chemical Engineers;* copyright 1984 by John Wiley and Sons, Inc., New York.

The author is a Professor of Chemical Engineering at the University of New Hampshire, Durham, NH 03824; (603) 862-3655.

Reprinted from *IEEE Trans. Prof. Comm.,* vol. PC-27, no. 1, pp. 14–19, March 1984.

```
                              Outline
                           Chapter Nine
                       REPORT PREPARATION

      I. Introduction
            (Attention) "Skills no better than before—abominable."
               A. Importance of communication skill.
                  1. More important than technical skills? Hogwash.
                  2. "All an engineer does is write (right?)."
               B. Philosophy of writing.
                  1. Honesty.
     II. Mechanics of report writing
               A. Outline.
                  1. Reader identification.
                  2. Who, What, When, Where, Why, How?
                  3. Write conclusions first.
                  4. Review literature or calculations.
                  5. Write thoughts on sheets of paper.
               B. Structure.
                  1. Cinnamon roll.
                  2. Dangers of rigid format.
                  3. Sample outline.
                     a. Purpose of section.
                         i. Present technical information.
                        ii. Define, recommend, encourage, promote action.
                       iii. Data repository.
                     b. Sample format (see Table I)
               C. Length.
                  1. Long enough to reach the ground.
                  2. 50-mile hike.
    III. Style and technique.
            (Interest) Hydrochloric acid to clean pipes.
               A. First person, humor, informal versus formal.
               B. Fog Index.
               C. How to improve.
                  1. Practice, practice, practice.
                  2. Invite criticism.
                  3. Read good writing appreciatively.
                  4. Read bad writing critically.
```

Some suggestions by Bolmer [4] for preparing a speech are also appropriate for prose. At each juncture, ask the magic questions: Who? What? When? Where? Why? How? The answers will usually lead you to the next step. Bolmer also suggests writing or identifying the conclusions first (asking the same questions) to provide focus in the outline. Next, review your notes and write prominent thoughts, quotations, and ideas on slips of paper. Do not then cast them into the air and pick them up randomly from the floor. Instead, organize them as your mind directs. In the shuffle, some ideas might appropriately land in the wastebasket.

Report Structure

Composing a report is much like baking cinnamon rolls. A cook does not put dough in one pile, raisins in another, cinnamon and sugar in a third, and then bake the ingredients separately. Neither does one place all the materials in a blender and atomize them into a uniform mass. Instead, individual elements are assembled wisely and in proper proportion to yield an interesting, attractive, and tasty result. So is a report organized to provide mental nourishment, impetus, and satisfaction.

I see three primary purposes of a design report:

1. To present technical information.
2. To serve as a repository of data.
3. To promote or define action.

The first two might be viewed as the dough, the third as cinnamon and raisins. Unfortunately, I cannot give you an exact recipe for composing a report. A rigid outline for all reports and situations is stifling. However, for a beginner, a skeleton format may be helpful. The format illustrated in

TABLE I
SAMPLE FORMAT FOR A TECHNICAL REPORT

Division	Section	Purpose		
		Present Information	Data Repository	Promote or Define Action
I. Beginning procedural section (front matter)	Letter of transmittal Title page Table of contents Abstract			
II. Summary	Summary			
III. Body	Introduction (background, literature survey, theory, etc.) Method of approach (procedure) Results Discussion of results Conclusions Recommendations			
IV. Concluding procedural section (back matter)	References Appendix			

Table I is discussed in detail. As you read about each section, think which of the foregoing purposes is satisfied. (I have provided space in Table I to keep score; I divulge my ratings later.)

I like to think of a report as containing four divisions: a beginning procedural segment, the summary, a body, and an end procedural segment.

• *Front Matter*
The beginning procedural segment usually contains a *letter of transmittal, title page, table of contents,* and *abstract.* It is much like the pages at the beginning of this book numbered in lowercase roman numerals. (This section is known as "front matter" in the publishing business.) In many reports, especially brief ones, some of these components are unnecessary. In a short or letter report, the title, abstract, and beginning of body may appear on the first page.

• *Summary*
The *summary* is an isolated section because it is often circulated separately to a wider audience that includes managers and nontechnical readers who are concerned with action and recommendations rather than computational detail. Because of its political impact and importance in decision making, the summary should be written most carefully, emphasizing vital conclusions and recommendations. Supporting data must be summarized and presented clearly and

interestingly to a less sophisticated reader. Illustrations should be used effectively but sparingly. Since the summary is based on the broader report, it is, of course, written last. It appears, however, near the front of the finished document as shown in Table I.

• *Body*
Asking who, what, when, where, why, and how leads smoothly to an efficient outline for the report body. An *introduction* of some sort is necessary to bring the reader "up to speed." Historical or chronological structure is oft-times effective in this section. If appropriate, *literature survey, theory,* and other topics may be folded into an introduction or inserted as separate sections.

To evaluate your report, a technical reader must understand how you derived the results. A section on *approach* or *procedure* serves this need. It should be written in a way that permits a reader to duplicate experiments or calculations independently if necessary. In a design project, pivotal assumptions and bases should be included and, where appropriate, explained. More common assumptions are listed in the appendix or not at all. Detailed calculations should not be placed here or in the appendix. They belong in your files. Representative sample calculations should be in the appendix.

A key structural role is played by the *results* section.

Information vital to the final conclusions and recommendations is found here. Peripheral data should be in the appendix or in your files. Inclusion of unnecessary detail obscures vital results.

The results section is followed by the *discussion* or *discussion of results*. This is where logical conclusions are exposed. Many authors fail to develop and manipulate their data enough. One table in my book, for example, was assembled from ten sources. I spent hours arriving at a format, days defining the details. This single table required more than a week's hard labor. The original ten sources could easily have been reprinted directly but I wanted focused data, not diffuse data. Many engineers do not invest enough energy in massaging results. They are satisfied with detailed tabulations of numbers when refined charts or curves would tell the story better.

The *conclusions* and *recommendations* sections represent the apex of your report. As you outline these sections, think, analyze, and ask the magic questions. Skilled technical writers, not unlike popular authors, often use suspense to create a climax. Since preceding sections have created a focusing effect, this segment can be concentrated and brief. Often conclusions and recommendations are combined into a single section. Sometimes recommendations are presented as a numbered list of statements.

How you say it *does* make a difference. We could imagine someone walking down a corridor, stopping at each door, knocking, and politely stating, "My senses perceive a conflagration at the extremes of this structure. I advise you to depart with haste." A real messenger would, of course, race up and down the hall screaming "Fire! Fire!" Provocations and emotions created by screaming "Fire!" in a technical report sometimes cause regret. On the other hand, we want readers to sit up, take notice, and, in many cases, act. Of the two approaches illustrated, a tone nearer "Fire!" is suggested.

• *Back Matter*
The final procedural section will not be opened by many readers, yet it serves a fundamental role in supporting the report. Not only should we be considerate of our more technical readers who will read this section but also want to help them find any mistakes that might be present.

References can be presented in any logical, consistent format so long as they are clear and unambiguous. The format used in this book should be acceptable in most reports. As a reader, I find article titles informative and recommend their inclusion. Sometimes, authors try to impress readers by citing exhaustive lists of nonpertinent references. This creates the same result as unnecessary detail in the text—foggy and misleading communication.

Efficiency and clarity are traits of an effective *appendix*. Sometimes, students dump their raw calculations here to prove the work was done and to impress the teacher. As a reader, I am confused, discouraged, and angered by this strategy. Writers often fail to separate the wheat from the chaff. In almost every case I have seen, computer printout is chaff and should not be included in the report. Raw calculations are also chaff and should remain in your files. They do serve nicely, nonetheless, as a useful outline for preparing the appendix. Illustrative and sample calculations selected critically from your work provide effective support to more focused information found in the report body. An effective appendix demands the same kind of creativity as any other part of the report. Sometimes even good authors are careless with this section.

By the way, in my opinion, purpose 1 (to present information) applies to the letter of transmittal, title page, and table of contents in Table I. The abstract, summary, discussion of results, and conclusions accomplish the same end and promote or define action (purpose 3) as well. Purposes 1 and 2 generally suit the introduction and approach sections. Action is promoted and defined primarily in recommendations. References and the appendix serve as data repositories.

I reemphasize that the outline is only a suggestion; the nature of the project—and your personality—shape the structure of the report. This reminds me of my first 50-mile backpacking trip. I had listened to a man who frequently hiked in California's Sierra Mountains. He stressed the importance of lightweight packing and illustrated it by telling how he took only three pair of socks. He wore two pair and carried the other. When camping for the evening, he changed socks and washed out the sweaty ones; laying them on a warm stone. The next morning, they were dry and ready for the day's hike.

I tried the same technique on a trip in the Appalachian Mountains in New England. What succeeded in dry California failed miserably and odorously in the Northeast. (Where does one find warm rocks in the rain?) Consider the situation in designing a report. Is yours a three-sock or a nine-sock project?

Report Length

The question of report length might be answered the same way Abraham Lincoln answered a query about a man's legs. He said they should be long enough to reach the ground. A report should be long enough to tell the story.

Length is also somewhat dependent on audience and other circumstances. Many of us, infatuated with our own writing, tend to inflate its length. The old saying "Length of a graduate thesis is inversely proportional to the data it contains" is boringly valid at times. It's as though there was a minimum weight limit. Even though I am considered sparing with words, a ruthless but respected critic eliminated about 20 percent of what was originally drafted for this chapter. The improvement was worth the pain.

STYLE AND TECHNIQUE

Some years ago, a New York plumber discovered that hydrochloric acid was dandy for cleaning clogged drains. He sent his suggestion to the National Bureau of Standards.

''The efficacy of hydrochloric acid is indisputable,'' the Bureau wrote back, ''but the ionic residues are incompatible with metallic permanence.''

''Thank you,'' replied the plumber. ''I thought it was a good idea too.''

Finally, someone at the Bureau wrote, ''Don't use hydrochloric acid! It eats hell out of the pipes!''

No doubt, crisp language communicates ideas efficiently. No one knows how many years scientific and technological progress has been retarded by foggy writing. Communication professionals have been criticizing the characteristically formal impersonal language of science for years. Yet, we still encounter unpleasant examples in our professional literature. Fortunately, promising trends are evident and we find more humor and use of first person in modern technical prose. Van Ness and Abbott wrote [3]:

> For many years the dominant attitude with respect to scientific and technical writing was that it should be impersonal, because science and technology were said to be impersonal. This forced adoption of the passive voice and promoted the lifeless syntax, the witless style, to say nothing of the grammatical mistakes of technical prose. We repudiate the whole of it. Not only does habitual use of the passive voice make for dull writing, it forces a convoluted style almost impossible for an engineer to make concise, precise, and grammatical. *I* and *we* are not four-letter words; they are entirely acceptable in technical reports and publications. We do not suggest that every sentence start with *I* or *we*; one seeks variety. If you are too humble or shy to bring yourself to write *I*, use *we*, in the sense of you, the reader, and I, the writer. *One* also has its place. Do not think you can avoid responsibility for what you write by adopting an impersonal style. No way; your name is on the title page. Take some pride in it; you are the expert.

The entire article is a useful guide for engineers.

I remember speaking, not long ago, with a student who went to work at DuPont. As a new recruit, he spent his first month on the job in a writing course. Instructors emphasized informal personal style because it makes written communication so much more effective. If the largest U.S. chemical corporation believes in it, we should feel free to promote it.

In the following example, the information about a project is written at different levels of formality:

> This experiment was designed to define the relationship between temperature, time, and location in the curing of a polyurethane automobile bumper. It was initiated because of failures in certain applications.

> About five percent of the bumpers we manufacture for the new Z cars are dropping from the vehicles at subfreezing temperatures. In a crash program to salvage our contract with Studebaker, Jean Doe assigned Dan Jordan and me to analyze the curing process and isolate any flaws.

Unfortunately, not all organizations tolerate informal technical documents. You may find the need to aim your language somewhere between that befitting an automobile purchase agreement and that in a letter to an intimate friend. However, grammar, punctuation, spelling, clarity, and precision of writing at any level of formality can be improved. As guides to the technical rules for good English expression, references 5 and 6 are recommended.

A recent article in *Science 82* [7] discusses the Fog Index used by Douglas Mueller, a writing consultant. It is a measure of writing clarity. As reported in that article, big words and long sentences are the two major culprits. The Fog Index puts these factors into a simple formula that tells how many years of schooling are needed to read a sample easily. The first letter to the plumber has a Fog Index of 26. To understand it requires a Ph.D. and seven years of postdoctoral study. The second letter, with a Fog Index of 6, should be clear to a sixth grader.

The article continues to describe how to calculate a Fog Index. My 12-year-old son Thatcher, intrigued with the challenge, computed indices for two important recent documents created in our family. A selection from the preface of this book scored 15, low enough for students with 12 years of grammar and high school and three years of college. My wife's recent book on colonial history rated 11. According to *Science 82*, she wins.

> At what Fog Index should a writer write? ''A low one,'' says Mueller. The nation's largest daily newspaper, *The Wall Street Journal*, got that way by lowering its Fog Index to 11. *Time* and *Newsweek* also average 11. *The New Yorker* usually comes in under 12. Technical journals range a lot higher, but most are notoriously hard reading, even for specialists. Good technical memos, according to a recent study at Bell Laboratories, average only 14. ''The truth is,'' says Mueller, ''no matter what Fog Index your readers can tolerate, they prefer to get their information without strain.'' Mueller says he's never met anyone, in any field, who couldn't lower his Fog Index to 15. ''Einstein could. It's easy. Just keep your average sentence length under 20, cross out every useless word, and never use a Big Word unless you absolutely need to. Remember: The less energy your reader wastes on decoding your language, the more he'll have left for your brilliant ideas'' [7].

Some examples, prominent and otherwise, were also given. From a business letter:

> We might further mention that we would be glad to furnish

any one of these whistles on a trial basis, to the extent that if the smaller size was not adequate enough, it could be returned in lieu of the purchase of a larger size, depending upon actual operation and suitability of your requirements for signal distance and audibility. (Fog Index: 28)

Translation:

If your whistle isn't loud enough, send it back and we'll give you a bigger one. (Fog Index: 6)

From the scientific journal *Nature*:

The current fashion for environmental impact assessment (EIA) is partly explained by the continuing force of the environmental protection movement in Western countries. That movement is now under severe pressure from economic recession, and there are signs that impact assessments themselves will play a decreasing role in planning and development. Certainly, this is the message that emerges from the U.S.A., where the emphasis is switching back to the costs of environmental protection. (Fog Index: 17)

Opening of the Gettysburg Address:

Fourscore and seven years ago our fathers brought forth on this continent a new nation, conceived in liberty and dedicated to the proposition that all men are created equal. Now we are engaged in a great civil war, testing whether that nation or any nation so conceived and so dedicated can long endure. We are met on a great battlefield of that war. We have come to dedicate a portion of that field as a final resting place for those who here gave their lives that that nation might live. It is altogether fitting and proper that we should do this. (Fog Index: 10)

Matthew 6:9–13 (King James version):

Our Father which art in heaven, Hallowed be thy name. Thy kingdom come. Thy will be done in earth, as it is in heaven. Give us this day our daily bread. And forgive us our debts, as we forgive our debtors. And lead us not into temptation, but deliver us from evil: For thine is the kingdom, and the power, and the glory, for ever. Amen. (Fog Index: 4)

Knowing the facts of good style does not necessarily create good writing. A reporter is said to have asked a famous football coach the secret of his success. He said there were three reasons: (1) practice, (2) practice, (3) practice. (A bystander added, "But it helps if the players are big and fast.") By analogy, to improve writing skills, you should write, write, write. (But it helps if you have grown up in an articulate family, studied debate for eight years, and taken a minor in English.)

Not only must you write, but also you should swallow your ego and invite expert criticism. In a less threatening vein, read quality writing by others and try to understand why it is good. When it is necessary to read bad writing, read it critically, noting errors and problems in margins as you observe them. Rewrite passages to see if you can improve them. (If the writer is your professor or a corporate vice president, it might be wise to destroy the marked copy.)

REFERENCES

1. Leesley, M. E.; Williams, M. L., Jr. "All a Chemical Engineer Does Is Write," *Chemical Engineering Education*. Fall 1978; 12(4): 188–192.
2. Ricci, L. J. "Chemical Engineers' Education Goes Downhill," *Chemical Engineering*. April 2, 1979; 86(9): 94–98.
3. Van Ness, H.C.; Abbott, M. M. "Technical Prose: English or Techlish?" *Chemical Engineering Education*. Fall 1977; 11(4): 154–159.
4. Bolmer, J. "Tips on Talking in Public." *Chemical Engineering*. Sept. 21, 1981; 88(19): 143–146. Also in *IEEE Transactions on Professional Communication*. March 1982; PC-25(1); 40–42.
5. Hodges, John C.; Whitten, Mary E. *Harbrace College Handbook*. 7th ed. New York: Harcourt Brace Jovanovich; 1972.
6. Strunk, W., Jr.; White, E. B. *The Elements of Style*. 3rd ed. New York: Macmillan; 1978.
7. Dunkle, T. "Obfuscatory Scrivenery (Foggy Writing)." *Science 82*. April 1982; 3(3) 82–84.

Writing a Trip Report

Marcia A. Petty

When you go to a professional conference, your follow-up trip report should be planned and written as carefully as any of your daily writing assignments. Your managers like to know they're getting value for their money, and a well written report that includes new information and reflects your expertise can increase the chances that the company will continue to support future conference attendance and membership. This is true especially for professional writers and editors; if you can't write clearly about what you learned at a meeting, a manager may think twice about sending you again.

My company sent me to the 34th ITCC in Denver in 1987, and my manager asked that I write about the trip afterwards. He wanted to circulate the report within our department and to the higher Intel Corporation managers who approved the trip. With no standard format to follow, I could have written a simple one-page list of the major stem topics and the sessions I attended. Such a report would have given the bare facts, but it would not have communicated much about the value I received from being at the conference.

So my report became more than a one-page item. As I outlined and wrote, I wanted to tell about the conference experience in general and to show good communication techniques by example.

THE CONFERENCE EXPERIENCE

It was hard to present in a few pages the complexity of four days of intensive information exchange. After all, I could attend only a finite number of sessions and talk with a finite number of people before experiencing information overload. But I did come back with certain key ideas that seemed to

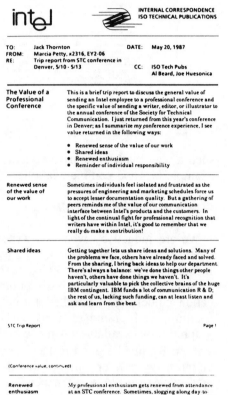

stand out. As I outlined my report, I did the following:

- Summarized the overall experience and its general value
- Synthesized the insights and examples I gained from different sources at the conference, including sessions, award-winning publications, papers in the *Proceedings*, and individual conversations
- Added practical examples of how conference insights might relate to the work in our own department.

I could have been off the report hook, so to speak, by writing the above information in two or three pages of tightly packed prose with no white space. I would have told my audience what I learned and how the abstract knowledge related to our specific experience. That's enough, right?

Not necessarily. I also wanted to *show* my audience what I learned, and that meant presenting the report in

Figure 1a. Company trip report submitted after the 34th ITCC, shown as an example of format and style

Reprinted with permission from *Technical Communication,* published by the Society for Technical Communication, May 1988, vol. 35, no. 2, pp. 130–131.

(Usability, continued)

Commands organized alphabetically	Some communicators recommend that in operating system reference manuals, we list commands alphabetically from beginning to end, not in separate sections by type of command. This way, the person who doesn't know the type can look up the command by name.
Some redundancy is O.K., she repeated	Include some redundancy. Much as we try for brevity and conciseness sometimes it's better to include a piece of information twice rather than make the reader search for it.

Readability—is the item easy to comprehend?

For this report, I'll define "readability" as a measure of how easy it is for the reader to comprehend the material, both text and graphics.

Job aids

The easiest document for a user to understand may not be a manual. One speaker gave an excellent argument for "job aids" that are not full-blown manuals. Examples are keyboard templates, wall posters of important warnings, and quick reference, downsized guides. The point is that good readability means giving the magical right amount of information...not too much, not too little. A job aid may have just the right information for the task in hand; a manual may result in overload. (Note that for two years we've been writing Technical Updates and 6-8 page installation guides to supplement manuals, I guess you could call these job aids.)

One competition manual had a tear-out, cardstock reminder list for "Managing Your System" right after the front cover. The list included page number references.

Enough background information

Make sure you provide or assume the proper background for understanding. This was referred to by one speaker as the given new contract: If I provide or correctly assume the given information, then the reader can more easily comprehend the new information.

Audience analysis

Determining the correct level of readability depends on correct analysis of the audience. This will continue to be a shortcoming if our audience is never clearly defined. I think we wind up with a schizophrenic view of our integrated system products. How should we respond if our questions get the following answers?

Page 4   M. Petty, 5/20/87

(Readability, continued)

Readable forms and contracts

Two authors made an argument for applying readability standards to such mundane items as contracts and forms. An excellent example was a contract that went to a customer in "term paper format": 8½x11 page, text the full width of the page (excluding narrow margins), headers barely distinguished from text. The customer thought it was a draft and marked it up! Needless to say, the company that wanted the contract signed redid the format so that it became much more readable.

We have a "good" bad example right in house: we need to reorganize the disclaimer page in H/W manuals. The page gives too much information at once and is barely readable. Maybe we don't care if users actually read it, it's just legal backup; if so, let's put all that stuff at the back of the manual! For example, one competition manual had the FCC statement at the back, but listed in the contents...odd, but workable.

Miscellaneous

A session on "visual literacy" reminds us that having an advanced text and graphics tool does not automatically create an experienced communicator. Training is still needed to use the features of any tool, and I don't mean just specific machine-dependent training. I refer you back to my paragraph headed "Individual Responsibility." The individual needs to learn the fundamentals of writing and graphics as well as the use of advanced electronic tools.

Summary

I hope Intel will continue to fund participation in professional conferences. The company, the department, and the individual all benefit. It may take time afterwards to put new ideas into place, but in the meantime, those ideas have a daily impact in our communication process.

Page 6   M. Petty, 5/20/87

(Readability, continued)

Question: Is this document for the novice or for the experienced person?
Answer: Yes.

Question: Do we sell to VARs or end users?
Answer: Yes.

One answer is to continue to provide multi-manual sets with different reader levels addressed in each manual. And to provide job aids.

Testing readability

The importance of testing manuals for readability is a common theme. IBM tests manuals all the time, with carefully selected subjects, one-way mirrors, and videocameras. The rest of us try to benefit from their general results. For specific results, we have to rely on our reviewers before a product is released and on customer comments after a product is released. Heck of a way to run a railroad! But our schedules don't allow the extra time needed for detailed readability tests.

Physical organization can increase readability

Although this may sound more relevant to a manual's usability level, physical structure can increase or decrease readability. This refers to general and specific levels:

- At the level of overall manual organization, for example, make sure the material appears in the correct sequence. Make sure the learning chunks aren't too big

- At the level of specific page format, if we help readers gain a sense of organization, we help them find what they need more quickly. I quote, "...typographic formats and white space create a sense of organization for a reader..." For example, headers are signposts to organization, and that's why we use different typefaces and weights to indicate hierarchy. This argument also supports hanging heads, which help you skim the organizational elements more quickly. Other organizational devices include rules, tables, boxes, and bulleted lists.

STC Trip Report   Page 5

Figure 1b. Company trip report submitted after the 34th ITCC, shown as an example of format and style, continued

such a way that it would serve as an example of good communication.

COMMUNICATION BY EXAMPLE

We are judged by what we write, whether a short memo or a lengthy manual. To give a lesson by example in my report, I followed basic publication guidelines. I checked format, spelling, and grammar. I used the graphic tools available on my workstation. I printed a good master copy on our network laser printer. I distributed copies on 8½ × 11 paper, stapled in the upper left corner, printed front and back so the readers would see a layout with planned facing pages.

I wanted a report that busy people could read and understand quickly, so I used a format with headers in the left-hand column to serve as quick-reference key words ("hanging heads"). To an engineering manager used to wordy reports, maybe the report would be a refreshing visual change. To a marketing manager, maybe the style of the report would emphasize what we've been doing in our manuals (in case they hadn't read our manuals lately).

Figure 1 on two pages shows a reduced copy of the report. I don't expect you to be able to read this reduced version, although it is legible even in small size. Rather, I want to show you the overall appearance of the report: the use of hanging heads, header subordination, rules of various widths to indicate topic changes, and page headers and footers. We've known for years that design influences communication; during the sessions in Denver, we heard the old principles reaffirmed. It never hurts to hear them again, and it helps to be reminded to use these principles in our daily work.

CONSISTENT QUALITY AND MANAGERIAL SUPPORT

Consistent quality is one path to a good professional reputation. If my report serves as an example of good communication, then on behalf of my peers I gain increased credibility for us all. The next time I or others in this department ask to attend a professional conference, I like to think I will have helped support a continuing, positive response from our managers.

Marcia A. Petty writes trip reports—and other documents—as a senior technical writer for Intel Corporation in Hillsboro, OR.

Clarifying Specifications

Tim Whalen

THE DEVELOPMENT OF SPECIFICATIONS, *which is of primary importance in many projects, is often a complex and difficult task. Both engineers and technical writers often encounter difficulty in developing correct and coherent specifications. For the consistent production of clear and authoritative specifications, this process must be clarified and reduced to manageable, step-by-step units.*

An initial characteristic of specifications that may cause problems for the specification developer is legal language. Such language is a basic trait of any specification. For example, "The subcontractor shall provide all necessary personnel, facilities, equipment, materials, supplies, and services capable of performing the tasks set forth in the remainder of this specification." The use of "shall" and "set forth" and the listing of what the subcontractor will provide are typical of legal phraseology.

Specifications contain such language because of their dual role as contractual documents and as technical descriptions. The specification is a legal document about technical material (see Figure 1). It is a lower-echelon *contractual document*, but it is in the upper echelon of technical documentation. This mixture of technical and legal content is often confusing to those who write and edit specifications.

SPECIFICATION TYPES

As a contractual document, the specification has all the formal language that many people associate with legal rhetoric, yet simultaneously it contains hard technical data such as electrical set points, assembly instructions, and equipment descriptions. One problem in dealing with specifications, then, is *reading through the legal language to get to the technical content.*

Many specifications are overwritten to legal formats that interfere with understanding. "Boilerplate" is a good title for the typical specification framework, which is full of obscure phraseol-ogy and unnecessary verbiage. The verbosity often present in specifications is caused in part by the repetitious nature of specification production. For example, in many architectural and engineering firms, as much as 80% of the company's specifications may remain unchanged *despite changing clients.* This accumulation of so much old data can obscure the new, meaningful 20% of the specification.

It is the responsibility of the specification writer to focus the language, remove the obscurity, and re-design the arbitrary formats that some specifications devolve into. A question that the engineer or technical editor should ask when involved in specification work is "Where are we in the stages of development of this specification?" The answer will provide much clarity to the writing since, almost universally, specifications tend to fall into one of four distinct categories: (1) Design, (2) Procurement, (3) Production/Construction, and (4) Operations (see Figure 2).

Each type of specification has unique properties. A design specification cannot be expected to satisfy an operational criterion, since they have radically different audiences: one, the engineers who create and prove a design; the other, operators who deal only with day-to-day operations (see Table 1). The specification writer can then tailor the data to the exact readership.

For the technical writer or engineer, the legalistic framework is a contract accommodation that must be dealt with in each of the four stages. After the technical information is defined and assembled, the transition into specifica-tion language can then be made by judicious use of legal language to state the technical requirements. The addition of the legalistic tone brings an advantage that many specification writers overlook—the legal tone is slow and cautious, and can be used to slow the pace and momentum of accumulated technical data. To some extent, the drawings and tabular information also supplement the legalistic language, but to clarify a specification, the writer must turn to "yes-no logic."

THE YES-NO LOGIC PATTERN

The key aspect of a technical specification is its ability to eliminate obscure and unclear ideas from the technology of the project. Any specification that allows "gray areas" has failed. The wording should be as clear and specific as possible, using every available linguistic technique to define precisely the qualities and types of materials and workmanship that are desired. Examples of some good limiting language are shown in the following pipe refurbishing specification from a commercial oil and gas firm.

Contractor shall accomplish at least the

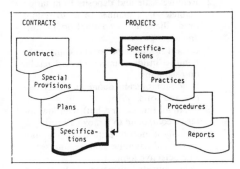

Figure 1—Role of specifications in contracts and projects. While each part of the contractual agreement is essential and complementary, specifications are lowest in order of importance. However, from a project perspective, the specifications are crucial to the technology. This paradox of values can obscure the role of the specification, i.e., to support the acquisition of goods and services that must be purchased.

Reprinted with permission from *Technical Communication,* published by the Society for Technical Communication. May 1982, vol. 29, no. 2, pp. 8–10.

Table 1— Goals of Specifications

Engineering	Procurement	Production/ Construction	Operational
1. Produce optimum design regardless of barriers	1. Meet engineering specifications	1. Convert the engineering specifications into detailed assembly procedures, in step-by-step sequences	1. Provide user with readable instructions for normal operations
2. Find means to integrate theory with practice	2. Obtain firm dates for delivery, and get commitments for quantities	2. Establish test points and parameters	2. Provide user with maintenance instructions
3. Apply known techniques to the design	3. Provide feedback to engineering on materials and services that do not meet the specifications and the delivery dates	3. Make provisions to inspect the materials and processes independently	3. Provide user with troubleshooting guidelines, parts lists
4. Research the unknown quantities	4. Aid engineering in surveying the materials and services that may meet the specifications and the necessary delivery dates	4. Provide feedback to engineering on performance of design	4. Provide feedback to engineering and production on operability and maintainability on user comment sheets (or equivalent)
5. Design around the true barriers			
6. Make design judgments and modifications from feedback from Procurement, Production, and Operational information			

services per Test Instruction T-38, in sequence, as set forth below:

1. Recondition 850 joints of 7-inch 23 psi/ft. K-55 LT&C casing, plus 1,050 joints of the same pipe to be pulled from wells in calendar year 1981;
2. Remove scale and deposits from inside diameter in a manner so as to restore inside diameter to nominal original diameter;
3. Inspect inside diameter and compare to nominal inside diameter, record results on each joint;
4. Sandblast outside diameter to remove rust and scale;
5. Clean all threads and buff with low abrasive compound to remove all rust;
6. Visually inspect threads for galling, tool marks and handling damage, and segregate defective joints;
7. Coat outside diameter with (brand name) or equivalent oil coating;
8. Apply API approved pipe dope to threads;
9. Provide and install (brand name) or equivalent thread protectors;
10. Stack pipe according to (company storage specification).

Briefly, the example limits its flow and individual contents to a yes-no logic pattern. This writing technique

eliminates all other approaches to the work and all other materials. In short, the yes-no step either makes the reader accomplish the step completely, or puts the reader on a "failed" or return cycle

to try the step again. Wording that contributes to the yes-no logic is specific to the point of excluding all other meanings: for example, numbering the joints of pipe precludes misinterpretation; using "shall" denotes *mandatory* compliance; naming the Test Instruction excludes all others; the word "sandblast" eliminates other methods of cleaning. Also, the presentation of the steps in a chronological order prevents the reader from ignoring or overlooking the technical intent of the specification.

In addition to precluding misinterpretation of contractual services caused by inexact language, the yes-no technique offers the specification writer other control points. The yes-no technique protects the company from getting charged for goods and services that are unwanted and unneeded, and precludes equivocation on the part of unethical or careless vendors. When vendors routinely seek clarification and direction after reading a firm's specifications, it is a sure sign that the specification does not follow the yes-no logic flow. The yes-no document provides a standardized, noninterchangeable vocabulary that clearly states the intent of each requirement. Lastly, following the logic path in the specification model ensures that the range of contractual operations is detailed.

The yes-no pattern may be visualized on a logic flow basis. This visualization begins in the engineering design stage (see Figure 3), where decisions are made to go ahead and buy goods, parts,

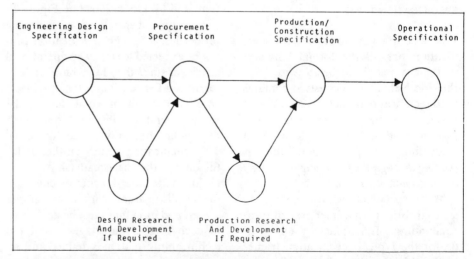

Figure 2—Types of specifications: stages of progress between design and use. An operational specification is generally written to a level that only operators would be concerned with, i.e., an operator's manual without the legalistic language and specificity that categorize the earlier stages of specification production.

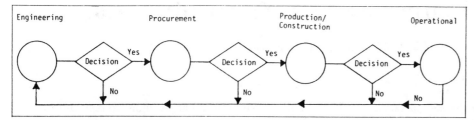

Figure 3—Logic flow of specifications.

and services for production of an item. Certitude in *how* to accomplish the work is shown in the exactitude of the specifications. At each stage in the logic flow, the specification is reworked to accommodate new data realistically. The information flow is always clearly yes-no. Introducing "maybe" into a specification is, therefore, a major error. Where equivalent goods or services may be used, carefully say so. (The conditional words *may, equivalent, equal, could, would*, or *should* can be useful to trade off goods, materials, or services that are *negotiable*, provided the variances introduced by such use will not affect the overall quality.) To reiterate: In yes-no logic, anything that doesn't meet the specification must be excluded.

Writers and engineers who work with projects requiring MIL-SPEC work are fortunate to have MIL-STD-961A as an aid to good specification writing. MIL-STD-961A is an excellent guide to yes-no logic since it precisely dictates the *only* form in which the Government will accept specifications. Furthermore, the familiar categorization of data into Scope, Applicable Documents, Requirements (the real technical focus of the document), Quality Assurance Provisions, Packaging, and Notes gives all MIL-SPEC specification writers a basis for understanding that commercial firms often lack. Engineers and technical writers who are unfamiliar with MIL-STD-961A will want to review it for tailoring into a commercial atmosphere.

SUPERCOMPRESSION

The correlative of yes-no logic is "supercompression," a linguistic phenomenon that occurs in specifications when the writer tries to pack too much into a section of the document. Supercompression goes beyond mere synopsis or summation. It results from accelerating the already telegraphic messages of a specification into a nearly incomprehensible jumble of technical statements. The following shows how a careless writer can produce an almost impenetrable document.

> Bus No. 2 shall be fed from WITR-2 IAW ABI Spec. 1.21.31 and adherent to approximate cycling characteristics across terminals 11-12 at about point A.12 to simulate permissive contacts from control panel 1-B12 then observe 480 ± 10% VAC on Bus. No. 1 to assure it is ready for checkout then proceed.

In this maze, supercompression created these flaws: (1) the intent of the specification—to describe operational startup—is lost, (2) the language is nearly undecipherable, (3) inexact language ("approximate," "ready," and "about") does not define the limitations that are desired, (4) the reader probably will not know all the acronyms, and (5) there is no numbering to place the data in sequence. Compare this "supercompressed" specification to its rewritten version:

Startup of Transformer for Cavern Operations

1. Bus No. 2 shall be energized first.
2. Bus No. 2 shall be fed from Transformer No. 2.
3. (Company Name) Specification 1.21.31 shall be used to cycle power per Table 1 of that document to readings shown for terminals 11-12 for point A-12 on Startup Drawing No. 1.21.32.
4. Simulate permissive contacts from control panel 1-B12.
5. Energize Bus No. 2 by closing circuit.
6. Observe 480 ± 10% VAC on control panel instrument.
7. Monitor for 5 minutes, noting fluctuations. Shutdown if out of tolerance with Step 6, above; go to Troubleshooting, if necessary.
8. Energize transformer by closing circuit.
9. Go to (company name) manual for transformer operation.

Reducing the specification to yes-no logic and using a standardized legalistic tone (including the formality of spelling out even the most simple acronyms) are positive steps to eliminate supercompression. Ω

REFERENCES

1. MIL-STD-961A, Military Specifications and Associated Documents.
2. Fred E. Eberlin, "What to Tell Your Engineers About Electronic Procurement Specifications," *Technical Communication*, First Quarter, 1975.
3. Martin Freir, "Effective Specification Writing," *Technical Communication*, Second Quarter, 1975, 14–16.
4. M. P. Jordan, "Elements of Efficient Communication," in *20th ITCC Proceedings* (Washington, D.C.: Society for Technical Communication, 1973).
5. J. D. Root, "Specification Writing: Precursor to Better Manuals," in *18th ITCC Proceedings* (Washington, D.C.: Society for Technical Communication, 1971).
6. Clarence W. Dunham, et al, *Contracts, Specifications and Law for Engineers* Second edition (New York: McGraw-Hill, 1971).
7. Irwin A. Page, "Engineering Communication and Professional Liability Claims," *Journal of Construction Specification Institute*, June, 1975.
8. Hans W. Meir, "Construction by Contract," *Construction Specification*, October, 1974.

TIM WHALEN *is a senior associate and a senior logistics engineer for CACI Inc.-Federal in New Orleans.*

Job Hunting: Sharpening Your Competitive Edge

RON S. BLICQ, SENIOR MEMBER, IEEE

Abstract—A job seeker who tailors each resume and application letter to capture the interest of a particular employer is far more likely to elicit a response than a job seeker who simply sends copies of a standard resume and letter to every employer. In a highly competitive job market, careful orchestration of the whole employment-seeking process is essential, from resume preparation to personal presentation during an interview.

UNTIL recently, engineers seeking a job or wanting a new challenge simply sought help from an employment agency or turned to the "Careers" section of a prominent newspaper or an engineering journal. They then applied to the employers who advertised positions of particular interest to them and expected the normal application-interview process to follow. But today the scenario has changed so greatly that scientists and engineers have to use entirely different tactics.

N. A. Macdougall, president of the Technical Services Council, writes [1]:

> In normal times, only 58 percent of jobs are advertised. Even fewer are advertised today, and some employers rely upon speculative applications. The vice-president of a chemical company says: "We never advertise any more. We rely upon marketing contacts and people who apply to us directly."

He also quotes the head of recruitment for a major oil company [2]:

> "Only 20 percent of jobs the company fills are from advertisements...we hire what we need today. If we lose someone, we will consider replacement. We no longer do long-range hiring.

> "Large employers get so many applications that they have to play the odds. Resumes which have obscure dates, don't explain summer jobs or miss data just get rejected. We can't afford the time to follow up. Little things can get you knocked out."

To see these remarks in the proper perspective, one needs to take a broad view of the whole employment process. Figure 1 shows the four stages a *successful* job applicant goes through, from making the initial contact to attending interviews and ultimately receiving a job offer.

If an employer has, say, 60 applications to consider for a particular vacancy, 59 of the applicants will be unsuccessful, some being eliminated at each stage. Most of them will be eliminated in the first two stages, since an employer will want to invite no more than eight to ten applicants for screening interviews and four or five for selection interviews. This means about 50 of the original 60 applicants are eliminated solely on the evidence they provide *on paper*, either in their resumes, letters of application, or application forms.

You cannot afford to be one of the 80 percent eliminated in this way. Consequently, *the impression you create when presenting your credentials to a prospective employer becomes critical if you are to be selected for an interview.*

This article examines the major factors one has to consider today when applying for a job. It emphasizes the written aspects, since applicants initially are judged almost entirely on the written documentation they provide. It also includes comments on presenting a confident, informative image during a job interview.

PREPARING A RESUME

For decades we have become accustomed to seeing and using the traditional format for our resumes (sometimes referred to as "curriculum vitae") in which the information we want to present is divided into five parts each preceded by an appropriate heading and listed in the following order:

1. Personal Information
2. Education
3. Experience
4. Extracurricular Activities
5. References.

This traditional arrangement is well recognized and still widely used. I do not describe it here because many books discuss the traditional resume in depth (e.g., [3]). Instead, I describe an alternative resume format which was highlighted at specials sessions of two technical communication conferences held in 1981 [4] and 1983 [5].

A major problem with the traditional resume is that its sequence tends to focus the reader's attention on the applicant's personal information and education which, although of interest to employers, often are not the factors they

Received April 12, 1984; revised July 11, 1984.

The author teaches technical communication at Red River Community College; Box 181–Postal Station C, Winnipeg, Manitoba, Canada R3M 3S7; (204) 632-2292.

Reprinted from *IEEE Trans. Prof. Comm.*, vol. PC-27, no. 4, pp. 201–210, December 1984.

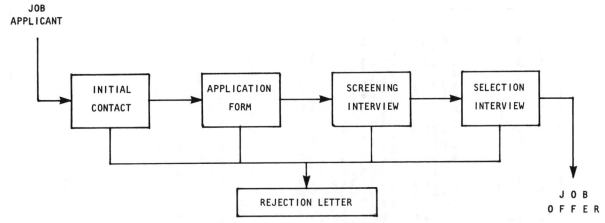

Fig. 1. The employee selection process.

most want to know. Today's technical employers are much more likely to be interested in *specifically what the applicant has done and special capabilities he or she has demonstrated that make him or her a particularly attractive choice.*

A contemporary resume directs its readers' attention to the skills and capabilities that will be of most value to a particular employer. It does this by using the "pyramid" method of writing described in the April 1982 issue of *Manitoba Technologist* [6], in which the resume opens with a summary statement that describes (a) what the applicant is particularly qualified to do or has extensive experience in doing; (b) what kind of work he or she wants to do; and, sometimes, (c) how the applicant's expertise can be used to the prospective employer's benefit. (Ideally, of course, there is a logical connection or development between these pieces of information.) This short paragraph is no longer than two or three sentences; it is titled Objective or Aim.

The second piece of information addresses the question a prospective employer is most likely to ask after reading the Objective, i.e., "What experience have you gained that specifically qualifies you to achieve this objective?"

The Experience section comes next and is divided into two compartments: first Related Experience and then Other Experience. This immediately focuses the reader's attention on those aspects of an applicant's work history that are especially relevant to the position being sought.

Moving the Education section much further down into the resume can be difficult for an applicant to accept if he or she has advanced degrees. Essentially it means taking a firm, objective look at all of one's qualifications *from the potential employer's point of view.* (The only times that education should be brought forward in a contemporary resume are when the job being sought calls for the applicant to have an extensive academic background and when the application is for a position in an academic institution.)

The remaining sections of the resume then follow, so that the complete list of headings becomes

1. Objective or Aim
2. Related Experience
3. Other Experience
4. Education
5. Extracurricular Activities
6. References

A two-page resume prepared in this contemporary format is shown in Fig. 2, in which the persons and circumstances are a composite designed to illustrate the methods for listing various information. The numbers in the margin are keyed to the following comments.

1. These essential details replace the Personal Information compartment of the traditional resume. Human rights legislation prevents job applicants from having to provide any more information about themselves (such as age, sex, weight, and ethnic background) than is shown here. However, if there are certain details they still feel they want to include, these can be inserted at the end of the Extracurricular Activities section.

2. The positions described with each Experience compartment should be listed in reverse order, the most recent experience being described first and the earliest experience described last. The most recent and most relevant experience should be described in considerably greater depth than early or unrelated experience (compare the descriptions of Dennis Hartley's Southcentral Contractors' experience with his Bowlands Stores' experience). The applicant should take great care to discriminate between factors he or she feels are most interesting and those an employer would particularly like to read about, and should concentrate on the latter.

3. The employer's name is listed first, underlined, and followed by the city and state. The person's position or job title is identified next, and then a description of

Dennis G. Hartley - page 2

(6) EDUCATION

Master of Science in Electronics Engineering, with major in fiber optics, University of Minnesota, 1984.
Bachelor of Science in Electrical Engineering, University of Montrose, Montrose, Ohio, 1982
Graduate Electrical Engineering Technician, Walter Halstadt Community College, Reece, Minnesota, 1973.
Graduate of Winona Collegiate, Duluth, Minnesota, 1967.

(7) ADDITIONAL ACTIVITIES/INFORMATION

Member, Institute of Electrical and Electronics Engineers Inc. (IEEE), 1973 to date. Secretary, St. Cloud, Minnesota Section, 1982-83.
Awarded Orton R. Smith Scholarship for proficiency in applied mathematics, Walter Halstadt Community College, 1972, and Ohio Power and Light Scholarship for achievement in communications engineering, University of Montrose, 1980.
Technical paper: "Accuracies of Computer Data Transmissions Attainable at High Baud Rates Over Fiber Optic Communication Links," in Communications Technology, 13:07, July 1984. (Paper based on thesis written as part of M.S. program, University of Minnesota.)
Military courses attended while in USAF:
    Transmission Line Installation Techniques, 1967.
    Supervisory Skills Development, 1969.
    First Aid and Safety Methods, various courses, 1968-71.
Junior Leader, Duluth, Minnesota, YMCA, 1962-67, teaching swimming and aquatic activities to boys age 9-15. Awarded Red Cross Bronze Medallion, 1965.

(8) REFERENCES

The following persons have agreed to provide information regarding my qualifications and work capabilities:

Martin F. Ebby, P.E.
Project Coordinator
Ebby, Little and Associates
360 Rosser Avenue
St. Cloud, Minnesota, 56302
Tel: (612) 544 1867

Philip M. Karlowsky
Contracts Manager
Southcentral Installation
    Contractors Inc.
1335 Westfair Drive
Lincoln, Nebraska, 68528
Tel: (402) 632 1450

Resume

DENNIS G. HARTLEY, P.E.
310 -- 408 Medwin Street
St. Cloud, Minnesota, 56301
Tel: (612) 548 1612

(1) OBJECTIVE

After four years supervising the installation and testing of wire and fiber optic telephone communication systems, I returned to college to obtain an M.S. in electronics engineering with a major in fiber optics. I am now seeking employment where I can apply my knowledge and experience in fiber optics engineering.

(2) RELATED WORK EXPERIENCE

June 1982 to
September 1983
and
May 1984 to date

Ebby, Little and Associates, Engineering Consultants, St. Cloud, Minnesota. Supervising engineer, responsible for installation, testing and analysis of tandem wire and fiber optic telephone communication links between Brainerd and Little Falls, Minnesota. Currently carrying out performance tests on installed links.

(3) September 1973
to August 1979

Southcentral Installation Contractors Inc., Lincoln, Nebraska. For first four years, member of team installing high voltage transmission lines and transformer stations along power grid between Weekaskasing Falls, Nebraska and Bismarck, North Dakota. After 18 months appointed crew chief in charge of team installing interconnecting and distribution systems to townsites along the route; responsible for: hiring, training and supervising local labor; ordering and monitoring delivery of parts and materials; arranging and supervising subcontract work; and preparing progress and job completion reports.
From June 1977 to August 1979, assigned as supervisor of team working under contract to Ohio Utilities Corporation, installing and testing fiber optic links between towns up to 28 miles apart.

(4) OTHER WORK EXPERIENCE

January 1967 to
February 1971

United States Air Force. Enlisted serviceman with Construction and Maintenance Directorate. For first two years, member of crew installing basic antenna systems and associated structures. For final two years, site technician responsible for maintenance of transmission lines and antennas at a midwestern USAF base. Attained rank of corporal.

(5) June 1963 to
December 1966

Bowlands Stores Inc., Duluth, Minnesota. Stock clerk in grocery store No. 16. Full time for two summers and June to December 1966; part time while attending high school.

2/...

Fig. 2. A contemporary resume.

what the job involved. If several positions have been held within the same firm, each is named and its duration stated so that the applicant's progress within the firm is clear.

4. Each position should draw attention to the personal responsibilities and supervisory aspects of the job, rather than just list specific duties. Verbs should be chosen carefully so they make the position sound as comprehensive and self-directed as possible. Verbs such as these are effective choices:

> coordinated
> monitored
> presented
> planned
> directed
> implemented
> supervised
> organized

If a paragraph grows too long (and this paragraph is rather long), it can be broken into subparagraphs:

...appointed crew chief responsible for
- installing interconnecting and distribution systems
- hiring, training, and supervising local labor
- ordering and monitoring delivery of parts and materials
- arranging and supervising subcontract work
- preparing progress and job completion reports.

5. Single-spaced typing should be used as much as possible to keep the resume compact. At the same time there should be a reasonable amount of white space on each side and between major paragraphs to avoid a crowded effect. A resume preferably should not exceed two typewritten pages (employers want to find the facts quickly; they do not want to wade through a mass of details) although it is acceptable to attach an extra page or pages containing, for example, a list of publications one has authored

6. Education can be listed either in chronological or reverse sequence. If a resume is to be sent out of the state, or if the applicant was educated out of state, it is best to identify the city and state of each educational institution attended.

7. Employers *are* interested in a job applicant's accomplishments and extracurricular activities, particularly those describing community involvement and awards or commendations.

8. When choosing people to supply references, always ensure that (a) the person's relevance is readily apparent (there should be a connection between the referee and the job applicant's previous work experience or community involvement); (b) the referee's full address and telephone number are listed (most requests for a refer-

ence are made by telephone); and (c) the referee knows the applicant is naming him or her in the resume (so that the referee will be prepared if a prospective employer calls).

Never overlook the importance of a resume's appearance. To a prospective employer the immediate impression created by the resume says a lot about the quality of work the applicant does. It may cost a bit more to have your resume typed by a professional stenographer on a high-quality typewriter and then duplicated properly, but the result can pay handsome dividends in the form of invitations to attend interviews.

And do not be afraid to use a display technique that enhances the professional quality of your resume. A soon-to-be-published textbook [7] describes how an engineer printed the two pages of his resume side by side on an 11″ × 17″ sheet and then folded the sheet so that the resume was on the inside. On the outside front he printed only his name and the single word "Resume" (Fig. 3). On the outside back he printed a cross-reference chart showing along one axis specific tasks in which he had experience and along the other axis projects he had been involved in. At the intersection point for each pair of entries he drew a small circle and then blacked it in to show his degree of involvement, as shown in the example in Fig. 4.

Another engineer, hearing that a sudden resignation had created a vacancy in a local company and was to be advertised the following day, prepared a one-page resume overnight (which he typed himself) and took it to the company shortly after door-opening time the following morning. He asked for, and gained, a two-minute interview with the manager of the department where the new employee would work. When he faced the manager he announced that he had all the capabilities to fill the position, that he could save the company X hours per month by using a special technique he had developed and used on previous tasks,

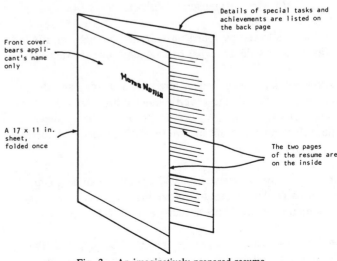

Front cover bears applicant's name only

A 17 × 11 in. sheet, folded once

Details of special tasks and achievements are listed on the back page

The two pages of the resume are on the inside

Fig. 3. An imaginatively prepared resume.

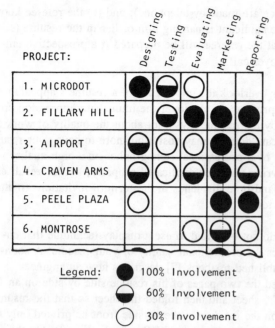

PROJECT:	Designing	Testing	Evaluating	Marketing	Reporting
1. MICRODOT	●	◐	○		
2. FILLARY HILL		◐	●	●	●
3. AIRPORT	◐	○	○	●	◐
4. CRAVEN ARMS	●	○	◐	●	◐
5. PEELE PLAZA	○			●	●
6. MONTROSE			○	◐	○

Legend:
● 100% Involvement
◐ 60% Involvement
○ 30% Involvement

Fig. 4. A cross-reference chart showing the applicant's involvement in various activities.

and that he had summarized what he could do for the company on the resume he had handed to the manager.

His resume, like his opening remarks, was oriented totally to the company's requirements. It was titled

RESUME PREPARED FOR ABC COMPANY

Its opening paragraph (the Objective) was introduced by the heading

How I Can Improve ABC's...

and then he listed three areas in which he would apply his expertise.

The next heading was

Previous Experience/Achievements in...

and was followed by four short paragraphs that identified with facts and figures what he had achieved elsewhere.

After two interviews he was hired the same day and the company withdrew the advertisement it had planned to place. His job-hunting tactics, and those of the engineer described earlier, point up two important aspects of current resume writing:

- Originality of presentation and approach can help capture a reader's attention, providing they focus on the potential employer's needs and are not brash.

- The days when a single resume could be prepared, duplicated, and mailed to numerous employers are over.

Now it is much wiser—indeed, essential—to prepare an individual resume for each employer, in which the emphasis and focus shift depending on the particular employer's requirements. These changes are most evident in the Objective and Related Experience sections of the resume.

WRITING A LETTER OF APPLICATION

Although some resumes may be delivered personally, the majority are mailed with a covering letter. Because potential employers will probably read the letter first, it must do much more than simply introduce the resume. It needs to state your purpose for writing (that you are applying for a job) and demonstrate that you have some very useful qualifications that the reader should take the time to consider. As such it becomes a letter of application to which you have attached a resume containing more definitive information.

A strong, interesting, well-planned application letter can prompt an employer to place yours among those whose authors he would like to see in person. On the other hand, a dull, unemphatic letter may cause the same employer to drop it onto the pile of "also rans" because its style and approach seem to imply you are a dull, unemphatic person. In today's highly competitive employment environment you cannot afford to let your letter be dropped onto the "also ran" pile.

Every business letter—and a letter of application essentially is a business letter—should follow the "pyramid" method of writing. That is, it should open with a brief summary that defines the purpose of the letter and then offer strong, positive details to suport the opening statement. Finally, it should close with a brief remark that identifies what action is to be taken next.

This means that an application letter can be divided into three parts:

- An *initial contact,* which states that you are applying for a job and briefly identifies what special qualifications you have that make you a particularly suitable candidate. (The intent should be to capture the reader's interest in the first one or two sentences.)

- An *evidence* section, which provides details and solid facts to support your contention that you are well qualified to hold the position. The facts you quote should be selected for their relevance to the position sought and their pertinence to the reader. (Remember that the most interesting experiences from your point of view may not be the most interesting to the reader.)

- A *closing statement,* which, rather than just closing the letter with a polite remark, *opens the door* to the next step (the employment interview).

There are two types of application letter: Those written in response to an advertisement for a job that is known to be open, or at the employer's specific invitation, are called "solicited" letters. Those written without an advertisement or invitation, on the chance that the employer might be interested in your background and experience even though no job is known to be open, are referred to as "unsolicited" letters. The overall approach and shape of both letters are similar, but the unsolicited letter generally is more difficult to write.

The Solicited Letter of Application

The main advantage in responding to an advertisement, or applying for a position that you know to be open, is that you can focus your letter on facts that specifically meet the employer's requirements. This has been done by Dennis Hartley in the letter illustrated in Fig. 5, which he has written in response to an advertisement for a project group leader with experience in installing, maintaining, and testing fiber optic transmission lines.

The following comments are keyed to the circled letters in this application letter.

A. For a letter that will have a personal address at the top, it is better to adopt the modified block style shown here rather than the full block style in which every line starts at the left-hand margin. Because the modified block style helps balance a personal letter on the page, it provides a more pleasant initial impres-

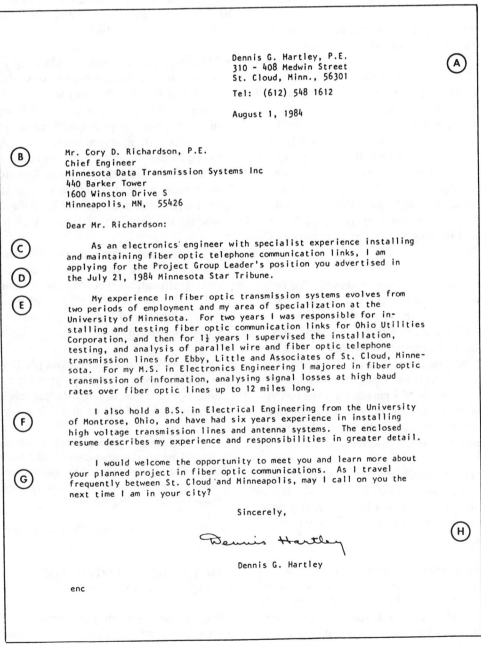

Fig. 5. A *solicited* application letter.

sion. Each line of the applicant's name, address, and telephone number, and the signature block at the end of the letter, should start at the page centerline.

B. Whenever possible, personalize an application letter by addressing it by name to the personnel manager or the person named in the advertisement. This gives you an edge over applicants who address theirs impersonally to the "Personnel Manager" or "Chief Engineer." If the job advertisement does not give the person's name, invest in a telephone call to the advertiser and ask the receptionist for the person's name and complete title. (You may have to decide whether it is better to send your letter and resume to someone in the personnel department or to a technical manager who is more likely to be aware of the quality of your qualifications and how you could fit into his or her organization.)

C. The first line of each paragraph may be indented about five spaces or may start flush with the margin.

D. This is the *initial contact*, in which Dennis summarizes the key points about himself that he believes will most interest his reader and states that he is applying for the advertised position. Note particularly that he creates a purposeful image by stating confidently "I am applying... ." This is much better than writing "I wish to apply... ," "I would like to apply... ," or "I am interested in applying... ," all of which create weak, wishy-washy images because they only imply interest rather than purposefully apply for a job. An equally confident opening is "Please accept my application for... ."

E. The *evidence* section starts here. It should offer facts drawn from the resume and expand on the statements made in the first paragraph. Broad generalizations such as "I have 13 years experience in a metrology laboratory" should be replaced with shorter-term descriptions that indicate the applicant's exact role and responsibilities and stress the supervisory aspects of each position. The name of a person for whom an applicant worked on a particular project can be usefully inserted here because it adds credibility to the role and responsibilities the applicant describes.

F. The *evidence* section should cover the key points an employer is likely to be interested in and draw the reader's attention to the attached resume. It may be divided into two paragraphs if a single paragraph seems to be too long.

G. This paragraph is Dennis's *closing statement*, in which he effectively draws attention to his interest in the position by referring to facts about the company's operations. He avoids using dull, routine remarks such as "I look forward to hearing from you at your earliest convenience" or "I would appreciate an interview

in the near future," both of which tend to close rather than open the door to the next step.

H. Contemporary usage suggests that most business letters should end with a single-word complimentary close such as "Regards," "Sincerely," or "Cordially," rather than the more formal but less meaningful "Yours very truly."

The Unsolicited Letter of Application

An unsolicited application letter has the same three main parts as a solicited letter and looks very much the same to the reader. To the writer, however, there is a subtle but important difference, in that it cannot be focused to fit the requirements of a particular position an employer needs to fill. This means the job applicant has to take particular care to make the letter sound both positive and directed. Here are some guidelines to help you shape a letter that you are submitting "blind."

• Make a particular point of addressing your letter by name and title to the person who would most likely be interested in you. This may mean selecting a particular department or project head, who will immediately recognize the quality of your qualifications and how you would fit into the organization, rather than the personnel manager. Never address an unsolicited letter to a general title such as "Manager, Human Resources," because, if the company does not use such a title and you have not used a personal name, it will likely be the mail clerk who decides who should receive your precious letter.

• Try to find out enough information about a firm that you can visualize the type of work it does and how you and your qualifications would fit the company's needs. This will enable you to focus your letter on factors likely to be of most interest to the employment manager or selected department head.

• Try to make your initial contact positive and interesting even though you are not applying for a particular position. For example, Dennis Hartley might modify the opening he wrote for his unsolicited letter to this:

As an electronics engineer with an advanced degree in fiber optics, I am seeking a position where I can use my ten years' experience in installing, testing, and maintaining both wire and fiber optic transmission lines. Have you any openings in your Project or Design Departments?

My experience in fiber optic transmission systems evolves from... .

COMPLETING AN EMPLOYMENT APPLICATION FORM

Filling in a series of company application forms can become a boring and repetitive task, yet any carelessness on an applicant's part can create a negative reaction from readers. Each company or organization usually uses its own specially designed form which, although it asks for generally the same basic information, may vary in detail. Consequently these suggestions apply primarily to the *approach* you should take rather than suggest what you should write.

- Always carry a personal data file with you so that you can search for accurate details such as dates, telephone numbers, and names of supervisors.

- Treat every application form as though it is the *first* one you are completing—write carefully, neatly, and legibly. Never let an untidy application form subconsciously prepare an employer to meet an untidy worker.

- Use words that describe the responsibility and supervisory aspects of each job you have held rather than list only the duties you performed.

- Particularly describe extracurricular activities that show your involvement in the community or in which you held a teaching or coaching role.

- Pay particular attention if there is a section on the form that asks you to comment on how your education and past experience have especially prepared you for the position. Think this through very carefully before you write so that what you say shows a natural progression from past experience to the job you are applying for. If you can, and if they fit naturally, add a few words to demonstrate how the position fits your overall career plan. This can be a particularly difficult section to write so do not be afraid to obtain an opinion of its effectiveness from an objective third party.

ATTENDING A JOB INTERVIEW

The key to a good interview is thorough preparation, by both the applicant and the interviewer. We tend to take it for granted that an interviewer comes well prepared—that he or she will have read the applicant's letter and resume thoroughly, will have clearly identified the type of person needed for the particular position, and will have prepared certain questions to ask. A good interviewer frames questions so that they elicit properly developed answers from the applicant, rather than laconic "yes" and "no" answers. Without adequate preparation by the interviewer, the interview may stumble along with no apparent sense of direction and have awkward pauses, resulting in a generally weak exchange of information.

Your role as a job applicant is to prepare thoroughly for the interview. This means researching information about the company, preparing answers to questions you anticipate will be asked, and developing a short list of questions that you want to ask.

Pre-Interview Preparation

Learning something about the employer means identifying the size of the company and where other offices exist, its principal products and services, specific projects or jobs it has handled, the number of people it employs, its involvement in community activities, and so on. Armed with this knowledge you will be able to ask intelligent questions during the interview and subtly demonstrate that you are a knowledgeable individual who is not treating the interview casually.

Preparing answers to questions you may be asked is more difficult because you have no way of knowing what types of questions the interviewer is likely to throw at you. There are, however, certain questions which frequently surface that you can prepare for:

1. What makes you want to work for our organization?
2. How do you think you can contribute to our company?
3. Why do you want to leave your present employer?
4. What do you like (or dislike) most about your present job?
5. What prompted you to leave (name of company) on (date)?
6. What salary are you looking for?
7. What do you expect to be doing in five years? ten years?

It is important to think about such questions before the interview so that you will be able to answer them directly and confidently. An unprepared job applicant who hesitates before answering them may create the unfortunate impression that he or she is a hesitant, unsure person.

Many job applicants ask "Why would an interviewer want to know what salary I want when he or she already knows the company has a fixed salary range that I would have to fit into?"

The answer is simply that the interviewer uses the question to assess how you value yourself and how well you have prepared for the interview. An ill-prepared job applicant may reply hesitantly, "Oh, between $22,000 and $26,000, I guess," whereas a well-prepared applicant will reply confidently, "With my experience I believe I should be getting $26,500." If you fear that the salary you want to quote may be too high, you can always add the qualification "...depending, of course, on the opportunities for advancement and the fringe benefits your company offers."

If you are well prepared you will attend every interview with certain questions in mind. Your interviewers expect this and so, toward the end of the interview, will ask, "Now, do *you* have any questions?"

Often you will find that most of your questions have already been covered during the discussion and that you have only one or two more you want to ask. But at this point, under the pressure of the moment, you may find you cannot remember what they were!

To overcome such a mental block you should jot down brief notes defining the questions onto a small card and carry it with you in a pocket or purse. Then, if you are not sure what you need to ask, you can produce the card and quickly check what remains to be covered. If you feel that a card is too much like a crib sheet and looks unprofessional, remember that to the interviewer the reverse is more true: It shows that you have done your homework and have prepared properly for the interview.

Creating a Good Initial Impression

You are being evaluated from the moment you step into the interview room. Perhaps the interviewer considers that the interview does not begin until you are both sitting down and have exchanged opening remarks, but subconsciously he or she is forming an opinion of you right from the start. Consequently you should

- Walk in briskly and cheerfully.

- Shake hands firmly, because a limp handshake creates an image of a limp, indefinite applicant (an image you cannot afford to create).

- Repeat the person's name as you are introduced and look him or her directly in the eye.

- Sit when invited to so do, pushing yourself well back in the chair, making yourself comfortable, and avoiding folding your arms across your chest (which psychologically suggests that you resist questioning).

Participating Throughout the Interview

The interviewer will want to set you at ease and establish a comfortable atmosphere conducive to a good exchange of information and so, during the initial part of the interview, often will ask questions about you and your personal interests which you can answer easily and readily.

To learn as much as possible about how you think and react, an interviewer expects you to give thoroughly developed answers to questions. An effective interviewer will pose questions and subsequent prompts in such a way that you are carried easily from one discussion point to the next and are automatically encouraged to provide comprehensive answers. But if you find yourself face to face with an inexperienced or inadequately prepared interviewer, the responsibility is solely yours to develop your answers in greater depth than the questions seem to call for.

For example, the interviewer may ask, "How long did you work in a mobile calibration lab?"

You might be tempted to reply "Three years," and then sit back and wait for the next question. You would do much better to reply, "For three years total. The first year and a half I was one of four technicians on the Minneapolis-to-Sioux City circuit. And then for the next year and a half I was the lab supervisor on the Fort Westin-to-Manomonee route."

An answer developed in this depth often provides the prompt (i.e., piece of information) from which the interviewer can frame the next question.

Sometimes you will face a single interviewer, while at other times you may face an interview board of two to five people. In a single-interviewer situation you will naturally direct your replies to the interviewer and should make a point of establishing eye contact from time to time. (To maintain continuous eye contact would be uncomfortable for both you and the interviewer.) In a multiple-interviewer situation you should

1. Direct most of your questions, and your responses to general questions, to the person who apparently is the chairperson. (But if an answer is long you should occasionally take time to look briefly at and talk momentarily to other board members.)

2. If a particular board member asks you a specific question, address your response to that person.

3. If a board member has been identified as a specialist in a particular discipline, direct questions to that board member if they especially apply to that field.

In certain interviews—often when applicants are being interviewed for a high-stress position—you may be presented with a "stress" question. A stress question is designed to place you in a predicament to which there may be two or even more answers or courses of action that could be taken. You are expected to think *briefly* about the situation presented to you and then to select what you believe is the best answer or course of action. Often you will be challenged and expected to defend the position you have taken.

The secret is not to let yourself be rattled and to defend your answer rationally and reasonably even though the questioner's challenging may seem harsh or unreasonable. Remember that the interviewer is probably more interested in seeing how you cope in the stress situation than in hearing you identify the correct answer.

Other points you should consider during the interview:

- If you do not know the answer to a question, it is far better to say you don't know than to try bluffing your way through.

- If you do not understand the question, again don't bluff. Either say you do not quite understand or, if you think you know what the interviewer is driving at, rephrase the question and ask if you have interpreted it correctly. (Never imply that the interviewer posed the question poorly!)

- Use humor with great care. What to you may be extremely funny may not match the interviewer's sense of humor.

- Bring demonstration materials to the interview if you wish (such as a technical proposal or report you authored, or a drawing of a complex circuit you designed) but be aware that you may not have an opportunity to display them. If the topic they support comes up during the interview, introduce them naturally into your response to a question. But remember that the interviewer does not have time to read your work, so the point you are trying to make should be readily identifiable simply by viewing the demonstration item. Never force demonstration materials on an interviewer.

- Do not smoke unless the interviewer also smokes and invites you to do so.

Finally, the most important thing to do during an interview is to *be yourself*. If you try to be the kind of person you think the prospective employer is looking for, you are likely to create a false impression which somehow rings hollow to the interviewer. An interviewer gains a much better impression of you if you answer questions firmly and confidently because you know the answers represent what you truly believe.

REFERENCES

1. Macdougall, N. A. "Job Hunting: Getting In the Door." *Design Engineering* (Canada). January 1983; 29(1): 28.
2. Macdougall, N. A. "An Employer's View of the Job Market." *Design Engineering* (Canada). January 1984; 30(1): 28.
3. Blicq, Ron S. *Technically-Write!: Communicating in a Technological Era.* 2nd ed. Englewood Cliffs, NJ: Prentice-Hall Inc.; 1981: p. 54.
4. The IEEE Professional Communication Society Conference on Communications with/for/by Government, Arlington, VA, September 16–18, 1981.
5. The IEEE Professional Communication Society Conference on The Many Facets of Computer Communications, Atlanta, GA, October 19–21, 1983.
6. Blicq, Ron S. "Tips for Writing Sharper Letters and Reports." *Manitoba Technologist.* April 1982; 12: 2.
7. Blicq, Ron S. *Administratively-Write!: Communicating in a Business Environment.* Scarborough, Ontario: Prentice-Hall Canada Inc.; 1985.

How to Write a Recommendation

ALAN D. WILCOX, MEMBER, IEEE

Abstract—Writing an effective recommendation requires an understanding of the reader and his specific needs. A "recommendation outline" is presented here which will meet those needs and also ease the burden of preparing the recommendation itself. In addition to the outline, two sample letters of recommendation are presented.

MANY times in your professional career you will be asked to write a recommendation for someone. Perhaps a good friend, a colleague, or a business associate is changing jobs and would appreciate your written evaluation of his or her performance. In the academic milieu especially, many students ask for recommendation letters.

Faced with this situation, and after agonizing over a number of letters, I located some of the standard "form letter" recommendation sheets used by various schools and companies. Unfortunately, these letters seemed inadequate to express my assessment of the candidate, and I ultimately abandoned them entirely. They were useful in one sense though: I gathered an inkling of what employers and personnel managers consider important. What I really needed was my own general-purpose outline to easily write an effective recommendation.

THE READER

Who is going to read the recommendation? The personnel manager, the department head, and the direct supervisor will probably see the recommendation, and it should be written with their needs clearly in mind. They need to see your friend as you do: as a real person with real strengths and weaknesses. After an interview they already have a "best manners" impression, but they need more substance. Perhaps they need to verify facts or clarify some uncertain points from the interview itself.

Well, if they need "strengths and weaknesses," it should be no trouble to put together a recommendation, even a recommendation with substance. The agony begins here. Consider what is really important to an employer: He needs to know whether your friend would not only be able to do the required work but also whether he or she would fit into the company and succeed as a long-term addition to the staff.

Several key factors seem to be of great interest in evaluating the potential employee and whether he or she will suc-

ceed: social competence, work competence, and a suitable blend of character attributes. These factors are shown as part of Table I. My impression is that a recommendation addressing these issues will be of great value. This is especially true if you have been candid and accurate in your presentation of the facts.

SOME POINTS TO REMEMBER

To help you write a recommendation that will help both your friend and the personnel manager, keep a few points

TABLE I
Outline of a Recommendation

Introduction	...is applying for a position as...and I would like to recommend him to you.
Background	Have known him for... In a relationship as...
Description	Career goals and interest in the field Relevant hobbies, activities, talents

	Social Competence
	  Works well with others
	  Is cooperative, congenial
	  Is understanding, open minded
	  Communicates well with others

	Work Competence
	  Able to see and solve problems; gets results
	  Speaks and writes effectively
	  Understands technical fundamentals
	  Has specific related experience
	  Is up to date in field of expertise
	  Can plan and organize work
	  Does quality work: accurate and thorough

	Character Attributes
	  Honest, sincere
	  Industrious, willing to work hard
	  Enthusiastic, motivated, self-starting
	  Initiative, ambitious
	  Sound, practical judgment
	  Imagination, vision, creativity
	  Independent thinker, problem solver
	  Logical
	  Perceptive
	  Positive attitude, desire to be effective
	  Emotionally stable
	  Intelligent, quick learner

Conclusion	Because of (all the above), ...would be suitable for this job. Would welcome phone call to discuss further.

Received April 16, 1984; revised July 3, 1984.

The author is an Assistant Professor of Electrical Engineering at Bucknell University, Lewisburg, PA 17837; (717) 523-0777.

Reprinted from *IEEE Trans. Prof. Comm.*, vol. PC-27, no. 4, pp. 211–214, December 1984.

in mind:

1. You are writing to a person, not to "whom it may concern." Ask your friend to give you the name and address of the person you should write. Then, when you write, keep him in mind and what he needs to know about your friend so he can make an informed hiring decision.

2. The hiring decision is based on the skills and character of your friend. He or she has various qualities, and the personnel manager needs to know what they are. Perfection is neither required nor expected! The relevant issue for the reader is to put the right person in the job.

3. Ask your friend what you should include in the letter. For example, he or she might be particularly competent

---

Bucknell University
Lewisburg, PA 17837
March 29, 1984

Prof. John Smith, Chairman
Biomedical Engineering Department
Xxxx University
Xxxxxx, Xxxxx

Dear Professor Smith:

Susan Xxxx is applying for summer employment and has asked me to write to you about her job qualifications. She is interested in two openings: Analysis of Heart Wall Motion and Computed Tomography. I can highly recommend her to you as an excellent selection to add to your summer staff for work in either of these two positions.

Sue attended my Introduction to Digital Systems class last fall and was one of my top students. I found her to be hard working and diligent to do the work assigned. Although her background is computer science, she did as well in this electrical engineering course as the EE students themselves. This semester she is in my Advanced Digital Systems class, and she is doing just as well as before.

Computer software is one of her strong areas, particularly in being able to relate it to the processing hardware. For example, she did a special project for me last semester in which she built a hardware multiplier circuit. In addition to doing the circuit design, she wrote the system software to interface the computer to the multiplier. This semester she's designing and building a small computer system using her own processor architecture.

I have no doubt that she can do application programming for you in FORTRAN. She has already programmed in that and other high-level languages in her studies at Bucknell. She learns quickly, and I know that she can successfully relate what she already knows to your job.

You will enjoy having her work with you this summer.

Very truly yours,

(signed)
Alan D. Wilcox, P.E.
Assistant Professor

Fig. 1.   Recommendation letter for a student summer job.

in a special field, and your discussion of this skill would reinforce what others might describe. Also, if the interview falls somewhat short, your clarification of some of the weak points will be beneficial.

4. Be honest about the qualities of your friend. If you write only in glittering superlatives, the recommendation says much less than if you are candid and down to earth. Give specific examples of what your friend has

---

Micro Resources, Inc.
Lewisburg, PA 17837
June 29, 1984

Mr. George Sherman, Engineering Manager
Computer Devices and Circuits, Inc.
Philadelphia, PA 19019

Dear Mr. Sherman:

I am writing in response to your June 4th inquiry about Ted Land, one of my former engineers. Based on my knowledge of his work, I am sure that he is fully qualified to be employed as a Senior Engineer, and I recommend him highly.

During the three years he was with our company, from 1978 to 1981, I worked closely with him when he developed a new, highly efficient, miniature switching power supply for our line of computer products. He was very methodical in planning his work and in his approach to solving the technical problems of the power supply. His attention to engineering detail helped us get the supply into production in near-record time with a minimum of changes from the prototype model he designed and built. Two novel ideas he had when working on the project also resulted in patents.

In addition to the power supply, he designed a water-level measuring system which went into production in 1980. He planned the entire project and did all the design and development of the system components. He was quite enthusiastic about the product and was here at the plant many evenings investigating new design approaches.

He is a very intense person and prefers to work alone rather than in a team effort. For that reason, when he was working for me with four other engineers, I made a special effort to give him assignments that required little direct involvement with others. I was pleased to discover, though, that he did get along well and went out of his way to assist the less-experienced engineers.

From what I have seen of Ted, I believe that he would be an ideal person for the job you described in your letter. Feel free to call me if you would like to discuss any points in greater detail.

Yours truly,

(signed)
Edward S. Johnson
Engineering Group Leader

Fig. 2. Recommendation letter for an engineer.

done that might be relevant to the new job. Avoid vague and unsubstantiated generalities.

HOW TO WRITE THE RECOMMENDATION

Once you feel comfortable with an understanding of the needs of the reader, the recommendation can be written by following the general outline shown in Table I. Certainly not all the items in the figure apply to everyone, but they should help give direction to your writing so it meets the needs of the reader.

For example, Fig. 1 is a letter I wrote for one of my students who was seeking summer employment. The letter follows the general thrust of the outline and provides as much information as possible to the reader. Hardware and software expertise were job requirements so I illustrated Sue's experience by examples of her work for me.

Figure 2 is a recommendation letter an engineering group leader wrote for one of his former team members. During the time Ted was working for him, Ted successfully designed a number of products, and the description of these achievements can help the prospective employer make a decision. Note that Ted does need some special attention: he's inclined to work alone. Balanced by the other information given, that is not really a negative factor in the recommendation. Overall, Ted comes across as a highly competent engineer who is a real person.

SUMMARY

Writing an effective recommendation requires an understanding of the reader and what specific information he or she needs to find in your letter. The recommendation outline in Table I can help meet those needs and also ease the burden of writing the letter. In addition, you can complete the task knowing you've made a positive contribution to the success and future of your friend.

Eight Steps to Better Newsletters

JANET H. POTVIN

Abstract—To the new or prospective editor, publishing a professional society newsletter can seem a formidable task. It need not be, however, with proper planning and organization. This paper describes a systematic approach to editing: (1) analyzing the editing task and assessing current publications; (2) considering costs and production alternatives; (3) formulating publication goals and an editorial philosophy; (4) planning a yearly calendar of issues and setting publication deadlines; (5) identifying critical issues and sources of information; (6) establishing methods for obtaining information; (7) creating a distinctive image; and (8) following through. Using a systematic approach simplifies the task of editing and makes it possible to publish newsletters that are accurate, attractive, authoritative, interesting, and readable as well as timely and informative.

ONE of the most important channels of communication between a professional society and its members is the newsletter. More timely than the society journal, which provides a permanent record of activity, the newsletter informs and keeps members abreast of current issues and trends, policies and procedures, pending legislation, society business transactions, upcoming conferences and symposia, competitions, calls for papers, nominations for awards, job openings, and the like—in short, the news. To serve its function well, a newsletter must be accurate, attractive, authoritative, interesting, and readable as well as timely and informative. But what specifically should the newsletter contain? What are the best sources of information? How should the newsletter be formatted? How should it be produced? How often should it be published? What does the process of editing actually involve? To the new or prospective editor, these questions and others can make editing a newsletter seem an awesome task. Limited published information is available to answer such questions. While there are many useful resources on copy editing (for example, [1-5]) and on publication design (for example, [6-9]), few sources specifically address the task of newsletter editing, especially the planning and organization required. (Exceptions are [10-12].) This paper describes a systematic approach to newsletter editing based on the engineering design process and provides planning guidelines and suggestions to aid those new to the task.

ANALYZING THE EDITING TASK AND ASSESSING CURRENT PUBLICATIONS

The process of editing a newsletter consists of eight steps. The first is to establish a framework for making decisions by

Manuscript received June 21, 1982.

The author is Director of Technical Writing at The University of Texas at Arlington, Department of English, P. O. Box 19035, Arlington, TX 76019, (817) 273-2692.

1. Analyzing

analyzing the task of editing and developing a profile of current publications. Unless the newsletter is a new publication, precedents for content, format, style, and quality will have been set, and knowledge of them is essential before formulating goals and an editorial perspective. Among the factors to be considered are

- Objectives of the newsletter
- Relation to other society publications
- Nature of the audience
- Content and news coverage
- Approach/tone/style
- Quality
- Image conveyed
- Frequency of publication
- Format (page size; columns per page; type style; type size)
- Number of pages per issue (or per year)
- Production mode (typewritten or typeset composition; printed or photocopied reproduction)
- Costs and available budget.

Specific questions concerning the existing publication include

- Are the format, design, and length satisfactory?
- Is the frequency of publication appropriate?
- Is the publication well received by members and officers?
- Is the news coverage adequate?
- Is the coverage authoritative?
- What specifically does the newsletter now include?
- What might it also include?
- What improvements (major or minor) in form or content might be made?

To aid in planning, collect newsletters published by others and evaluate the content, format, and appearance of individual issues. What are the strong points? The weak points? What features make a newsletter really outstanding? Average? Mediocre? It is also helpful to inventory the current publication and to compare its contents with a list of the possible inclusions in a newsletter (see Table I). The results of such analyses can be

Reprinted from *IEEE Trans. Prof. Comm.*, vol. PC-25, no. 4, pp. 204-210, December 1982.

TABLE I
AN INVENTORY OF COMPONENTS OFTEN INCLUDED IN NEWSLETTERS

Abstracts	Legislative reports
Advertisements	Letters to the editor
Annotations	List of society officers
Announcements	Logo
book releases	
competitions	Masthead
conferences	Messages (from society officers)
elections	
grant application deadlines	Opinions
upcoming meetings and symposia	Order forms
Awards	
	News
Bibliographic listings	articles
Book reviews	briefs
Calendars of coming events	Patents
Calls for	Photographs
applications for fellowships and grants	Puzzles
nominations for awards	
papers	Queries
program proposals	Questionnaires
Cartoons	Quotes
Classifieds	
	Readers'
Deadlines	corner
Departments	exchange
	forum
Editor's column/corner/viewpoint	Regular columns/features
	Reports
Feature articles	Requests
Forecasts	
	Standards
Grants	Statistics
application information	Subscription forms
awards	Surveys
	questionnaires
How-to-do-it articles	reports
Interviews	
	Table of contents
Job listings	Tutorials
openings	
placements	

used to advantage in formulating goals and an editorial philosophy; in determining what to include and what to improve in the newsletter.

CONSIDERING COSTS AND PRODUCTION ALTERNATIVES

Quite likely, the publication task is well defined: Convey important news to an audience of professionals in the most timely manner at the lowest reasonable cost. The challenges are how to accomplish the task: how to cover the major issues fully; how to obtain the news to print; how to provide an interesting, attractive publication; how to involve society members in reading the issues; and how to maintain interest once it is created. Many of the decisions are affected by cost and available budget, including the mode of production, page size, number of pages per issue, and number of pages per year—so alternatives must be considered.

Mode of Production

Newsletters can be composed by typesetting, typewriting, or word processing and can be reproduced by printing or photocopying. Typesetting gives a professional appearance, allows greater flexibility in design and thus more attractive pages, and permits more information to be published in fewer pages. Having a newsletter professionally typeset and designed is more expensive than having it typewritten, however, and additional lead time may be needed for copy preparation and production.

2. Production Cost

Typewriting is less expensive and can be done in-house, but the appearance of typewritten newsletters is not as attractive and more pages are required. As an alternative it may be possible to have headings set in display type, which greatly enhances the appearance of the typewritten page. If available, the use of a word processor with capabilities of producing boldface subheadings in the text and left- and right-justified margins further improves the appearance of the typewritten newsletter and simplifies copy preparation. To maximize the amount of information contained in a typewritten newsletter, pages can be prepared on oversized sheets and photoreduced to the finished page size.

Newsletters can be reproduced by photocopying or printing; the number of copies to be produced and the number of pages per copy are determining factors. Before making production decisions, cost estimates for the alternatives for composition and printing should be obtained from one or more publishers in order to consider the tradeoffs of composition, printing, and mailing costs.

Page Size

A related consideration is page size. Nonstandard sizes and special weight papers are expensive. Use of the standard 8.5" × 11" page is convenient for both printed and photocopied newsletters and is generally cost-effective. The base-size paper used for offset printing is 17" × 22"; it produces eight 8.5" × 11" pages when printed two-sided and is convenient not only for design and production but also for mailing. The size is familiar and comfortable to readers [7] and issues can easily be retained in file folders or binders for later reference [13].

Number of Issues per Year; Number of Pages per Issue

The newsletter can be published weekly, monthly, bi-monthly, quarterly, or with any regular frequency. The amount of information to be published and the publication budget determine both frequency and number of pages per issue. A regular frequency should be established and, ideally, each issue should contain the same number of pages. A newsletter should not be less than four pages but it should not be excessively long. Those who are busy tend not to read more than eight pages at a time [13]. If consistently more material is obtained for each issue than can be accommodated, increasing the total number of pages for each issue or changing to a more frequent publication of the same or smaller size should be considered.

Mailing costs are a factor in this decision. For example, adding four pages to a newsletter might increase the weight enough to effectively double the mailing cost. When changes in the number of pages are contemplated, it is helpful to have a dummy or mockup prepared (with allowance for the additional weight of ink and mailing label) and have the weight checked in advance. If the weight change is significant, the paper stock can be changed and the weight rechecked, or an alternative selected. When the newsletter approaches 20-24 pages per issue, changing to a magazine instead of continuing to publish a newsletter might be considered. Time, staffing, and funding are the critical factors in this decision.

Advertisements

A quality newsletter is expensive—it may be necessary to consider alternatives for offsetting costs. Accepting advertisements can be a good way for a society to recover part of the costs, society publication guidelines permitting. If not handled carefully, however, advertisements can cause problems without generating sufficient additional funds. Layout and design of a newsletter are more complex when advertisements are included, more pages are needed, and a more sophisticated (and expensive) printing process may be required.

If the decision is made to include advertisements, it is important that the rates be set high enough to ensure that the increased revenue actually offsets the increased costs. To alleviate potential difficulties, specifications for advertisements should include not only available sizes and prices but also precise descriptions of the kinds of artwork acceptable [14].

3. Goals and Philosophy

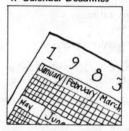

4. Calendar Deadlines

FORMULATING PUBLICATION GOALS AND AN EDITORIAL PHILOSOPHY

Once cost information has been obtained and a budget established, production decisions can be made, size and frequency of publication determined, and editorial goals set. The goals may be to continue the present level of effort or to improve publication content, design, quality, and image. As part of the planning, the editorial perspective needs to be considered: Is the newsletter to be conservative and noncontroversial in its presentation of the news? Influential in determining policies and procedures of the society? Provocative? Thought provoking? Controversy stimulating? Should readers be encouraged to respond to issues by exchanging ideas in a reader's forum or by sending letters to the editor? The answers to these and related questions shape the editorial philosophy or perspective of the newsletter.

PLANNING A YEARLY CALENDAR OF ISSUES AND SETTING PUBLICATION DEADLINES

A next step is to plan the calendar of issues for the year, including a general theme for each, and to set the deadlines for each issue. Depending on the size and content of the newsletter and the manner in which it is mailed, the length of time required to accomplish various tasks will vary; however, in scheduling, time must be allowed for

- Receipt of articles
- Editing articles and preparing copy for composition (typing or typesetting)
- Composing text and setting headings
- Proofreading
- Layout of pages
- Pasteup
- Printing or photocopying
- Applying mailing labels and sorting
- Mailing.

A useful procedure for scheduling is first to set the target date on which the newsletter should be received by readers and then to work backwards to determine

- Date on which newsletter must be mailed

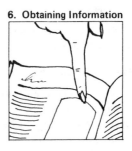

- Date on which newsletter must be ready for printing
- Date on which contributions must be received.

Estimates must be made of how much time it will actually take to perform each of the necessary tasks and for delivery of the newsletters after they are mailed. How mailing is to be done and how far the newsletter must travel are factors in scheduling. For example, three to four weeks are required to bulk-mail from east coast to west in the U.S. If newsletters are to be bulk-mailed, deadlines must be set much earlier than if they are to be mailed first class.

The deadlines must be set realistically but also with a view to obtaining and conveying the maximum amount of timely information within the constraints of printing and mailing schedules. To aid potential contributors, the calendar of issues and the deadlines for each should be published as a standard part of the newsletter so that anyone who wishes to contribute can conveniently do so.

IDENTIFYING CRITICAL ISSUES AND SOURCES OF INFORMATION

Next determine the specific content to be included in individual issues. Identify critical issues and list potential sources of information for the articles and features to be included. The process involves five steps: (1) brainstorming for ideas; (2) listing sources of information (authorities in the field and published sources); (3) consulting with knowledgeable advisors; (4) reviewing relevant publications; and (5) attending society business meetings and annual conferences.

Brainstorming

Drawing upon the group problem-solving technique of brainstorming is helpful for developing a list of topics to include in the newsletter. Sitting down with associates and thinking of all the possible ideas for contributions and articles and writing them down as they occur provides a starting point for planning the content of individual issues. What are the major issues currently of interest in the field? What are the major problems? What are the major controversies? What pending legislation could have an effect on members? Which of these would be of sufficient interest to members to cover in the newsletter.

After developing a list of topics, identify individuals who might develop articles in the various areas or provide advice on whom to consult. Among the possibilities are society officers, experts and authorities in each of the subfields of the society speciality, and those who regularly participate in the annual conferences.

Third, develop a list of the relevant published sources that might provide information. Among these are newsletters of other societies; newspapers, magazines, and trade journals; so-

ciety proceedings and journals; news releases; reports; catalogs; and calendars.

Consulting Knowledgeable Advisors

Through letters, telephone conversations, personal contacts, and discussions at society business meetings and annual conferences, knowledgeable advisors can provide vital information on topics to be covered, information resources, and ideas for expanding and improving the newsletter. The newsletter editor should be in continual communication with society officers, especially the president or chairperson, to exchange ideas about policies and information to include. It is important for the newsletter editor to recognize that the society runs the newsletter and not vice versa.

Reviewing Relevant Literature

Regularly reviewing the journals, reports, and newsletters relevant to the interests of society members provides both a source of ideas for articles and information to reprint with credit given the original source.

Attending Society Business Meetings and Annual Conferences

The most valuable sources of information about what is relevant to professional society members are the business meetings and the annual conferences. The newsletter editor should attend such meetings regularly and actively seek ideas for contributions and content.

ESTABLISHING METHODS FOR OBTAINING INFORMATION

One of the most difficult aspects of newsletter editing is obtaining information of sufficient quality to fill each issue. It has been estimated that the newsletter editor will have to generate approximately 50 percent of the copy for each issue [13], a considerable amount if the newsletter is large. Several methods can be used to obtain information: (1) Establish a staff of contributing editors; (2) inviting authorities in the field to prepare key articles; (3) select significant articles and proceedings papers to reprint; (4) publishing requests for contributions in the newsletter; (5) subscribing to other publications; and (6) subscribing to a clipping service.

Establishing a Staff of Contributing Editors

One of the most effective ways to obtain interesting material for the newsletter is to establish a staff of contributing editors who will provide regular feature articles or columns in the areas of major concern to society members. This involves inviting the authorities in each of the major interest areas to join the newsletter staff. Contributing editors do not have to write an article for each issue but merely to arrange for an article for their regular columns. The presence of a staff of contributing editors who are technical experts provides substantial, authoritative coverage of the items of major interest to readers.

Inviting Authorities to Contribute Feature Articles

A related method is to invite authorities in the field to write a single article (or a series if they desire) on a topic of key interest. Not everyone is willing to become a regular contributor but many individuals are willing to write a single article, if given sufficient lead time.

Selecting Significant Articles and Proceedings Papers to Reprint

Many papers on timely topics are written for conference presentation but unless members attend the conference, they may not be aware of significant developments and findings. Selecting key conference papers for reprinting can provide an additional source of articles for the newsletter and also a service to society members. (Permission of both author and proceedings editor is required for reprinting.) Reprinting selected papers from the proceedings can serve to stimulate interest in that publication as well—a supplemental benefit.

Publishing Requests for Contributions in the Newsletter

A standard method of obtaining contributions is to publish an announcement of the types of information the newsletter covers, then to publish requests for contributions along with a calendar of deadlines. Although ordinarily a limited amount of information is obtained by this method, it should not be neglected.

Subscribing to Other Publications

Much of the information to be included in the newsletter will come either directly or indirectly from materials published by others. To obtain access to information directly or peripherally related to readers' interests, it is helpful to subscribe to magazines and journals in appropriate areas and ask to be put on the mailing list for newsletters, reports, and news releases of relevant organizations.

Subscribing to a Clipping Service

If the budget permits, it is often convenient to subscribe to a clipping service to receive copies of key articles on topics of interest.

CREATING A DISTINCTIVE IMAGE

The society's publications are not only the major channel of communication between the members but also the basis on which members form their impression of the society. Creating a distinctive image is of utmost importance. It can be done by (1) developing a functional format; (2) creating an attractive design; (3) editing articles for grammatical correctness and consistency in style and form; (4) writing eye-catching headlines; (5) proofreading meticulously and double-checking accuracy of all information; and (6) consistently publishing on time.

Developing a Functional Format

Newsletters can be styled with either one, two, or three columns per page. Double-column format is preferred by many:

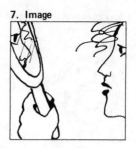

7. Image

It has the most attractive appearance in typewritten newsletters, especially if copy is prepared on a standard typewriter without the capability of justifying both right and left margins. The format is flexible and pages are easy to lay out. Headings can be set across one or two columns to distinguish major and minor articles, and announcements and filler can be set in wider measure for variety and emphasis.

A three-column format has a more professional appearance but more sophisticated typesetting is required and layout is more complex [14]. The three-column format works well only if copy is typeset and both right and left margins are justified. It is extremely difficult to produce an attractive-looking typewritten newsletter using three-column format.

Using a single-column format is generally ineffective, especially for typewritten newsletters, because the finished product too closely resembles a typewritten report. The format selected should be functional for the production mode used; it should be flexible; and it should be easy to produce.

Along with a standard format for the newsletter, a standardized organizational plan should be developed to create a sense of order and continuity. There should be a standard sequence of sections and departments in each issue, readily identifiable by major headings. In structuring the organizational framework of the newsletter, consider using an interesting photograph or feature article on the front page of each issue, along with the society logo and a table of contents, "News at a Glance," or "In this Issue" section.

The table of contents at the beginning is especially valuable because it highlights important articles and enables readers to locate information of interest quickly and easily. Either the second or the last page should contain the masthead: purpose and scope of the publication, frequency of publication and deadlines for contributions, and identification of the editorial staff. If the newsletter staff includes contributing editors, the masthead should include not only the name but also the address of each so that readers who wish to contact them can do so easily. This is especially important if a forum for reader exchange of ideas is part of the newsletter. The newsletter should also list the society officers in a standard place each time. Major articles should follow the first page, interspersed with regular features, announcements, and news briefs. Articles should be grouped with regard to content.

Not all important information should be put at the front, however. Studies have shown that some readers start at the back and move forward; thus information to attract reader interest should be placed at the end as well as at the beginning [7]. Major sections of the newsletter (for example, messages from officers, contributing editors' regular columns, minutes, awards, calls for papers, calendar of coming events, classifieds) should be arranged in standard sequence and format in each issue. Part of the last page should be reserved for mailing information.

Creating an Attractive Design

It is possible to print articles exactly as they are submitted by contributors, especially if an 8.5" X 11" page size is used, but doing so does not make a very attractive newsletter. With different typefaces and margins, pages resemble excerpts of a manuscript or report more than a unified publication. Not only is good design lacking but space is wasted.

A newsletter combines features of newspapers (transferral of specific facts), magazines (transferral of concepts), and advertising (transferral of concepts and stimulation of response); as a result, visual presentation is especially important. The appearance of a newsletter influences a reader's response and conveys an impression of the society to the reader. Type, headlining, graphics, and white space should be arranged to create reader interest and convey a favorable image. The design has other functions, too. Like the makeup of pages in a newspaper, it should help the reader get the maximum amount of information in the shortest possible time; it should grade the news—letting the readers know which articles are more important and which are less important, which are light and which are serious; it should help the reader move easily from one item to another and provide quick and handy summaries of the news at a glance [7].

Creating an attractive design involves decisions about body and display type, page layout, and use of illustrations. The typeface selected for a newsletter contributes greatly to the overall image of the publication. Serif types are conventional and conservative; sans serif types suggest a modern or avant garde approach. Readers generally prefer serif types; they are familiar, legible, and readable [7]. One typeface should be chosen for the body copy and one for headings; type styles should not be mixed on a page: A more harmonious appearance arises from setting all articles in the same body type and all headings in the same display type, with variations only in size and boldness to designate major, secondary, minor, and special articles. Using a different type style for each article and a different display type for each heading creates a jumbled, hodge-podge effect.

Columns of type, headlines, and graphics should be arranged attractively on the page and should create an effect of balance and symmetry. Pages should have some contrast as well as balance and symmetry. If there is excess white space at the bottom of a page or if columns end unevenly, the impression is one of sloppiness or carelessness. Balanced pages are created by alternating long and short articles and selecting and editing filler and announcements to take up extra space. At low cost, rules or boxes can be placed around articles and announcements for emphasis and added interest. Shading can be used, although at greater cost.

If the budget allows, photographs are an excellent way to add interest to a newsletter and to improve the appearance. Photographs should supplement the articles and draw the reader's attention to the text rather than being merely window dressing. Photographs of society officers can accompany messages and photographs of contributing editors can be used on a staff page or as an accompaniment to their regular columns. Photographs of meeting sites can be used to stimulate interest in upcoming conferences; photographs of participating society members (not officers) can accompany reports of conference activities. Photographs can be used to introduce candidates for office and to identify and honor those who receive awards. Artwork or cartoons can add interest to newsletters and enhance the attractiveness of the design. Graphics should be used judiciously, however; too much detracts rather than adds interest.

In summary, a good design should not call undue attention to itself; it should create interest and draw the reader's attention to the articles themselves. Thus, in designing a newsletter, it is helpful to

- Select a single typeface for body copy and a single display type for headings, varying size and boldness of headings rather than type style.
- Fit as much information on a page as possible without crowding.
- Arrange columns on pages to provide symmetry and balance. For newsletters printed on two sides of the paper, consider facing pages as two-page spreads rather than as individual sheets.
- Keep as much of an article together as possible. Avoid awkward or excessive continuations.
- Alternate long and short articles and edit filler to complete pages.
- Use extra white space to advantage by adding announcements, queries, reminders, requests, and the like, delineated by rules or boxes.
- Use graphics to add interest.
- Be guided by moderation. Excess or garish graphics destroy the image of professionalism.
- When the layout of the newsletter is finished, apply a storyboarding technique: Lay out all the pages in order on a large conference table (or pin them to a storyboard if one is available) and view the story the newsletter tells. Is the sequence of articles visually interesting? Is there a balance of long and short articles? Do the pages create reader interest? Are announcements, queries, requests, and other filler properly interspersed? Is the overall design attractive? Modifications can be made as needed until the organization and design are pleasing.

Editing Articles for Grammatical Correctness and Consistency in Style and Form

The quality and style of contributed articles vary widely. Therefore, to establish high quality in the publication as a whole, it is necessary to establish standards of style and form. Some articles will be well written and carefully proofread; others may be characterized by any number of major or minor grammatical and stylistic errors. Editing for grammatical correctness and stylistic consistency increases reader interest and the readability of articles without destroying individual writing style. Attention to the quality of writing enhances the overall readability and attractiveness of the newsletter.

Writing Eye-Catching Headlines

In a newspaper, headlines help sell the paper, index the news, grade stories according to importance, summarize the news for hasty readers, and make the paper more attractive [7]. In a newsletter also, the headings serve to create reader in-

8. Following Through

terest and to induce readers to read the articles. The headings or headlines should be indicative of the contents of the articles, relatively brief, and phrased to catch the eye.

Proofreading Meticulously and Doublechecking Accuracy of Information

Once articles and headlines are set, they should be proofread extremely carefully. Particular attention should be paid to names and affiliations and to numbers: statistics, addresses, telephone numbers, and dates. The accuracy of all information to be included should be double-checked before publication.

Publishing on Time

A high quality newsletter is produced by setting high standards for content, writing quality, style, and design; by maintaining them; and by consistently publishing on time.

FOLLOWING THROUGH

The final step in the process of editing a newsletter is following through. This involves brainstorming each issue to develop topics and sources of information; coordinating and following up until all material is received, edited, composed, proofread, designed, pasted up, published, and mailed; and meeting deadlines.

Well before contributions are due, each issue should be carefully planned; possible ideas for contributions and articles to be included should be brainstormed and possible contributors listed. Each issue should have a major theme, whether it is the announcement of an upcoming conference, discussion of a landmark court decision, or results of a major report or survey. Once the issue has been brainstormed and logical sources identified, decisions should be made concerning ideas to pursue, and they should be followed up. Letters should be sent to regular contributors reminding them of the deadline and to the special contributors inviting them to prepare articles. Followup through additional telephone calls or letters may be required.

Contributions are usually received slowly; not all who promise to contribute will actually send material. The contributions received and the production tasks must be coordinated. Keep a record of articles requested, sources contacted, articles received, production status of each article, tasks completed, and those remaining to be done. A simple checklist is quite helpful. As articles are received, they should be edited and prepared for composition; afterwards, they should be carefully proofread. Pages should be designed for maximum in-

terest, readability, and attractiveness. Each phase should be completed on schedule to ensure timely publication.

The process of editing is not a difficult one, but it requires planning, organization, coordination, and attention to detail.

CONCLUSION

In engineering design, the quality of a product depends on accurate and extensive analysis, complete and thorough investigation of alternatives, detailed and purposeful design, and careful and ordered application of the design [15]. The process is equally applicable to newsletter editing. A quality newsletter should be timely, interesting, informative, eye-catching, readable, accurate, and authoritative. More specifically, it should contain accurate, timely, well written articles on topics of interest to members; authoritative coverage of issues; and accurate, carefully researched, carefully proofread information. By using a systematic approach and following each step through, it is possible to publish an interesting and informative newsletter on time.

ACKNOWLEDGMENT

A number of individuals have provided helpful suggestions about newsletter editing and I appreciate their assistance: Joseph Bordogna, Lyle D. Feisel, Allen W. Hahn, Edwin C. Jones, Jr., Donald E. Kirk, Robert E. McIntosh, James R. Roland, Ronald J. Schmitz, and, especially, Floyd L. Cash, co-editor of the joint newsletter of the IEEE Education Society and the American Society for Engineering Education, EE Division.

REFERENCES

[1] J. B. Bennett, *Editing for Engineers*, New York: Wiley-Interscience, John Wiley and Sons, 1970.

[2] L. M. Zook, Ed., *Technical Editing: Principles and Practices*, Washington, DC: Society for Technical Communication, 1975.

[3] R. Van Buren and M. F. Buehler, *The Levels of Edit*, 2nd ed., JPL 80-1, Pasadena, CA: Jet Propulsion Laboratory, California Institute of Technology, January 1980.

[4] *A Manual of Style*, 12th ed., Chicago: The University of Chicago Press, 1969.

[5] *U.S. Government Printing Office Style Manual*, rev. ed., Washington, DC: Government Printing Office, 1973.

[6] W. J. Bowman, *Graphic Communication*, New York: John Wiley and Sons, 1968.

[7] A. T. Turnbull and R. N. Baird, *The Graphics of Communication*, 3rd ed., New York: Holt, Rinehart, and Winston, 1975.

[8] M. E. Skillin and R. M. Gay, *Words Into Type*, New York: Appleton-Century-Crofts, 1964.

[9] J. V. White, *Editing by Design: Word and Picture Communication for Editors and Designers*, New York: R. R. Bowker Co., 1974.

[10] L. H. Wales, *A Practical Guide to Newsletter Editing and Design*, 2nd ed., Ames, IA: Iowa State University Press, 1976.

[11] *How to Produce Newsletters*, Washington, DC: Association for Educational Communications and Technology, 1979.

[12] M. Beach, *Editing Your Newsletter*, Portland, OR: Coast to Coast Books, 1980.

[13] *Helpful Hints for Newsletter Editors*, Washington, DC: The American Society for Engineering Education.

[14] J. A. Steeh and E. S. Maxey, "Newsletters for Fun and Profit," *Proceedings of the 29th International Technical Communication Conference*, May 1982, W95-98; Washington, DC: Society for Technical Communication.

[15] J. W. Souther and M. L. White, *Technical Report Writing*, 2nd ed., New York; John Wiley and Sons, 1977.

Writing to Persuade

J. W. GILSDORF

Abstract – Technical communication may be chiefly informational in intent, but there are persuasive elements in almost all communication. Sometimes, even in technical matters, persuasive communication skills are very important. The author provides arguments for improving these skills, discusses a number of "mini-strategies" that can be used to persuade the recipients of written communication, and touches on the ethics of using persuasion tools.

Technical writing would seem, on the surface, a simple matter of transfer of information. And, in fact, much of what is written about technical writing deals with informing. However, most engineers and managers would agree that the ability to persuade is a necessary skill. Although numerous articles on persuasion appear in the business and professional press, they typically address the "hotter" topics of oral and nonverbal persuasion.

Ironically, as the functions of an enterprise become both more specialized and more interdependent, written persuasion becomes even more essential. Written persuasion is needed more than ever to coordinate activity within the technical organization. A few examples include justification reports, explanations of delays or changes in plans, requests for increased staff or capital outlay, and position papers. Even more numerous are the daily routine persuasive memos, written, for example, by engineer to engineer, arguing for closer cooperation, or by manager to technicians, motivating them to meet a shortened timeline on a project.

The subjects of pieces published in business magazines – on negotiating, silent command, motivating, influencing, wielding power, one-minute management, and even power dining – do not add up to the skill set needed for persuasive writing. Oral persuasion skills, though valuable, do not substitute for the equally important ability to persuade in writing.

Oral persuasion shares numerous elements with written persuasion. In both, for instance, audience analysis is essential. But many factors are different. Some of the tools of oral communication (facial expression, posture, tone of voice, and interpersonal space) are unavailable for written persuasion. On the other hand, the receiver of written persuasion has more opportunity to examine the message closely. Many of the quick-compliance triggers used in oral persuasion will not work in writing. And, unlike oral persuasion, written persuasion is often one-way and one-shot.

Let us look at some aspects of persuasion that underlie any written persuasive task.

WHAT IS PERSUASION?

Persuasion is a conscious effort by a sender to change the opinions, attitudes, beliefs, or behaviors of a receiver through the transmission of a message. Based on thorough audience analysis, persuasive writers select, order, and word the elements of their message so as to encourage desired responses in their receivers. Persuasion does not coerce; the receiver is free to choose.

Persuasive writing is usually tougher than informative writing. To persuade, the writer has to cause a reader not only to pay attention initially but also to read through to the end: the move to action is usually last. Because they are multisided, issues and situations requiring persuasion are often more complex than those requiring just information. Then, too, everyone is short of time. The longer and more complex a message, the greater its potential for putting readers off.

Selection of words, evidence, and many other creative resources becomes more crucial in persuasion. The person reading information generally wants the message – indeed, has often requested it. The person reading persuasion often resists it, along with its efforts to work change. Any ill-chosen element can cause such a reader to stop reading.

Dr. Gilsdorf is Associate Professor of General Business at Arizona State University, Tempe, AZ.

Reprinted from *IEEE Trans. Prof. Comm.*, vol. PC-30, no. 2, pp. 68–73, June 1987.

Persuasive messages contain subtle emotional as well as rational components in an effort to establish credibility. If the rational evidence "feels right," and if the emotional appeals do not offend the receiver's rational sense, then the message tends to sound believable.

Credible, ethical persuaders say nothing that is untrue. Untruth is neither wise nor necessary. Rather, effective persuasion strategy begins with statements the audience finds agreeable, and gradually adds new material that will move the audience toward the new position of agreement.

INFORMING AND PERSUADING – A CONTINUUM

Engineers and technical writers who have been asked for objective information strive to provide material that is as free of personal bias as possible. The communicators are careful not to persuade accidentally when the assigned task is to inform. They monitor their intentions toward the receiver and keep the purpose of a message clear in their own minds.

On close examination, however, it is clear that few human communications are entirely free of persuasiveness. Informing and persuading differ more in degree than in kind. Even in objective reports, the writers present their information in such a way that the readers will accept the facts at the author's word. Establishing credibility for an objective message is, at least minimally, persuasion.

In all messages, informative as well as persuasive, a communicator selects among all possible words and ideas that could be presented. The mere act of selection precludes complete objectivity.

Most messages, however, have a primary focus, a primary intent that is either to inform or to persuade. A persuasive message might ask the reader to set up a committee, to support an untried idea, to allocate funds for training, or to grant an extension on a deadline. All these work to create a change in the reader's attitudes or behavior. In contrast, an informational message, such as the write-up of an experiment, primarily conveys data.

Most written messages in organizations fall somewhere on the information-persuasion continuum sketched in Fig. 1.

Near the extreme persuasive end of the scale, one finds the simple, single, repetitious message, containing very little information but aiming at almost a conditioned response. Near the extreme informational end of the scale is the empirical scientific research report, where even the selecting of data to present must be done with as nearly complete impartiality as the researcher can achieve. Most of what we think of as technical written communication falls at a considerable distance from the persuasive extreme. Yet many messages in a technical organizations will have some persuasive intent.

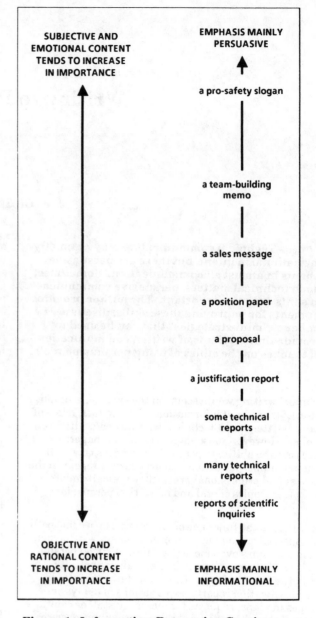

Figure 1. Information-Persuasion Continuum

Most persuasive messages, of course, inform as well as persuade. Typically, as the persuasive content of a message increases, the subjective component increases in importance – although it often remains quite subtle. In either informing or persuading, the impression of credibility must never wane. No matter how true or well-reasoned a message, if it stops seeming true, the receiver stops accepting it.

CREDIBILITY, RECEIVER'S EXPECTATIONS, AND TONE

To be perceived as credible by the receiver, the sender first must be credible. Within an organization, people accumulate a personal history. Persuasive efforts are not accepted from sources who have not shown themselves reliable, competent, and trustworthy.

To be perceived as credible, the sender must also meet the receiver's expectations. The writer must assess, for instance, the reader's knowledge level, sensitivity to language, and style preferences, and adapt the message accordingly. Not all arguments carry the same weight for all readers.

Important advice to all persuasive writers is, "Know thyself." If writers analyze themselves as well as their intended receivers, their own mental sets and emotions can work for them – suggesting a useful word connotation, for example – rather than against them, impeding their effectiveness. A person not fully aware of his own attitudes on a subject will have difficulty getting far enough outside those attitudes to influence others.

In part, and particularly when writers send messages to persons who do not know them, credibility is established by control of tone. If a persuasive message strikes the receiver as sly, evasive, exaggerated, crafty, or secretive, for example, the writer is unlikely to be believed.

Tone must be sincere, courteous, and professional. Keeping the "you" attitude at the mind's forefront helps a writer maintain an appropriate tone. The writer needs constantly to ask, "How would this sentence strike me if I were in the reader's place?" A persuasive message must never seem to scold, preach, shove, lecture, talk down, beg, threaten, whine, blame, pontificate, insult, or flatter. Rather, it invites the reader to think along with the writer, as the writer pursues a reasonable, easy-to-follow, reader-centered discourse.

Always, a persuader needs to intend, and seem to intend, good toward the persuadee. Mutual benefit is acceptable to the receiver of the communication, but the persuader must not appear to be promoting his or her own gain over that of the persuadee. The persuader can present evidence and appeal to emotion but must not force an issue or psychically corner the receiver. If the receiver picks up any signal meaning, "Aha! Now I've got you!," the persuader has just lost the cause.

PERSUASION AND THE HUMAN PSYCHE

Several elements of the human mind take on added importance when persuasion becomes a primary goal.

For one thing, human beings are influenced by emotion, to a greater or lesser degree. Most technical readers feel that they keep emotion out of their thinking. By emotion, however, they may mean wild grief, fury, scorn, ecstasy, or weepy sentimentality.

Most emotions that actually move people to action are on a much smaller scale. Quiet pride, for example, is important to most people, engineers included, as is self-esteem. There is an emotional component in people's commitment to ethics and responsibility, their role as a behavior model, or their protective feelings toward subordinates.

Fear, guilt, and uneasiness are other emotions that move people, perhaps at the subconscious level.

The persuasive writer's appeal to emotions such as these must be subtle, not obvious. A nuance, a word connotation may tap into an emotion successfully where a blatant appeal to emotion would surely fail.

Another influential element is "dissonance" or "lack of closure," and it works to a persuader's advantage. Human beings dislike being undecided or confused about events they perceive as meaningful to them. They are eager to resolve problems and to make decisions, proceeding on the basis of the best information they have available to them. Good persuasive communication helps them do this.

Third, people value what is hard to get. This applies to more than possessions. The contract won through planning and persistence means more than the one that was a foregone conclusion. Hard-to-obtain information or new ideas are more desirable if offered to the reader exclusively or as one of a select few.

People generally want to be nice guys. A writer can mention a past kindness the reader performed and express appreciation again. The reminder tends to revive the pleasant, generous feeling the reader experienced and may thus move him to look more kindly on the present topic. The desire to be nice also improves the success of horsetrading (see below) and related strategies.

Human beings often share opinions, attitudes, and beliefs with reference groups, that is, with others whom they respect and wish to resemble. They make a great many decisions mainly on the basis of what they see one or more of their respected peers doing. Sometimes one of the receiver's reference groups, cited as supporting the persuader's position, can be influential.

In a related vein, a persuader can suggest ways in which writer and reader are similar. When the writer can create the impression of common interest, the underlying message is, "I'm like you. I understand what you need." The receiver of that subtle message will be more likely to follow the sender's leads and arguments than one who feels that they are very different.

People do not like to seem inconsistent. Sometimes a writer can show that what he recommends is very similar to something the reader already believes or does. Or he can win a receiver's agreement on one idea and then use the consistency principle to win agreement on a related idea. People redefine themselves constantly in small ways, but they need to feel consistent in doing so.

Knowing the receiver's frame of reference can sometimes tell a writer what to stress and what to avoid. Say, for example, that a department supervisor has three separate projects, all threatening to go over budget. An engineer has developed a proposal for cross-training several subordinates. If it is offered to the supervisor as an employee morale-builder, the proposal does not answer the supervisor's immediate needs. If, however, the proposal shows clearly how the cross-training plan integrates with one of those critical projects to cut down time or cost, the supervisor has a strong motive for reading and for adopting the plan.

Facts are essential. Facts are what we begin with. Yet people's attitudes toward those facts determine their decisions. Attitudes can change, and be changed.

PATTERN AND STRUCTURE

As the form of a building should reflect the function it is to serve, so structure is important in written persuasion. Depending on the nature of the task, the writer can select from a number of different organizational patterns. Although the basic patterns for persuasion are mostly inductive, he might have good reasons sometimes to write deductively, that is, to "bottom-line it." In addition, many tasks contain persuasive elements but must follow a preset, expected order of development. Proposals, for instance, are not usually structured inductively.

One useful persuasive pattern, called the motivated sequence, can help persuade when the reader is initially uninterested.

The motivated sequence has five steps – attention, need, satisfaction, visualization, and action:

- An attention-getting opener hooks the receiver into beginning to read.

- The writer works to establish and develop the reader's need for his plan, idea, service, or product.

- The writer demonstrates that his plan, idea, service, or product will satisfy the need.

- The writer encourages the reader to visualize himself having solved the problem, enjoying the product, or putting the service to use.

- The writer moves the reader to whatever action he has in mind: authorizing something, for example, or placing an order.

The well-known AIDA pattern (Attention, Interest, Desire, Action) roughly follows this sequence.

A familiar (and inductive) structure for an argumentative or advocacy message arranges arguments and evidence according to relative strength. The following is an order that works well:

- Opener, where subject is introduced using ideas both reader and writer know and agree on

- Strong supporting argument(s)

- Weaker supporting argument(s)

- Citing and refutation of one or more opposing arguments

- Strong supporting argument(s)

- Summary

- Move to action

The placement of the strongest arguments first and last makes use of the principles of primacy and recency. An audience remembers best what it hears or reads first (prime position) and what it hears or reads last (most recently). The less important things are placed in the middle, where an audience's attention tends to lessen.

Some of the opposing arguments need to be mentioned and refuted. The persuader chooses opposing arguments he can deal with, and shows how they are weaker or less important than the reader might think. Sometimes an argument against one's position can be turned into a weak argument in favor of it.

The opposition's strongest counterarguments are best ignored. The reasoning here is that a sound refutation of any argument will demonstrate that the opposing position has its weaknesses. If a strong opposing argument is acknowledged but not effectively attacked, the opposing position seems stronger.

Within the larger structure of a persuasive message, a number of lesser elements can be utilized to make a message more effective. These "mini-strategies" are discussed below. Technical readers would not be persuaded solely on the basis of any one of these, and some of them are appropriate only to certain kinds of messages. Any persuasive messages would contain many other elements to fill out and balance emotional and rational appeals.

Word Connotation

In strictly informative writing, writers generally stay with low-connotation words, that is, words without strong emotional associations. In persuasive writing, because people generally make a decision based at least in part on how they feel, sometimes words with stronger connotations work better.

However, whereas highly flavored words will move people more, it is harder to control the direction and distance they are moved. A writer needs to analyze the audience with great care before choosing the words. Too emotional a word, or the wrong kind of word, can misfire and move a reader a great deal but in an unwanted direction.

To illustrate the difference, the words *fairness* and *reciprocity* are both favorable (not neutral) words for persuading a person to give something for something. The saying "You scratch my back, and I'll scratch yours," carries roughly the same idea, but has stronger connotations and the smack of collusion and underhandedness about it. Except in unusual circumstances, a persuader probably would not use it.

Offering Alternatives

People are often more willing to go along when a persuader tacitly assumes they will do as asked and gives them a choice of ways to do it. For example, a persuasive memo might take a page and a half explaining to the receiver why his help is needed on a new project and then ask, "Would you rather work on component modification with Pam Craig or on fabrication planning with Gregg Baxter?" The receiver is fairly likely to choose one of these responsibilities, either of which the persuader likes, rather than choose the unspoken, unoffered, but very real third option: "No, I don't want to help on this project at all."

Another way a persuader can use alternatives is by setting up a good one and one or more bad ones. From the hearer's or reader's point of view, the alternatives are about equally attractive. The persuader, though, presents attractively the one he wants chosen and presents unattractively the ones he does not want chosen.

Reward and Punishment

A persuader can tell the receiver "what's in it for him" if he decides rightly and what unattractive consequences might follow an adverse decision:

- "You'll feel relieved when ...."
- "If our trailer rigs have to survive Dakota winters without this undercoating .. ."

A persuader can subtly praise a receiver for having the right attitudes or doing the right thing.

Indebtedness

A small favor, even something like a minor tangible gift, can arouse in the receiver the feeling that he ought to reciprocate. For example, a national journal enclosed a quarter in a letter persuading subscribers to provide some demographic information. The coin taped to the short questionnaire created in the receiver a feeling of obligation. The quarter was not offered as payment for time, for few people would consider a quarter good wages for 15 minutes of effort. Yet the response to the query was high. Even when the receiver cares little or nothing for the favor, the feeling of indebtedness tends to arise and to exert some motive force.

Participation

Sometimes persuasive writers can get readers to perform a requested activity or at least to think through a process with them. Participation, especially if it involves actual motion, reinforces what the readers think or feel as well as the likelihood of remembering the message. Promotional material sent to sales prospects often invites the reader to "take this simple test," which will show him how much he needs whatever the seller has to offer.

A writer who needs to demonstrate that a certain cost overrun is inevitable might discuss a critical set of figures and request that the reader check the last and most significant calculation, actually leaving space to do it. Working the figures himself would tend to reduce the reader's skepticism and irritation with the unwelcome news.

Another use of participation involves getting readers to visualize themselves in the shoes of the person whose po-

sition is being advocated. Say that an engineer needs to persuade a manager to change glass tubing suppliers. Among his arguments, he might induce the manager to imagine how it felt, at 3:30 PM last Saturday, to hear a hiss from the vacuum rack and see a crack develop, letting air into the system in which he was measuring adsorption of gas molecules on an experimental compound.

Horsetrading

To horsetrade, a persuader makes an original request bigger than the one he hopes to gain, then further along in the persuasion names a lesser request he could get by with. The reader experiences relief from the original discomfort of the larger request and gets the chance to be a nice guy as well. He might think, "Well, sure, I guess we can let him have *that* much."

Christmas-Treeing

The Christmas-tree strategy works on reader perception of the relative size of things. If a persuader is already fairly sure that a persuadee will agree to a large request, the persuader can sometimes hang one or more considerably smaller requests onto the big one because they seem minor alongside it.

The Yes Mode

This technique is related to audience participation. If the writer has been persuading well, and if the reader has been going along agreeably, the persuader may use a series of four or five questions that the reader is likely to answer with yes. The last question poses the new point on which the persuader wants agreement.

For example, a few years ago a law was being considered that would oblige banks and brokerages to withhold 10 percent of any interest that their customers earned. The financial institutions, not wanting to bear the expense or trouble of doing the tax man's job, sent letters to their customers urging them to protest to their congressmen. Near the end was this sequence:

> Wouldn't you like to keep the government from getting any farther into your affairs? Wouldn't you like to avoid having your hard-earned interest taxed the second it's earned? Wouldn't you like to show how you feel about Uncle $$$am not trusting you? Then isn't it worth a few seconds of your time to sign and mail the enclosed card to your congressman?

Use this technique sparingly and cautiously. A series of questions like these tends to carry high emotion. And, as mentioned above, high emotion can backfire.

Learning

People like to learn new things. They often respond favorably to "Did you know ....?" All messages are partial and selective. (Just consider how Mobil's institutional advertising and Greenpeace would address the same set of facts.) "Did you know" messages can be selected as well, to inform and educate hearers on the aspects of a subject that work to a persuader's benefit.

Easy Agreement

Persuaders should mentally assume the receiver's position and complete this sentence: "I'd do it in a minute except that ...." What does the reader need? A resource who can answer questions on a continuing basis? Some training for subordinates? A reliable, steady supplier? Line things up for him.

One point deserves repeated mention. Persuaders must not say things that are false. But they can select and emphasize what promotes their position and select out or deemphasize what militates against their position.

CAN WE PERSUADE AND BE ETHICAL?

Since persuasion manipulates a reader's emotions, ignores unhelpful facts, and permits persuaders to get their own way, is it therefore unethical?

The answer lies in the persuader's intent. If persuasive skill is used to cheat or otherwise harm the recipient or others, it is unethical. If the persuader is fair and does no harm, the persuasion – so long as it does not contain untruth – is ethical. Persuasion is a tool which can be used in many ways; ethical use depends on the user.

We are all consumers as well as providers of persuasion. Daily, we expose ourselves to the persuasion of others. We need to do this; we need others' arguments and evidence because we are busy. We cannot and do not want to discover and reason out everything for ourselves.

If a persuasive message offers us something we want or can use, and if we trust the sender, we are willing to let the communicators select, order, word, and all the rest.

Persuasion is a fact of life and a fact of organizational life. Persuaders who understand their task can bring much greater skill and success to that task.

BIBLIOGRAPHY

1. Bettinghaus, E. P., *Persuasive Communication*, 3rd ed., Holt, Rinehart and Winston, New York, 1980.

2. Cialdini, R. B., *Influence*, William Morrow, New York, 1984.

3. Clark, R. A., *Persuasive Messages*, Harper & Row, New York, 1984.

4. Fisher, R., and Ury, W., *Getting to YES*, Houghton Mifflin, Boston, 1981.

5. Karlins, M., and Abelson, H. I., *Persuasion*, Springer, New York, 1970.

6. MacLachlan, J., "Making a Message Memorable and Persuasive," *Journal of Advertising Research 23*, 6 (1983/1984), 51-59.

7. Simons, H. W., *Persuasion: Understanding, Practice and Analysis*, 2nd ed., Random House, New York, 1986.

Some Guidance On Preparing Technical Articles For Publication

RICHARD MANLEY

JUDITH GRAHAM

RALPH BAXTER

Abstract—Writing for publication in the professional literature brings prestige to the author and recognition to his employer. Yet in many organizations, few technical articles are written for publication in the professional literature. This paper discusses some of the elements involved with writing technical papers for publication within a technical corporate environment. It outlines some of the major incentives and barriers to publishing and offers suggestions for addressing the barriers from both the author's viewpoint and that of management. Those elements composed the foundation of a workshop designed to encourage technical professionals to write for publication. The workshop included exposition of the reasons to publish, addressed barriers to publication, discussed skill building, and contained company-specific briefings on internal release procedures and management support for writing.

MANY ENGINEERS AND SCIENTISTS carry out their research and analysis projects in the corporate world, on internal development projects or on contract projects awarded by some outside agency or sponsor. Often these project professionals write documents describing the solution to a technical problem or recommending some action based on their analysis or study. Yet few of the technical professionals who write these business documents take the next step and prepare articles for the technical community at large.

Business documents have a specific purpose: to present the results of the research, study, or analysis, cast in terms of a solution to a particular problem. Although the research and analysis itself may prove exceptionally worthy, the documents are usually the means to satisfy the initiating inquiry. In many organizations, detailed discussion of the substance of technical analyses is often relegated to an appendix of the business report.

Engineers and scientists often record results informally in briefing charts, technical memoranda, working papers, or occasionally a technical report. These documents see limited distribution, often only within the sponsoring organization, and concern themselves with procedure and analysis as an adjunct to reporting results.

Documents like these may contain answers that reveal insights developed by the researchers or point to some new area of analysis that might fall beyond the scope of the current investigation. Unfortunately, because the documents themselves are often company confidential or even classified in a national security sense, their circulation is further limited. Yet the researchers have done good work and the investigators take pride in their accomplishments. Most would be pleased that the technical community is interested in the paths they followed and in sharing the insights they gained.

Clearly, the investigators can benefit professionally by disseminating their work to as large an audience as possible. [1]. Publication demonstrates command of a specific area, exhibits the company's involvement in the particular field, and shows active participation in the technical community. In many cases, technical publications provide a means for the ideas and work of the staff to gain exposure to the management of their own organizations. Yet few authors of in-house technical publications take the next step. That is, few extract the seminal ideas from their work and prepare an article for publication in the professional literature of the field.

MITRE Corporation's in-house technical training and education wing is the MITRE Institute. The staff of the Institute recognized that barriers within the corporate environment can prevent disseminating technical work in a form suitable for communication to the larger community. Our work led us to design a workshop to guide aspiring authors

Richard Manley is Technical Program Manager for the MITRE Institute, the educational and training unit of the MITRE Corporation.

Judith H. Graham is President of Graham Associates, a Virginia consulting firm specializing in technical and business writing instruction.

Ralph C. Baxter is President of New Dominion Services and Professor at George Mason University (Fairfax, VA).

Reprinted from *IEEE Trans. Prof. Comm.*, vol. 32, no. 1, pp. 5–11, March 1989.

in surmounting those barriers over which they have some control and in managing those over which they have less.

This paper discusses some of the elements involved in writing technical papers for publication within the corporate environment. It outlines some of the major incentives and barriers to publishing and offers some suggestions for addressing them from the viewpoints of both authors and management. These elements then became the foundation of a workshop designed to encourage technical professionals to write for publication within our company's specific environment.

SOME MAJOR ELEMENTS IN PREPARING TECHNICAL ARTICLES

Writing technical papers for publication involves a number of activities that are strongly influenced by many factors both within and beyond the direct control of authors. We considered the following list of topics to be a comprehensive and useful one for the purposes of our workshop:

- Understanding why/why not to write professional papers
- Identifying an appropriate conference or publication
- Classifying publications
- Analyzing the elements of journal articles
- Using research resources
- Gaining public release authority for the work
- Clarifying management support

WHY PUBLISH

In an intense technical environment such as ours, deadlines, briefings, and new projects all have high priority. In the face of these demands, it is often necessary to remind the technical staff of the benefits that can accrue to those who write about their work in the literature of a professional field. Our staff members have found the following benefits in varying degrees when they write for publication:

- New insights
- Expert feedback on the reported work
- Professional interest in sharing results
- Prestige
- Personal/professional gratification
- Renewed research funding
- Management awareness of the value of individual work to the field

Among the most compelling reasons to publish is that writing a succinct, clear paper aimed at one's colleagues helps the author generate new insights into his work. Putting one's thoughts down on paper seems to discipline the thought process. Reconstructing the steps that were taken to achieve some set of analytical results and developing those novel or innovative portions of the analysis in greater depth allow an author to revisit work with the benefit of hindsight. Thinking about the facts and drawing them to-

gether in a logical arrangement—which often differs from the chronological order—and then setting them down tends to yield a fresh perspective and new understanding [2].

When the manuscript is submitted to a publication for consideration, many publications and all scholarly journals have it carefully reviewed by experts in the area. The author clearly benefits from the feedback his colleagues can offer regarding the context of this writing. They often point out new information, ask questions that lead to new insights, and offer recommendations that can clarify thinking.

Of course, once the author has completed his paper and finally views his work in print, he feels a sense of personal accomplishment and professional gratification. Adding to this is the fact that colleagues and associates are now aware of the author's contribution to a particular field and tend to hold the author in higher regard for the capabilities which the paper demonstrates.

In some cases, renewed funding for research projects depends at least in part on reporting results to the community. Many research organizations consider it a part of their responsibility to disseminate their findings once the material has been released for public distribution. Papers written about the company's research results demonstrate its presence and capabilities in a field that the company considers important enough to fund with research dollars.

Management awareness of an individual's contribution to a field on behalf of the company is still another important benefit. Most professional staff in large organizations have only limited exposure to their upper management. But publishing distinguishes the individual author and sets him apart as one who may be furthering corporate goals by writing on topics that the management considers worthy.

Some organizations have formal programs that recognize authors of technical papers. In our organization, management strongly encourages publication of technical work. In an annual "Best Paper Contest," MITRE awards a substantial cash award for the year's best contribution to the literature of a field; papers published in refereed journals or proceedings earn lesser but considerable incentive awards. The writers are featured in the company newsletter and are usually photographed with company officers presenting the award and offering their appreciation. The newsletter usually briefly describes the winning paper and lists authors who received the incentive awards. The Corporation's publishing awards program has features similar to that introduced some years ago at Motorola and reported by Marsh [3], and to that described by Anderson [4].

WHY NOT PUBLISH

The path that leads from an idea for a paper to an actual publication of an article is usually a long, unclear, often confusing one. It demands patience and clear communication with one's management and company organizations as well as sponsors and with the publication's editorial staff. Technical people fail to write technical papers on work-related topics due to a host of difficulties. Reasons given for failing to write for publication seem to fall into two main categories:

- Lack of information about writing opportunities and procedures
  - Difficulty in choosing a topic
  - Difficulty in identifying an appropriate forum for the article
  - Difficulty in keying the article to the forum
  - Low odds of acceptance squelching motivation
  - Critical reviewers
- Lack of skills and resources for formal writing
  - Lack of basic writing skills
  - Little business time for writing
  - Minimal financial incentive preventing use of personal time on such endeavors
  - Uncertainty about management policy, difficulty of security review, controls denying discussion of seminal results, uncertainty about release of data
  - Lack of support resources (management, secretarial, editorial, research)

ADDRESSING OBJECTIONS

Writing Opportunities

Most staff members write in association with their jobs. Nearly all staff have some writing responsibilities for documenting their work and the results that they achieve. Most papers written by the technical staff can be classified as technical business papers; they represent a body of knowledge upon which the author can draw for a technical article.

The general format of a technical business paper opens with a brief executive summary aimed at decision makers, focusing on the results and how they impact the business. Other sections of the business report develop the procedure, elaborate on the nature of the problem, discuss results in relation to other work or the limitations of the current work, and sometimes include appendixes that explain technical details. But the primary purposes of the business paper are to report some results in a business context and to provide some recommendations for action based on the results. The procedures and methods used, the study's relation to other work, and the technical details mainly provide little interest compared with the results.

Usually, readers of the literature of a professional field are most interested in those very parts that are subordinate in the business paper. So, in using a business report as a source document for a technical article, a major step is to change its emphasis. The author must often recast the article to clarify and focus on those aspects of the work that are of primary interest to technical colleagues who read a particular publication and are the likely audience for the article. It is also important to keep the publication's editorial policy in mind and target one's writing to accommodate it.

Selecting a Specific Topic

Each project or related work usually contains dozens of possible ideas for technical papers if evaluated carefully. The individual author and his management should review his work assignments together and glean a set of promising topics; the set can be narrowed down further as each is better understood. A major criterion for choosing one topic over another may very well be related to the corporation's overall strategic goals.

The company management usually has a clearly articulated strategic plan for the company's growth in the near future. Often, a component of the plan involves establishing some credentials in new fields of interest to the company. If an aspiring author is familiar with, or working in these fields, he can further his firm's reputation and distinguish himself by writing on such specific topics. Technical papers may be strategic for their subject matter, but good timing is also very important. Papers that appear late are usually of little value to the corporate strategic plan.

Even if no new fields are to be conquered, writing about one's work in the technical disciplines is important to advancement of the field. Authors should work closely with management and technical experts to select topics that are both interesting to the members of the profession and yet likely to pass corporate release approval, declassification procedures, and the review process. A case study, a novel application of a well-known approach, or a description of a new model for a process may all be appropriate topics. The author must keep in mind the operative constraints—the audience's fields of interest, the editorial policy of the targeted publication, corporate goals and policies, sponsor goals and policies, and the availability of resources to deal adequately with the chosen topic.

Outlet Opportunities

According to the New York Times, more than 40,000 professional journals and business/trade publications document work in nearly every field imaginable. Although the literature of electrical engineering and computer science represents a small portion of that total, it offers many opportunities to recast the business report as a technical paper or article, including the following:

- Professional journals
  - National societies (e.g., *Proceedings of the IEEE, Journal of the Association of Computing Machinery*)
  - Subsocieties (e.g., *IEEE Transactions on Computers*)
  - Special interest groups (e.g., *IEEE Transactions on*

Software Engineering, ACM SIGSOFT Notices, IEEE Computer magazine)
- Conference proceedings
  - National conferences
  - Regional conferences
  - Special interest group/subsociety meetings
- Broader interest publications (e.g., *BYTE, Defense Electronics*)
- Trade journals (e.g., *PC Week, Electronic Business*)

A comprehensive list of IEEE journals appears in *IEEE Spectrum* and in the membership materials distributed by IEEE. Similar lists appear in ACM publications.

SOME CHARACTERISTICS OF AN ARTICLE

Journal Article

Understanding the characteristics that various classes of professional publications expect for articles helps to target the writing. The journals and transactions of the various professional societies are the most scholarly. Journal articles convey information to other practitioners in a particular field, people already familiar with the subject matter of the technical material. The introduction to a journal article typically includes a rationale for the work based on a set of technical objectives and features a literature review of related work. The journal article differs from the business paper in that it contains not only the results, but also details on how they were reached. The discussion section focuses on the results and their significance to other work in the area. Detailed technical aspects of the work inappropriate for the body of the paper often appear in an appendix. Michaelson [5] describes several formats that journal articles might take.

A journal's editorial policy may appear in the publication itself. It describes the subjects covered by the publication, acceptable length, writing style, and format expected. Although reviewers' comments and correspondence with the professional journal offer expert feedback, authors face difficulty in gaining acceptance for their papers in such publications. Because space in them is at a premium, stiff competition and delays in publication are commonplace. The length of the manuscript is crucial for a journal. Usually, only those that meet strict length limits are accepted.

The author and editor of the journal have a critical relationship in preparing an article for a technical journal. The editor judges whether an article should be included in the publication. Normally, the editor assigns the article to a group of peer reviewers for comment and critique. When they complete the review, the editor returns the article to the author with suggestions and reviewers' comments. Because most authors dislike criticism of their work, they often take it personally even when the legitimate comments are helpful. Experienced authors, however, view such feedback as valuable free advice from leading figures in the field. Not only does the critique lead to a better paper;

it sometimes points out other relevant technical issues or resource materials that can shed further light on the topic.

Special Interest Group Publication

Special interest group (SIG) publications offer another outlet for technical writing. Usually less scholarly than the transactions of the parent society, SIGs are likely to be more timely and relevant to a specialized readership. Often the SIG dedicates a particular issue of its publication to a special theme. Articles in theme-related issues include tutorials on the basic technology of the topic, a review of the state of the art in one or several aspects of the field, and reports of likely new directions the field might take based on recent results. For example, in 1988, the IEEE *Software* magazine published separate issues focusing on the special topics of parallel processing, CASE, and fourth-generation languages, respectively. Most publications promote their special theme issues well in advance. Their advance promotion efforts should give authors enough time to find an issue that relates to their work. Articles appearing in IEEE special interest magazines are peer-reviewed and edited.

Conference Proceedings

Conferences are an excellent forum for technical authors to begin the process of writing for publication. The oral conference presentation is usually accompanied by a paper that presents the material and makes the presenter's points in writing. The conference paper provides an opportunity for the author to assemble thoughts in an organized way and to prepare a paper describing the work with somewhat less demanding editorial standards than the journals call for. Conference papers often report on work in progress or tentative results.

Most conferences issue a call for papers well in advance of the conference itself and usually request that only an outline or an abstract be submitted to the program committee for consideration. On the basis of this material, the program committee determines whether the paper relates to the theme of the conference and decides to reject or accept the author's submission for the conference session and publication in its proceedings. However, the review and inclusion policies vary greatly from conference to conference. An author can benefit from having the points of the paper discussed during the presentation at the conference. A paper appearing in the conference proceedings can then incorporate elements of the discussion, be revised, rewritten formally, and submitted to a journal for consideration.*

In nearly every area of electrical engineering and computer science, conferences take place on a more or less regular

---

* *Ed. note*: IEEE policy expressly prohibits ''double publication''— that is, publication of the same paper in both the proceedings of a conference and in a journal. It is acceptable, however, as Dr. Manley suggests to enhance the paper with additional material and submit it for journal consideration.

basis, usually annually. The professional societies almost always sponsor an annual meeting at the national level that focuses on the special interests of the subsociety as distinguished from those of the profession as a whole. Regional and local chapters of the subsocieties hold annual conferences with a topical theme that usually changes from year to year.

Notices announcing upcoming conferences usually appear in the last sections of IEEE publications under the heading "Call for Papers." Some conferences may have more extensive promotional campaigns, particularly if the likely participants come from different disciplines. Additional promotional techniques include mailing announcements to targeted mailing lists and running extended advertising copy in society publications or in the publications of related societies. Some conferences have such an impact, either by their sheer size or by the topic they deal with, that even the popular media will report on the plans (for instance, Comdex).

Other Outlets

Technical writers can turn to other outlets as well. The magazines published by the various societies of the IEEE (such as *Computer, MICRO,* or *Software*), special interest group letters of the ACM (such as SIGADA, or SIGPLAN), other special interest journals such as *BYTE,* and trade journals like *PC Magazine* all appeal to various audiences. Articles in the IEEE magazine format and ACM letter format publications are peer-reviewed. Each publication maintains its own editorial policy but, by and large, articles that report research results from a practical viewpoint or describe applications of new technologies or procedures to existing problems merit publication. Of course, the audiences of such articles vary widely from publication to publication. They vary in education, interest focus, and expectations of style, currency, and level of technical detail; the author who is aware of these characteristics can successfully target an article for a particular publication.

SKILLS AND RESOURCES

Writing Skills

For most researchers, the memory of college writing assignments has faded dimly into the past. Most of their current work is documented in briefings and short information papers with a decidedly business thrust. As Eisenberg [6] points out, the technical writing professional needs technical writing skills.

Many corporations have recognized that technical writing skills are vital for their staff. They often run courses that deal with technical writing principles and offer practice for the attendees. These courses can be developed internally by the company or can be offered as professional development courses outside the office. Technical writing courses are offered by local community colleges as well as by the extension services of state universities. Specialized consultants are also in business to teach technical writing and overall communication skills to the technical professional. The IEEE Educational Activities Department offers self-study courses in writing and communication; the "Put It In Writing" Program (7) offers a videotape and workbook designed to build technical writing skills. There are many other quality courses available as well.

Research Resources

Many organizations maintain a library which houses a collection of reference material relevant to their work. Most industrial libraries maintain a close working relationship with local public or university libraries or have access to documents from distant sources through an interlibrary loan system. Most staff members realize that the library exists and maintains a collection of reference works, but many don't known that modern computer technology has made the library a formidable reference identification and retrieval system. Many industrial libraries maintain online data base services designed to match researchers' topic areas and keywords to documents that address them in the literature. A particularly powerful system is the online citation index, designed to track the citation history of technical papers as they are themselves cited in the literature. This allows the researcher to follow a seminal idea first discussed in a technical paper into succeeding years, and to find those paths of research that built upon the original idea. Also, most libraries have a collection of "how-to-write" books, the corporate writing style standard and other style manuals, journal guides, as well as many journals themselves. Corporate libraries can also obtain copies of citations that an author finds of interest.

In recent years, authors have become increasingly aware of copyright issues. Before a copyrighted article can be excerpted or a figure used in an article, the copyright holder must grant permission. Libraries and information services departments usually encourage authors to secure copyright permission through them. Because they have blanket agreements with many publishers and the experience of seeking permission from many others, they can save the author time and effort.

TIME PROBLEMS

One way authors can deal with time constraints and delays is to write a technical paper as the project progresses, using the "incremental approach" outlined by Michaelson [5]. The writer constructs an outline from project plans and details it as the project progresses. Each writing increment takes only a short while because each time the writer visits the manuscript, he produces only small increments, written while the material is fresh in mind.

From a broader viewpoint, timing the completion and publication of a technical paper can be important to a company and its strategy for asserting its presence in a particular field. Articles and papers that fit a corporation's publication plan reflect well on the author. They provide additional exposure to management and enhance the author's standing in a field where the company has an interest.

COMPANY POLICY

Gaining Public Release

Before employees can offer to write about their work for the general community, document control authorities in the company must usually approve the original paper for public release. This is particularly important in research fields dealing with material that is classified in either a company confidential sense or a national security sense. Although this practice relates to the safeguarding of sensitive information, it also helps assure the company that its sponsoring agencies are aware of information being disseminated. Nothing is more embarrassing for an agency than to discover some new issue in the technical press and note its discovery attributed to work performed under its own sponsorship.

In our organization, each paper must go through a prescribed release process; a time lag of six weeks is usually required for approval. Our procedures require an author to file a standard form with the document control office. This office then acquires the approval signatures from the internal corporate management chain, the technical release office, and the sponsor's project officer. This public release step can cause some delay; it's always prudent to allow plenty of time for this process.

There are several ways to lessen the impact of the delay. First, co-authoring with a member of the sponsoring agency allows many potential obstacles to be dealt with by the co-author. Experience shows that sponsoring agencies usually expedite material that members of their own organization submit. Also, keeping one's own management apprised helps speed the signature process through the internal organization. Good planning and consultation with management early in the preparation process will shorten turnaround time and help eliminate potential snags that could send the author back for a major rewrite. Another tactic is to submit a suitably sanitized draft of the paper to the acceptance committee for review purposes only, noting that the complete paper will be forthcoming only after release approval. It is not uncommon for public release approval to be denied when a paper reports results in the context of a sensitive project or program. Another approach may be to obscure the sensitive nature of the project by reporting the elements in a hypothetical, unclassified context. The author could discuss only small parts of the work without revealing the motivation for it, or plan to discuss only those parts that are most likely to be declassified [8].

What Managers Can Do

Management can play a strong role in encouraging technical professionals to write about their work. By carefully structuring assignments and understanding the benefits of publishing, they can take several steps to enable and encourage the technical staff to prepare papers for publication:

- Recognize that writing about technical work enhances the company's reputation and prestige
- Include writing for publication in project planning and coordinate professional writing with business reports
- Encourage incremental preparation of manuscripts
- Brainstorm with staff to identify fertile topics for papers within the technical work program
- Realize that writing is an important part of professional development
- Include professional writing and publication as part of the corporate strategic plan
- Acknowledge and reward those who successfully publish technical papers
- Keep sponsoring organizations apprised of the professional writing agenda
- For classified or sensitive projects:
  - Plan to gain release of suitable results
  - Identify elements of the work that are not sensitive

Management is an important force for the technical author. By taking a proactive role, the manager can help guide the author's work and encourage preparation of a paper that meets corporate strategic goals, enhances the department's reputation and builds the author's esteem. A helpful manager can provide unique guidance on the author's choice of subject and approach and help prepare the way for release of technical data that the author may want to include.

MITRE'S PUBLISHING WORKSHOP

MITRE management was concerned that, although our work program continues to show steady growth, the staff has not kept up with contributing to the technical literature at a commensurate rate. Over the last two years, our Washington facility prepared nearly 1400 technical business papers, supported by a great deal of interesting and useful work. But it seems that little of this information reaches the wider technical community. Of the business-oriented documents produced during the last two years, staff members requested releases of only about 50 for public distribution; only 70 papers were published in technical journals or conference proceedings.

Why the discrepancy? As detailed above, new authors don't fully appreciate the benefits of technical publishing; they face uncertainty, obstacles, and setbacks. They wonder what to write about or how to locate an appropriate publication or conference, worry that their work may lack sufficient interest or quality (and thus be rejected), and struggle to find the time and resources to prepare articles suitable for publication. We felt that a short course or workshop that would address these areas would help eliminate some barriers and encourage more staff to publish their work.

Workshop Approach

The MITRE Institute collaborated with two local writing consultants to develop and deliver such a one-day workshop. The program emphasizes some of the incentives and barriers discussed in this paper. The course shows participants how, what, and where to publish. We also developed a notebook of resource materials to which prospective authors could refer. One goal of the workshop is to offer guidance on how to prepare the technical article from the body of knowledge on which a business report is based.

Our business reports are the product of a general process that is more or less consistently applied across the company. Although report topics vary greatly, within the electrical engineering, systems engineering, and computer science fields most of our business reports share a common format and expository style and their preparation usually begins as a response to a sponsor's question or expressed need. For this reason, the process of preparing technical articles for publication based on corporate business reports can be somewhat standardized for our organization as a whole, and a workshop that offers guidance on the process is generally applicable to the work of most of the technical staff.

We opened the workshop to the entire technical staff at our Washington facility. The workshop participants realized that attending the publishing workshop would benefit them professionally and personally. They perceived that the course materials would help them prepare for publication and they attended because they wanted to. Sanders [9] observes that many people receiving technical writing training these days are college-trained professionals. All the professionals attending our workshop were formally trained in a technical discipline and had many years of experience. The average experience of the class was nearly 17 years since the bachelor's degree; no one with fewer than four years of experience attended. As a key aspect of the workshop, we coordinated its offering date with a call for papers from a local ACM conference that dealt with an area of computer science in which many of our staff were experienced and qualified to contribute.

Several speakers were invited to share with the participants their experiences with publishing technical articles based on their work for the corporation. We also invited representatives of the Information Services (library) and Security/Document Control departments to discuss their respective services and how authors might make particular use of them.

Participants found the expository portions of the workshop to be of greatest benefit. Few were aware of the considerable number of outlets for publishing, or that targeting articles is a well-developed strategy, or that so many potential topics were hidden in their work, or that writing can forward both the organization's goals and their own. Most believed that if management would offer support in a more proactive way, they would be more inclined to support corporate goals by writing technical papers for publication.

CONCLUSION

Technical professionals can meet the challenge of publishing their work by familiarizing themselves with the process of preparing technical papers based on the work they perform in their jobs. By understanding that the business report can serve as a resource for the technical article, by recognizing the barriers to publishing and how to overcome them, and by becoming familiar with the nature of the various types of technical articles and their various outlets, technical authors have the tools they need to prepare an article for publication.

This familiarization process can be expedited through conducting a short workshop designed to expose aspiring authors to the elements of preparing a technical article for publication in the professional literature. The workshop helps to demystify a process that appears arcane and obscure to many technical professionals.

REFERENCES

1. Corbin, N. C., "The Importance of Competent Technical Communications to Career Advancement," *Proc. IEEE Professional Communications Conf.* (1985), pp. 17–20.
2. Michaelson, H. B., "How Writing Helps R&D Work," *IEEE PC-30*, 2 (June 1987).
3. Marsh, J., "Motorola's Silver Quill Program to Encourage Writing about Work for Publication," *IEEE Micro Magazine* (June 1985), pp. 53–57.
4. Anderson, D. T., "Company Incentive Program for Getting Engineers to Write," *IEEE PC-26*, 4 (December 1983), pp. 170–171.
5. Michaelson, H. B., *How to Write and Publish Engineering Papers and Reports*. Philadelphia: ISI Press, 1986.
6. Eisenberg, A., "The Importance of Writing Skills for the Engineer," *IEEE Potentials* (February 1984), pp. 24–28.
7. Joseph, A., *Put It In Writing*, Cleveland: International Writing Institute, 1986.
8. Michaelson, H. B., "Publishing Classified Papers," IEEE *Institute* (January 1987).
9. Sanders, S. P., "What Is Communication Education and Training?" *IEEE PC-30*, 1 (March 1987), pp. 45–46.

Part 4
Manuals and Instructions
Giving Directions That Work

THE SEVEN articles in this section deal with writing clear instructions. The instruction manual is perhaps the one technical document most of the public is familiar with, and there is no shortage of complaints about the frustration that poorly written instructions can cause. You will learn from these articles, however, that there is no mystery to writing clear instructions and procedures.

DETERMINING YOUR PURPOSE

As the opening article indicates, different kinds of instructions have different formats and names. What John Major emphasizes, however, is the necessity of thoroughly understanding the needs of your readers before writing instructions, and then selecting the right format for your purpose. The key word is *accessibility*—the ease with which your readers can get the information they want. What to include, what to leave out, and how to arrange information must all be considered when writing a manual, and the decisions you make will determine how easily and reliably your instructions are followed.

STYLE, CLARITY, ACCURACY

Donna Reddout's article contains tips on how style and organization determine whether your instructions or procedures are clear. Audience analysis is again the primary concern, and the author shows how you can first identify your readers and their needs. Essential directions and peripheral information should be separated, and specific items in your manual should be easily accessible through a table of contents and index. Some of Reddout's topics are further explored in the next article where Lidia Lopinto looks at conceiving, organizing, and producing a manual. Her eight steps for this process conclude with the need to review your completed document for accuracy, and her final words remind us that few manuals can be engraved in granite—they will always need to be revised and updated.

USER FRIENDLY

James Gleason and Joan Wackerman investigate the difference between effective and poor computer manuals, and from their findings derive the most successful ways for you to link machine and user. Dividing a manual into two parts, the operator manual and the reference manual, is yet another example of foreseeing your readers' needs. The authors' emphasis on testing and retesting instructional material opens a theme to be further developed later in this section.

MEETING READERS WHERE THEY READ

Your audience's reading level will always concern you as a manual writer. Gleason and Wackerman indicate you can safely assume no more than an eighth-grade reading level for most manual users. This could vary, of course—most engineers should have a higher level—but as Benjamin Meyer points out, a lot of manuals now have to be designed with a "new look" due to the poor reading skills of their users. Meyer discusses ways to design such manuals for nontechnical or semitechnical personnel who operate or maintain quite complex equipment. By imitating the comic books these readers are probably familiar with, you can still design workable instructions, if the additional expense is no barrier.

TESTING YOUR MANUAL

In the next-to-last article, Marshall Atlas describes how you should test your technical manuals once they are completed—the "user edit." This article reminds you that if instructions can possibly be misinterpreted, someone will do so. Two types of testing can be applied: formal and informal. By following Atlas's approach you will be able to expose the flaws and ambiguities often missed by a manual's authors—people who are already familiar with the process.

OVERSEAS READERS

Finally, we turn to a topic prevalent in our shrinking world: how to write instruction manuals for people whose first language is not English. Stacey Sanderlin not only describes the problems involved, but also surveys what some global companies have done to overcome them. Human and machine translations are useful up to a point, but are costly. However, Basic or Fundamental English as developed by the Caterpillar Tractor Company and others can help overcome some hurdles and can even pave the way for more effective and economical translation. Sanderlin (like Meyer) recommends extensive use of visuals in manuals, and her 29 references are a rich source of information on the development of this topic over the past two decades.

What Should You Write: A User's Guide, Tutorial, Reference Manual, or Standard Operating Procedure?

John H. Major

A USER'S GUIDE, A REFERENCE MANUAL, A TUTORIAL, AND A STANDARD OPERATING PROCEDURE—*each has its own function and design. Research has shown that on-the-job readers must glean, from mounds of printed material, crucial information specific to their tasks and that they must do this quickly to meet task-related deadlines. Technical writers must provide their readers with the information they need when they need it and at a level they can comprehend. The first step in making this information accessible to on-the-job readers is to determine what they need in order to perform specific tasks. Matching document type with the needs of users has been successful in a computer center with over 60 employees, most of them part-time, and for the external documentation needs of a clientele numbering well over 5000. This article tells how it was done.*

In *Zen and the Art of Motorcycle Maintenance*, Robert Pirsig wrote, "Writing and editing technical manuals is what I do for a living . . . and I [know] they [are] full of errors, ambiguities, omissions and information so completely screwed up you [have] to read them six times to make any sense out of them . . ." [1]. I too have written what Pirsig calls *spectator manuals,* isolated manuals unrelated to the task at hand or the need of specific users.

I work at Academic Computing Services (ACS) at the University of Wisconsin—Eau Claire. I've seen tutorials I had written being used as reference manuals. I've written user's guides attempting to combine reference material with tutorial material. And I expanded an inherited set of standard operating procedures that were merely a well-written collection of recipes; when asked how things were going, my week-end operator once told me that if you can cook, there could be no problem. What he didn't realize was that the recipes never mentioned *what* to cook *when*.

ACS serves a user community of over 5000 faculty, staff, and students with a staff of 10 full-time employees and over 60 relatively short-term, part-time undergraduate student employees. Necessity dictated some structure for our documentation needs. To provide that structure, I started by reviewing research in the areas of study skills, reading comprehension, and cognitive psychology.

Thomas Sticht, in his chapter on "Understanding Readers and Their Uses of Text," indicates two classes of reading tasks: (1) the *reading-to-do* reading task, in which information is looked up, used and forgotten; and (2) the *reading-to-learn* task, in which information is extracted and retained to be drawn upon at a later time. Writing text for the reading-to-do class concentrates on different reader skills than writing text for the reading-to-learn class. Sticht suggests that "students' experiences with textbooks and learning in school may lead to the development of school-like, topic-oriented manuals when, in fact, workers want job-related, performance-oriented guidance." [2].

In my previous writing, I had been producing manuals that were merely reading-to-learn textbooks! I realized that I had to learn about tutorials, standard operating procedures, reference manuals, and users' guides. I also had to understand how readers comprehend what they read and how they access and process the information available to them, especially on the job. And finally I had to determine which document types were appropriate to specific tasks.

Matching document type with purpose produced significant changes. It has turned standard operating procedures from a collection of well-written instructions into a schedule of production *backed up* by well-written instructions. It has removed confusing, esoteric options from tutorials, thereby almost entirely eliminating the use of tutorials as reference manuals and resulting in tutorials which indeed aid instruction. It has introduced reference manuals from which information can be applied to the task at hand. And it has resulted in a user's guide that the Chancellor of the University has called "not only

Reprinted with permission from *Technical Communication,* published by the Society for Technical Communication. May 1989, vol. 36, no. 2, pp. 130–135.

informative but also readable.''

DEFINITIONS

The first step in matching document type with purpose is to define the document types and their respective audiences.

Tutorial

A tutorial is a teaching tool whose intended audience is a group of readers who have never used the product being documented but who will need to use the product in the near future.

The tutorial starts with the assumption that the reader knows nothing about the topic, and then it explains the topic simply and logically, building on the reader's growing comprehension [3]. It does not detail every option; instead, it provides a foundation upon which the reader can build (through practice and the use of reference manuals). This journey, then, concept by concept, from simple to complex, dictates the structure of the tutorial. Readers are READING TO LEARN.

Standard Operating Procedures

A standard operating procedure is a step-by-step *when to* guide written for the technician or operator. The reader has not performed the procedure often enough to have it committed to memory.

Standard operating procedures (SOPs) are grouped together in logical clusters and designed to take the operator from 12:01 a.m. of the first day of the year to midnight of the last day of the year. SOPs also provide for handling impromptu demands of the job, which always seem to arise. The reader of an SOP is READING TO DO.

Reference Manual

In her book *How to Write Computer Manuals for Users,* Susan Grimm defines a reference manual as ''a collection of information organized so that specific information can be located as quickly and effortlessly as possible. It is not meant to be read from cover to cover.'' [3, 38].

The intended audience of the reference manual is a group of readers who are familiar with the topic and who have a working knowledge of at least the context of what is being documented. This audience needs specific information about some detail of their work.

Whereas the tutorial is structured to lay a foundation and then build upon that foundation, becoming increasingly complex, the reference manual is designed for ready access to a specific detail or group of details. The reader of a reference manual is READING TO DO. The key to success is the reader's ability to locate crucial information quickly.

User's Guide

A user's guide is a summary of all related resources, such as available services, software, and hardware, written for everyone who might use those resources. It serves as a system of pointers to more detailed information, such as reference manuals, tutorials, and standard operating procedures.

The audience of the user's guide needs an overview of related topics. Some of these topics will be of immediate interest; some will be of interest later.

AN IMPLEMENTATION

At ACS, documentation had developed along two separate paths: user documentation and ''internal operations documentation.''

User documentation developed in bursts and was driven by requests for information in response to isolated questions. How-to guides began popping up to meet these demands (e.g., *How To Log Onto The Mainframe Computer, How to Use the XYW Plotter*). Some of these were related to each other; others were not.

Some of the more extensive how-to guides began to have *tutorial* or *reference* added to their titles. There was no organization to the development of documentation. A ''user's guide'' was developed in an attempt to organize these documents. This guide merely arranged these documents into sections of hardware, software, and services.

Operations documentation was highly organized—from SOP 1.0 to SOP 12.4.1.13a—centered around like functions; for example, *daily saves, weekly saves, monthly saves, master saves, grandmaster saves,* and *disk pack saves* were separate entries under one main standard operating procedure. And these SOPs were how-to instructions, very good how-to instructions, but there was no way to figure out when you were supposed to do what.

In matching document type to user purpose, we scrapped most previous documentation (a major hardware conversion made the scrapping pretty painless) and we modified (or are modifying) the rest. The following is an implementation plan based upon what we have learned. It is presented as a blueprint that you can follow to do what we have done, and in some cases are still doing, with our documentation at UW–Eau Claire.

After determining which of the four major document types will meet your audience's needs, the first task is to organize these types to make the information they contain accessible.

Organizing the User's Guide

The user's guide (1) points to all available documentation, (2) guides the reader through a general background, and (3) lays the groundwork upon which assumptions can be made about the level of comprehension readers have when they reach a given point in the total documentation picture. Thus, it is the best place to start.

You should design the user's guide to be the lead document. It should be the first place readers go for information because it contains the necessary pointers to further information.

You cannot just document information; you must provide access to it—not just access to the documentation but also access to the specific information within the documentation. The well-planned user's guide is a big step in gaining access to that needed information.

For example, our User's Guide provides a common thread running through all the other, supporting documentation. This is where the reader will learn what resources are available, including documentation. As indicated in the preface to the *Academic Computing Services User's Guide:*

> This user's guide is general in nature, designed to introduce you to academic computing resources at UW—Eau Claire. It contains summary descriptions of all services, software, and hardware, as well as maps, sample forms, and the location of resources. Specific details on particular resources are covered in additional documentation, all of which is cited.

As you write the user's guide, keep in mind the KISS principle. When tempted to get into involved descriptions or instructions, leave a note citing a tutorial and/or a reference manual that currently exists or is to be written later and go on. When you have written all summary descriptions of the needs to be covered (don't forget a thorough glossary), go back and decide what to expand, either within the user's guide or through additional tutorials, reference manuals, and/or standard operating procedures. Ask yourself, "Will EVERYBODY need this information?" If the answer is "YES," expand the user's guide at that point. If you are honest with yourself and your users, you will find the answer to be "NO" much of the time.

Determine which type of additional document is needed for each kind of user task. For example, let us assume that most of your users will have to use an editing package called EDIT and that some of them have no experience with editing software. You will need to write a tutorial on the use of EDIT to get them started. For those who are familiar with and who must use some of the more esoteric aspects of EDIT, you will need to write a reference manual. If your vendor has already provided you with either one or both of these, and they meet with your approval, you will merely reference these—and their location—and go on.

The User's Guide entry for EDIT would look like this:

> EDIT is an editing package which assists you in creating and modifying text files using your microcomputer and printer. These files can be stored on diskette for future reference or further modification.
>
> For further information see the following documentation provided with the EDIT diskette (an extra copy is available in the conference room):
> *The EDIT Tutorial for New Users*
> *Advanced EDIT Reference Manual*

You look around, make a few phone calls, and discover that a tutorial on EDIT does not in fact exist and that nothing worthwhile is available in the way of reference material; you're going to have to write both.

Organizing the Tutorial

Tutorials, standard operating procedures, and reference manuals, all contain short, procedural *how to* instructions, written to explain a specific task. The structure of the how-to instruction is step-by-step: begin at step 1 and continue until step last. How-to instructions should not stand alone. Decisions made about the audience and purpose of each of the document types determine the context, and therefore the tone and language, for the how-to instructions.

The tutorial presents specific tasks that will take the reader from the known to the unknown, building concept upon concept. Make sure that the reader can complete the task of the first example; make it as fail-safe as possible. Nothing succeeds like success. Start with a task that is likely to occur in the reader's working environment. Make it real. There's nothing more boring and demeaning than being forced through some "cutesy" exercise in order to learn how to perform a task integral to making a living. A very small symptom of this is in the following statement:

COPY MYOLDFILE INTO
MYNEWFILE

Replacing that with

COPY BUDGET INTO FY87BUDGET

indicates a familiarity with the reader's environment and is more readily understood by the reader, especially if it is part of a larger example on working with budget files which a vast majority of your readers understand.

In the preface of the tutorial, tell the reader to use the tutorial, hands-on, once or twice to become familiar with the subject and then to put the tutorial aside.

Do not document alternatives that this specific example will not explore; leave those for the reference manual.

Organizing the Reference Manual

Reference manuals are not required to be long. It is quite possible that the tutorial will be longer than the reference manual, since the tutorial must provide explanations that the reference manual does not.

The key to a good reference manual is *accessibility*. Can the reader find the needed information and comprehend it in order to perform a job-related task now? Reference manuals are generally organized alphabetically. The reader needs more information on a specific detail and generally knows where to look or, at least, has several good guesses at some key words.

Tables of contents, indexes, appendixes, charts, tables, call-outs, icons—all can be used to enhance the accessibility of information. The reference manual is the final word on the topic. If it isn't here, the only thing left is trial and error. Since to get to this point the reader has gone through the user's guide and the tutorial, you, as writer, can assume that the reader is ready to comprehend the information contained in the reference manual. What you must do is make sure that that information can be found.

Organizing Standard Operating Procedures (SOPs)

Where do the standard operating procedures fit in? The audience of an SOP is generally quite limited. An SOP is related to production schedules. It is not enough just to have *how to* instructions, you must tell when these tasks are to be done; that's part of what makes SOPs *standard*. For an example, let us assume that a clerk must prepare a particular report every Monday at 10:00 am. An SOP might include an entry like this:

MONDAY 10:00 Prepare last week's Weekly Report
    EDIT the file PRELIM.WKnn where nn is the two-digit number representing last week; i.e., last week was the nnth week of the fiscal year. On the first line of the report, . . .

    For more information on EDIT, see the following documentation provided with the EDIT diskette (an extra copy is available at the console):
    The EDIT Tutorial for New Users
    Advanced EDIT Reference Manual

MONDAY 11:15 Et Cetera

Note the citation of additional information in the form of vendor-supplied (or locally written) documentation. You could also include such documentation as maintenance manuals, schematics, and engineering reports.

CONCLUSION

Technical writers must provide their readers with the information they need when they need it and at a level they can comprehend. A user's guide, a reference manual, a tutorial, and a standard operating procedure—each has its own function and design.

The reader of the user's guide needs a summary of all related resources, such as available services, software, and hardware, written for everyone who might use those resources. It serves as a system of pointers to more detailed information, such as tutorials, reference manuals, and standard operating procedures.

The reader of a tutorial wants to learn about a topic in order to do something later. The tutorial is bound by a distinct scope and sequence.

The reader of a reference manual needs information that is organized so that specific details can be located as quickly and effortlessly as possible. Accessibility to information is the key to reference manuals.

The technician using a standard operating procedure needs a step-by-step *when to* guide, bound by scope, sequence, *and* schedule.

After you define these major document types, your first task is to organize them. Make the information they contain accessible to individual readers who have specific, and often unique, needs.

Design the user's guide to be the lead document; it is the set of necessary pointers to further information. Determine what additional documentation, if any, is needed. And keep in mind the functional keyword of each type of additional documentation: tutorial—*sequence,* reference manual—*accessibility,* standard operating procedure—*schedule.*

This strategy will enable you to provide your readers with comprehensible information when they need it.

The following outline illustrates the transformation of the library of ACS documentation:

Old Library
User's Guide
- Non-existent

Tutorials
- Duplicated intro material
- Duplicated log-on sequence

Reference Manuals
- No cross references
- Unknown level of audience expertise

Standard Operating Procedures
- No schedule included

- Topical Arrangement
  - General Policies
  - Staffing & Personnel
  - Work Areas
  - Operations Utilities
  - Emergencies & Incidents
  - Configurations & Conventions

New Library
User's Guide
- Lead Document
- Cross references all other documents

Tutorials
- Intro material in User's Guide
- Log-on sequence in User's Guide

Reference Manuals
- User's Guide serves as cross reference
- Audience level of expertise developed through following the User's Guide

Standard Operating Procedures
- Personnel & General Policies
- Schedule
  - End of Year
  - End of Semester
  - End of Month
  - Daily
- Requests
- Emergencies

This matching of document type with the needs of the users and employees of Academic Computing Services has been successful. ACS users and employees can readily access the information they need when they need it. Ω

REFERENCES

1. Robert M. Pirsig, *Zen and the Art of Motorcycle Maintenance* (New York: William Morrow, 1974), pp. 34-35.
2. Thomas Sticht, "Understanding Readers and Their Uses of Texts," *Designing Usable Texts,* ed. T. Duffy and R. Waller (New York: Academic Press, 1985), pp. 315-340.
3. Susan J. Grimm, *How to Write Computer Manuals for Users* (Belmont, CA: Lifetime Learning, 1982), pp. 37-39.

JOHN H. MAJOR currently works for Academic Computing Services at the University of Wisconsin—Eau Claire. He also teaches for the University's English Department.

Manual Writing Made Easier

You *can* make your manuals easy to use and read. Just know your audience, define jargon, use a personal approach, organize well, and write clearly.

By DONNA J. REDDOUT

Technical training documentation in the late 1980s emphasizes a task-oriented, user-friendly manual format. After all, as manual writing expert Edmond Weiss points out, ". . .users don't want to know how the system works; they want to know how to work the system." But, before you can have a good manual, your writing must be clear, concise, and comprehensible. Use the following five writing tips to improve the usability of your manuals.

Know your audience

The more you know about your audience before you begin writing, the better your readers will understand what you write. Your audience determines style, use of technical jargon, content and structure, and the reading grade level.

Find out who will use your manual. List specific names, if you can, or specific groups of people. Then find out how much they know about the subject. Survey all or a representative sample of your audience to discover the extent of their knowledge. The ratio of novices to experts determines the type and amount of information you put into your manual and how that information is organized.

Identify and define language

Every discipline has specialized words, phrases, and acronyms that experts in that discipline use when they communicate

Reddout is the supervisor of Telos Federal Systems' Missile Training Group in Lawton, Oklahoma.

with each other. This language—or *jargon*—often accurately communicates ideas and concepts within a discipline but is often unfamiliar to those outside the discipline. Law and medicine are two common examples.

The manual's user does not need to know a discipline's complete vocabulary—only those terms necessary to perform the tasks. List the jargon your subject uses, then decide which terms you *must* use and which terms may be substituted for a more common word. Acronyms create special problems since technical fields tend to use them extensively.

Clarity is the most important goal of manual writing; therefore, identifying un-

familiar terms and defining them is crucial. Here are some other suggestions:
■ If a term is subject-specific and you must use it, assume the novice is unfamiliar with it and define it.
■ If the user will see the term used during the performance of the task, then define it.
■ Include a glossary of words, phrases, and acronyms in your manual and design the glossary to make reference quick and easy. Be sure the glossary is complete and accurate.

When you define terms remember the following suggestions:
■ Use plain English. Don't use jargon to define jargon.

Figure 1—Verbs

USE:	INSTEAD OF:
Recognize	Be cognizant of
Try	Make an attempt to
Gather	Accumulate
Begin	Commence
Show	Demonstrate

Figure 2—Word usage

USE:	INSTEAD OF:
Concrete/specific words	**Abstract/general words**
Typewriter	Office equipment
Truck	Vehicle
Six	Several
Pencil	Writing system
Common words	**Formal words**
Use	Utilize
Help	Facilitate
Best	Optimum
Rest	Remainder
Skills	Competencies
Single words	**Phrases**
Calculate	Perform a calculation
If	Should it prove to be the case
May	Have permission
Except	With the possible exception of
Because	For the reason that
Until	Until such time as
For	For the purpose of

Figure 3—Buzzwords

Thrust (of your report)	Reconfigure
Impact	Target
Parameter	Access
Time frame	Ongoing
Input	Downsize
Viable	Optimize
Overview	Prioritize
Quantify	Synergistic
Throughput	

■ Identify how the term is *like* something familiar, and then describe what makes it *different*.

■ Write the definition objectively. Identify what it is, not what you think of it.

Select a personal point of view

The best point of view to use is the personal point of view. The personal approach, which uses the pronoun *you*, emphasizes the reader. Using *you* in a manual has four advantages: you establish a rapport with your audience; the reader becomes personally involved; you will not be "talking down" to a novice; and you can express technical ideas and concepts more clearly than if you were using the impersonal point of view.

A note about humor: Some sources suggest using humor to make the novice user feel more at ease and to relieve anxiety, but use humor sparingly. Humor may be appealing at first, but it rapidly becomes irritating and even gets in the way of your message. Most users prefer concise, to-the-point instructions. They don't like wading through paragraphs that are entertaining when their purpose is to learn a process.

Organize for clarity

How you organize your manual determines not only its usefulness but also the user's attitude toward the document, the equipment, and the job. Nothing is more frustrating than having a task to perform and not being able to find the procedure in the manual. Consider both content and structure when organizing your document.

The content should emphasize critical tasks. These tasks are difficult to perform, and errors cause serious consequences. Do not overemphasize low-risk errors. The content should make a distinction between necessary information and nice-to-know information. The content should also give your readers realistic explanations of what to expect. (What *really* happens if I push this button?)

The structure is almost as important as the content. Structure the document so it can be used quickly and easily. Avoid jumps, skips, and loops that keep the user flipping pages. Organize the manual from simple to complex, by order of performance, or in any other order that suits your objective, audience, and subject.

Minimize the amount of conceptual information; i.e., how the machine works or descriptions of its components. If you include it at all, keep conceptual information short, simple, and easy to read.

Design a detailed table of contents as a ready reference for the user, and highlight the more frequently used or most critical tasks. Also consider having an index that keys on words, phrases, acronyms, and procedures. Anticipate your user's needs. What information might the user have to look up in a hurry? Consider having an outline at the beginning of each chapter.

Write clear, factual prose

After you have written your first draft,

concentrate on editing. When you edit, check verbs, word usage, and sentence structure.

If you use weak verbs, your writing will be weak. Use forceful, descriptive verbs. Minimize the use of the verb *to be*, and avoid sentences that begin with *there is* or *there are*. Use one-word and short verbs. Figure 1 gives some examples of good and bad verb forms. Search for verbs that precisely describe actions the user must perform: attach, center, crimp, disconnect, isolate, list, mix, press.

Avoid passive verbs. Passive verbs cannot express the authority or urgency necessary to explain step-by-step procedures. They do not clearly identify what the subject is doing. However, passive verbs *may* be used if the doer of the action is unknown or unimportant. For example, "The software was designed in 1983."

Substitute concrete and specific words for abstract and general words, common words for formal words, and single words for phrases. Figure 2 gives some examples.

Delete buzzwords—words that are currently popular. These words often lack precise meaning, and they usually fade out of use in a short time. Figure 3 shows examples of these.

Distinguish between suggestions, recommendations, and requirements. Match the helping verb to the level of action required. When you suggest an action, use low-force helping verbs, like *should* or *ought to*. When you recommend an action, use moderate-force helping verbs, like *must* or *have to*. When you require an action, use the most forceful helping verb, *shall*. Do this consistently and your reader will recognize the appropriate level of action. Using *shall* when you merely suggest an action confuses your reader.

As a general rule, keep sentence length short. Ideas and concepts are easier to grasp in short sentences. The length of the sentence and the number of syllables determine the reading difficulty or reading grade level (RGL). To have an RGL of 7 or 8, you must average 11 to 15 words per sentence and 1.5 syllables per word.

Put the main idea (subject and verb) first in the sentence: "Turn the lever clockwise until it stops." However, if a condition determines the action, state the condition first. For example, "If the green light goes off, press the panic button," not "Press the panic button if the green light goes off." Arrange the steps of a procedure in the order in which the user must consider them.

The finished product

According to Yale writing professor William Zinsser, well-done technical writing is ". . . the principle of leading a reader who knows nothing, step by step, to a grasp of the subject." If you use the writing tips discussed in this article, you will have a manual that the user will be glad to use.

Designing and Writing Operating Manuals

LIDIA LOPINTO

Abstract—**A manual should communicate the design engineers' intentions for the operation of a process. To do this, it must be accurate, detailed, and logical, and it must be completed in time for the startup. Steps in preparation are structuring the manual; gathering and organizing information; making an outline; writing a draft; checking for technical accuracy; reviewing for consistency, style, and format; cross referencing and making an index; and distributing and accumulating updates and corrections.**

A N operating manual is a key document for training personnel in the operation and maintenance of a new plant. If made available before the startup, it can help avoid costly operating errors. After the startup, it becomes a source of reference for experienced personnel and a source of instruction for replacement personnel.

To serve these purposes, the manual should cover startup, routine operations, trouble-shooting, normal and emergency shutdowns, and equipment maintenance and repair. For those learning the process, the information must be technically accurate, clearly written, and in logical sequence. For those familiar with the process, it should also be indexed and arranged for ready access to specific items. Its format should allow easy revision as the process is modified and new insights are gained about operations.

THE INFORMATION REQUIRED

Before the writing of an operating manual can begin, the process must be clearly conceived by the designers and be at least partially documented in process flow diagrams and engineering drawings.

Generally, the task of preparing the manual is assigned prior to the completion of the engineering work. This makes preparation difficult because not all of the information needed has been developed and finalized. However, there is an advantage to such scheduling: The missing information can be requested while there are still engineering labor-hours remaining in the budget.

The project engineer is usually responsible for the preparation of the manual. If the project is small, the project engineer may actually write it. If the project is large, another person may do the writing, with the project engineer coordinating the flow of information from the various engineering groups.

The author is editor and publisher of *The Engineering Software Exchange* newsletter, 41 Travers Avenue, Yonkers, NY 10705; (914) 963-3695.

Certain information must be available, at least in preliminary form, before preparation of the manual can begin, including process-flow and piping-and-instrument diagrams, control-panel design, a statement of the control logic for routine and emergency operations, and operating and maintenance manuals from equipment suppliers.

A systematic procedure for preparing the operating manual includes these steps:

- Designing the structure
- Gathering and organizing information
- Making an outline
- Writing a draft
- Checking for technical accuracy
- Reviewing for consistency, style, and format
- Cross referencing and making an index
- Distributing and accumulating corrections and updating.

DESIGNING THE MANUAL'S STRUCTURE

Many engineers make the mistake of plunging into the writing of a manual without a plan. Because a manual is in

Reprinted from *IEEE Trans. Prof. Comm.*, vol. PC-27, no. 1, pp. 29–31, March 1984.

many ways analogous to a complex computer program, it makes sense to first prepare a preliminary structure, i.e., a flowchart.

The structure is not an outline (which comes later) but a flowchart of the types of information that will be presented. This can be done early in the design stage because details are not yet required. Having a structure, the details can be filled in more easily later. Time and work will be saved by securing approval of the structure by those responsible for the design of the plant. If the structure must be rearranged at a later date, ensuring the consistency of the parts can be a burdensome task.

Although the structure of the manual depends on the type of process and the specifications and needs of the operator of the plant, a proven approach to this structuring might be called "modular design."

By this approach, the manual is divided into modules, each devoted to a specific group of equipment. Each module is self-contained, providing all the essential operating and maintenance information for each equipment group. The idea behind this approach is to divide the manual into manageable parts that can be easily referred to by those concerned only with portions of the process.

A module typically contains the following:

Summary A brief statement of content.
Introduction A discussion of the process design concept.
Description A detailed description of the construction and arrangement of the equipment.
Routine operation A detailed discussion of how the equipment is to be operated and controlled.
Troubleshooting A description of procedures for pinpointing and diagnosing equipment malfunctions.
Maintenance A compilation of procedures for disassembling equipment and replacing parts, plus testing and lubrication schedules.
Emergency operation Detailed procedures for handling emergencies such as a runaway reaction, loss of cooling water, fire, leaks, and overheating, and for shutting down the process, with a description of the operation of automatic controls.
Appendix Parts lists, spare-parts lists, drawings and figures, and manufacturers' literature illustrating equipment and describing its operation and maintenance.

An obvious way to organize the modules is according to their sequence in the process. The first module of the manual encapsulates all the other modules, generally taking the form:

• Process design philosophy—explains why the process was chosen and gives the rationale behind the design.
• Overall process description—describes, with the aid of an overall process flow diagram, the main process flows, the physical and chemical changes occurring, and the instrumentation.
• Following in order—the overall table of contents, general operating instructions, maintenance philosophy, summary of controls and alarms, cross-reference index, equipment list, instrument list, environmental records, and general emergency procedures.

GATHERING AND ORGANIZING INFORMATION

After the manual has been structured, the next tasks are to gather all the information and organize it under the major headings. Arranging the information alphabetically or according to the process sequence will help pinpoint what may be missing. Those responsible for this material should be made accountable for supplying it by a specified date, and systematically prompted via memorandums or phone calls.

SETTING UP THE OUTLINE

The next step is to make a list of subheadings for each of the major headings. This is more easily done if the information that has been gathered has been separated into file folders designated by the major headings. The subheadings, with the information belonging with each, should next be organized into a logical sequence.

STARTING THE WRITING

After the material has been organized and carefully reviewed (so that one becomes familiar with all the technical details), the writing can begin, and it will be easier because of the systematic approach that has been followed to this point.

REVIEWING FOR ACCURACY

Each section of the first draft should be reviewed by the engineers who designed that portion of the process. Because a lot of time can be wasted if all the reviewers act as literary critics, it should be pointed out to them that they are to check technical accuracy and not style. The manual should be clear, logically ordered, informative, and useful; it need not be a literary masterpiece.

REVIEWING THE FINAL DRAFT

After the results of the accuracy review have been incorporated into a final draft, the manual is ready for a final review. Now, items to be checked include consistency of units, terms, and nomenclature; correctness of crossreferences; completeness of sentences; logical structure of paragraphs; and proper spelling.

CROSS REFERENCING

Each listing in the table of contents should be keyed, by

means of numbers, to the section of the manual it covers. Generally, these numbers overhang the left-hand side of the table so they can be picked out easily. They should not be page numbers because such a system can become meaningless as the manual undergoes revision.

A cross-reference index will be helpful to the reader who is familiar with the process and who needs specific information. An alphabetically arranged index will enable this person to find the information wanted without going through the entire table of contents.

CORRECTING AND UPDATING

Upon distributing the manual, it is important to have a system for receiving corrections and changes. One way is to send out forms, along with the manual, on which reviewers can make entries.

Changes can be made easily if the manual has been written with, and stored in, a word processor. If the manual is available to a user via a word processor, that person can quickly gain access to any section of it by means of the search routine for key words.

ACKNOWLEDGMENT

I thank Dell Archer of Engelhard Industries for his contributions to this article.

Manual Dexterity—What Makes Instructional Manuals Usable

JAMES P. GLEASON AND JOAN P. WACKERMAN

Abstract—This paper discusses how properly designed instructional manuals can meet the needs of operators of home computers, office systems and word processing equipment. It also details several ways that structure and content presentation can help you produce more effective manuals.

PICTURE a woman returning to office work after a 25-year hiatus. She recalls her 20-page Smith-Corona typewriter manual and then looks at her new word processor manual, which is more than two inches thick. It's immediately obvious to her that things have changed in the past few years.

How much will this new manual help her? Because today's market offers greater function and flexibility to less sophisticated users, manuals like this must meet her needs, but must also work for a variety of users with different word processing backgrounds.

As a new product developer, how can you ensure that the documentation you develop is going to meet the users' requirements? How can you make sure that your manuals will be simple, easy to use, and appropriate for your target market?

WHAT MAKES AN EFFECTIVE MANUAL?

There are a number of factors that distinguish the best and most popular manuals from the ones that people throw in the corner and never look at. Consider the following:

- *Organization* Good manuals have a structured format, a complete index, sections set off by tabs, and a table of contents.
- *Content* The material focuses on operator tasks, contains practice exercises, provides clear illustrations, and is concise.
- *Appearance* The presentation is attractive and colorful, with plenty of white space, and is "packaged" in booklets that are small and easy to handle.
- *Language* The text is conversational and easy to read, geared toward an eighth-grade reading level.

On the other hand, operators do *not* like documentation that

- Is inaccurate
- Has too much information or detail
- Has a demeaning or childish tone
- Is formal, stiff-sounding, or full of jargon
- Has poor printing qualities (broken type and so on)
- Is poorly organized (the information is of no use if you can't find it).

In providing the best documentation, manuals must be as usable as possible. That is, they must enable your users to do what they want to do with the equipment. That usually means two different kinds of documents need to be available: an operator guide (for training) and a reference manual. Also, you might want some sort of job aid, such as a quick-reference card, to summarize keys, functions, and probably the most common procedures.

Why so many kinds of documents? First, the operators need a training manual to take them step-by-step through the equipment's uses. This kind of documentation needs practice exercises. On the other hand, the reference manual should be organized in a more traditional sense around system features and functions. As a rule, it should also provide a greater level of detail.

One recent trend that is becoming increasingly popular among equipment manufacturers is toward customer problem-solving and repair. In most cases, this means the manufacturer provides some sort of problem determination

Reprinted with minor changes from *Conference Record* of the IEEE Professional Communication Society Conference held in Atlanta, GA, October 19–21, 1983; Cat. 83CH19160-4, pp. 142–144, IEEE Service Center, 445 Hoes Lane, Piscataway, NJ 08854; copyright 1983 by the Institute of Electrical and Electronics Engineers, New York.

Jim Gleason is a technical writer for IBM Corporation, Dept. F98/962-3, 740 New Circle Road, Lexington, KY 40511; (606) 232-7967. Joan Wackerman is area manager of training systems for Courseware, Inc., 427 N. Lee St., Alexandria, VA 22314; (703) 684-1000.

Reprinted from *IEEE Trans. Prof. Comm.*, vol. PC-27, no. 2, pp. 59–61, June 1984.

guide and the operator uses it to isolate machine problems. On the positive side, this approach can significantly decrease system downtime for things like operator error. On the other hand, the guide itself may be difficult to use. The same criteria for usable manuals apply here—maybe more so, as these guides are usually very task-oriented and instructions must be followed exactly.

Further, if the operators are expected to repair some part of the machine, not only must the manual be explicit and usable, but also the task must be practical. Make sure that the operators are not expected to do anything beyond their technical ability.

KEEP THE CONTENT BRIEF

How do companies keep manuals small and friendly when there are complex features to cover? One way is to divide one manual into separate booklets, based on different procedures or tasks. These booklets can be contained in a ringed binder or in one box, similar to the reading kits available in schools.

Another, and perhaps better, way is to ensure that the manual spells out only the content that the user needs to know to do his or her own tasks. For example, in word processor documentation, users need to know how to enter text on the machine, edit the text, and then print it out. In this case, the first portion of the manual must address only those areas. (Often the hardest battle for the author is preventing the engineers from putting every detail about how the system operates into the manual.)

Users want only brief, boiled-down, step-by-step keystrokes for the operations they want to do. Thus, one way to keep your content slim is to identify the series of tasks your users will perform and then list the step-by-step procedures for doing these tasks on the equipment.

Avoid covering all functions of the machine in an operator's manual. As they become more experienced, operators will tailor their machines to their own specific applications and will learn how to do many of the functions by experimentation. The place for detail is in the reference manual.

Practice exercises should be short and interspersed systematically throughout the structured format. They should be clear and procedural and provide feedback to the operators as to what they did right or wrong.

Illustrations should be as simple as possible. Good technical artists know how to take a complex piece of equipment and represent it simply. Authors of good manuals use their artists' expertise (or ask for their advice) to produce these simplified illustrations. Remember that the reason for having illustrations is to clarify an idea and to avoid using a large number of words to explain a concept. All illustrations should reduce the number of words required and should

stand alone—that is, they should make sense together with the caption and the callouts.

The organizational structure should be very apparent to the user. The operator's manual should be sectioned according to common performed tasks or the ways people use the equipment. This is in contrast to organizing by functions or what the machine can do—an approach more appropriate to organizing engineering materials. Where possible, the task sections should be separated by tabs. Further, these sections should be broken into elements that are the same for each section. For example, you might want to have a preview, a system overview, the main content, practice exercises, and then a review.

By following this format with each section, the user quickly becomes accustomed to your manual and can use it more effectively. Further, users *like* a structured format. It makes them feel comfortable and they can skip sections they know they won't need.

Once they work through a few sections and know what to expect under each kind of heading, they can successfully skim for the information they need. Whether we like it or not, users will skim and will also skip sections. That's the way people use manuals; they are often impatient to do something and need information quickly. For this reason, it is important that a manual remain somewhat modular, especially for equipment with many and diverse functions. Every user will not use every function.

A structured format forces you as a writer to provide only information that fits the format. This prevents rambling, unnecessary content, and excessive detail.

The index must be complete and must be approached from several directions: features of the machine, functions of the keys, tasks the operator wants to do, and so on. Users must be able to find information for a variety of reasons. The most popular documents let the users quickly find the pages they need.

APPROPRIATE LANGUAGE

It is fairly well known that jargon should be avoided in technical documentation. Terms must be as simple as possible, and their use must be consistent; the reader should not find a variety of terms for the same concept. Further, when successful writers introduce a new term, they try to tie it to something that the user already knows. For example, you might contrast the way a particular key on a word processor works with a similar key on a typewriter.

There is some new documentation on the market now that attempts to be conversational, but ends up being demeaning. It's not easy to write a readable document without becoming patronizing—that is, using childish language or telling the customer to do something that sounds too elementary. For example, several users recently complained that, in a particular operator's manual, the text told them they were "doing a great job and should take a break." The users found this somewhat presumptuous and resented it.

A good rule of thumb is to read the manual aloud. If you feel it sounds patronizing or demeaning, others probably will too.

MANAGING DEVELOPMENT

If you are managing the writing or development process, you need to do several critical things to turn out a set of documents that really work.

First and foremost, you need to ensure that the materials are tested, retested, and tested again. Do not allow your writers or yourself to become ego-involved with the manuals to the extent that you cannot change them when the data say you should. The manuals must work.

As you begin the documentation process, use an instructional designer or an experienced document designer to give input to your format. It is worth hiring a consultant for a day or two for this. The format will act as a blueprint or specification for your writers. It will give you the tool you need to ensure that the documentation produced is relevant. The document should be designed not only for what the machine can do but, more important, for what the users will do with it.

Once you have your format designed, meet with your writers and show them a model section of a manual, using your format. Also show them some good examples of existing documentation that is similiar. Next, train your writers to write to the format. Make sure that the editor is part of your team all along the way. Involve him or her in the design of the format and the writing of the prototype section. Then ensure that as the authors begin to write, they frequently submit outlines and drafts to the editor.

Give writers immediate and frequent feedback on what they are doing right and wrong. The more specific you are about what is and is not working, the better your final product will be.

Stay close to your users. Interview them and find out how they use or intend to use the equipment.

We do not go into detail here about how or why to write at lower reading levels. There are many guidelines available on how to do this. Briefly, they recommend using short, simple, concrete words and writing in short sentences. Ask someone to calculate the readability of your document or send it through a computer program that analyzes reading level.

The important thing to emphasize is that it does not hurt to aim for an eighth-grade reading level for any document. Brighter people are not offended by simply written material. They just read it faster.

Arrange to have one editor, if possible, edit all your documents for a particular piece of equipment. This will help related documents appear more consistent. Ensure that this editor understands your desire to use terms consistently and to keep the language simple, clear, and readable. Again, make him or her part of your team.

MAKING IT LOOK GOOD

We all know that we are drawn to documentation that is attractive. But what makes it attractive? A variety of characteristics come into play, such as color, novelty, and simplicity.

How do you develop a creative, attractive package? If you are a writer or manager, let your artists be creative—brief them on your overall direction and give them a chance to select typography, layouts, and covers. Tell them what type of reaction you want the user to have upon seeing the package and let them come up with ideas. It is wise to have at least two artists working on ideas in parallel before you decide on one direction.

Take time to make sure there is plenty of white space and the layout is clean, simple, and useful.

The most recent finding is that users like documentation that is small. They like the smaller 6-in. × 9-in. size as opposed to the traditional 8.5-in × 11-in. manual. Why? It is less intimidating. Also, when they are operating a piece of equipment and using the documentation, they often put the manual on their laps. It is awkward to use the machine if the book is too large.

GETTING THE POINT ACROSS

Let the final test of how the material should look and read rest upon customers or users. Keep as close to the users as you can.

As a writer, you provide the most important (and, in many cases, the only) interface between the designers and engineers who developed the equipment and the operators who must use it. Unless the documentation you produce is the most informative and usable possible, you do both groups a grave disservice.

The ABCs of New-look Publications

B. D. Meyer

COMMUNICATING MAINTENANCE REQUIREMENTS *for complex weapon systems to recruits lacking in basic reading skills is a major concern of the military. The Army's response to this concern is the "new look" Technical Manual. The new-look format visually demonstrates operating and maintenance procedures with a minimum of words. This article describes the new-look manuals and offers advice on how to prepare them.*

The new-look format is a relatively new requirement being stipulated in technical publications contracts. This presentation provides an overview of the new-look concept and the reasons underlying its creation. A functional description of the new-look technical publications team as well as its interaction with the customer and support groups (such as logistics and training) is also provided.

WHY THE NEW-LOOK FORMAT?
Blame It On Television

Today's soldiers were nurtured on a steady diet of television: it was their entertainment, their babysitter, and their window to the world. Television encompasses action, rapidly changing scenes, and minimal narrative. How little narrative is actually used in television is demonstrated by the fact that an average news script, for a half-hour evening news program, would occupy less than one column on the front page of *The New York Times*. The military

has been increasingly concerned about recent recruits' lack of basic reading skills. This controversial issue was given national exposure by columnist Jack Anderson in February, 1982, when he noted that almost 40 percent of the Army's junior enlisted personnel read below the sixth grade level, which means that they are functionally *illiterate* by United Nations standards.

Getting the Soldier to Read

Traditional service literature, with its high ratio of text to illustrations, is not an effective communication vehicle for visually oriented people. Even the best technical publications are cumbersome to use, and therefore the operator or technician often does not make full use of the contents. The consequences are weapon system degradation, or possibly, non-availability in a combat environment. The soldier is more likely to use a manual that contains a short sentence that tells the soldier what to do and a simple illustration to demonstrate the procedure.

According to Government specifications, the reading level for equipment operators is eighth grade and for organizational and intermediate level personnel, no higher than ninth grade.

New Technology/New Approach

Achieving a high level of operational readiness with hardware that is functionally complex requires a new approach to maintenance. Today's weapon systems, and those now on the designers' drawing boards, include embedded diagnostics called BIT (built-in test). BIT enables the system to detect a fault and to identify the specific unit to be replaced. System components and modules are designed for both logical fault analysis and ease of replacement. In many instances, the nontechnical crew can make necessary repairs themselves, guided by BIT and the instructions in the new-look technical manual. The additional maintenance tasks rel-

Reprinted with permission from *Technical Communication,* published by the Society for Technical Communication. February 1986, vol. 33, no. 1, pp. 16–20.

egated to the crew level are one of the major reasons for the new-look format.

Nontechnical/Semitechnical Audience

The new-look format is intended for a nontechnical or semitechnical audience. The trained electronics technician has traditionally represented the target audience for technical manuals written on early-generation weapon systems. Using the theory of operation and logic diagrams, the technician exercises ingenuity, skills, and generic test equipment to keep the system operational. Unfortunately, this is not a viable approach for a crew lacking the training or *technical competence* to make qualitative decisions.

A new-look manual does not describe how the system operates: it does not contain logic diagrams, schematics, or wire lists. Instead, it provides concise, step-by-step instructions for locating a problem and replacing a part or assembly. Little or no deductive reasoning is required on the part of the crew, and nothing is taken for granted in generating the instructions. *Technical competence is not a prerequisite.*

WHAT IS THE NEW LOOK?

Modular Format/Information Rate

Interactive text and illustrations are the essential elements of new-look material. Words and pictures must work together to accurately portray the maintenance task to the reader. The basic new-look text/illustration unit is called a module. A module is similar to a storyboard panel, which is used to plan film or videotape production; both contain the narrative and the visual elements for a single scene or topic. The new-look publication finalizes the storyboard as a series of printed pages. Ideal new-look pages include no more than one or two modules.

Information rate can be defined as the amount of new information presented in a given page, which is controlled by both the writing and the graphic technique. The information rate must be tailored to the reader's comprehension level and the complexity of the tasks being communicated. Simple tasks can be presented in a single module, while complex instructions will require many modules of text and graphics that run for several consecutive pages. Too much information in a given module will confuse the reader, but too little will bore the reader and increase page count unnecessarily. Once a rate is selected, it should be maintained consistently throughout the publication.

Simplified Text

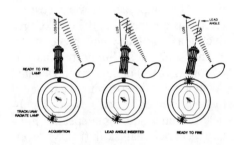

New-look instructions are simple, direct, and extremely precise. Each procedure must stand alone with a minimum of cross-referencing, and each step within the procedure must contain all the information necessary for the reader to perform a task without having to flip pages. This layout increases the page count, but page count can be minimized by carefully eliminating nonessential information and including only the amount of detail required. New-look publications are actively used during training; each procedural step is performed so that the student achieves the necessary degree of proficiency. New-look instructions assume that the crew has received hands-on training for each procedure.

Procedural steps for new-look manuals are written in the *second-person,* imperative mood. Narrative text is minimized because new look is an active, performance-oriented format. The reader doesn't have to be told how the system works, simply *how to fix it.*

Personal language is used to maintain reader interest. To accomplish this, writers must make the writing "you" oriented, putting the reader's interest and resultant comprehension first and placing themselves in the reader's circumstances. Writers must "speak" to the reader as if they were at the reader's side, talking the reader through the procedural steps.

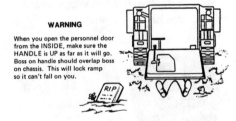

In certain circumstances, humor can be a valuable element when used at the right time to maintain interest and in the proper context. Select humor carefully: misused humor can lose its intended impact and detract from your writing. When used, humor should be directly related to the text material, typically highlighting the consequences of not following the given procedures.

Specific Illustrations

The illustrations accompanying each set of procedural steps permit the use of fewer words. *They demonstrate and visually verify the procedures explained in the text.*

Illustrations must show hardware in the same perspective as viewed by the reader, and should not show unnecessary detail for parts of the equipment not directly involved with the procedure. Operational scenarios, showing the soldier performing the procedures, are used extensively. Movement of mechanical parts is shown with arrows.

Two-color illustrations (green plus black or red plus black) are currently used; however, a specification for single-color new-look manuals is being developed. Whether color or shading is used in an illustration, it must make a contribution. Color or shading highlights complex mechanical parts when installed in a system or shows an adjustment point in a physical relationship with the assembly on which it is mounted.

Detailed views are another important illustration element. They show the location of the equipment being described in relationship to the overall system. They are also used to show movement through "before" and "after" scenarios. Proper use of illustrations can demonstrate movement even though the page is static.

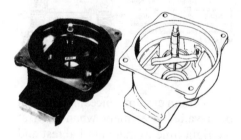

Line drawings are used exclusively in the new-look manual. Accurate perspective and isometric (exploded view) drawings create a more realistic "hands-on" view than engineering drawings. In situations where system hardware is available, an operational scenario can be "staged" and photographed. The illustrator then traces the desired information from the photograph, thereby creating a line drawing. This method requires less time than generating a perspective drawing and is therefore more cost-effective. The illustrator should be careful not to embellish the tracing with unnecessary detail from the photograph.

In contrast to traditional manuals, the new-look manual uses the human

form to enhance illustrations. The person shown performing a particular task must be correctly positioned in each module. Figures are well-proportioned to prevent them from appearing as caricatures. The same character, or set of characters, appears in a set of modules that make up one procedure. Drawings of the system and crew in a normal operating environment are used when appropriate. The format for these scenarios is flexible; however, the illustration must be realistic and functional. The hardware and the crew should stand out from the background.

Repetition of artwork is preferred over cross-referencing. Repeating the same view of the equipment does not necessarily create monotony. By adding a hand, or a human figure, a commonly recognizable image or view can be created to hold the reader's attention. This approach is very cost-effective as compared to the cost of producing a completely new piece of art.

The Logistics Support Analysis Record (LSAR) database is a valuable document for determining the material to be presented. Each new-look publication must address a specific military occupational skill (MOS) level. The MOS is a "job description" of the duties to be performed by the soldier in operating and maintaining the system. The MOS requirements for a particular system are listed in the "D" sheets of the LSAR together with the procedure.

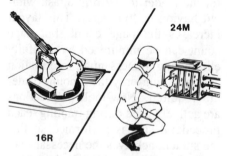

In the Product Improved Vulcan Air Defense System (PIVADS) Program, the operator level is the 16R gunner and the organizational-level person is a 24M. The 24M maintenance people are promoted from the ranks of the 16R gunners who demonstrate proficiency in their MOS.

A single BIT Program was devel-

oped for PIVADS in accordance with contract requirements. Most of the corrective actions required by the BIT Program were beyond the capability and training of the 16R. As a result, the technical writers reviewed the MOS requirements and interpreted the maintenance tasks differently for the crew and organizational-level manuals. When a specific repair task diagnosed by BIT is not authorized for the 16R, the manual reads "notify organizational maintenance." The organizational maintenance manual specifies a technical procedure for correcting the fault that is within the capabilities of the 24M.

Hands-on Experience

Hands-on experience with system hardware is a prerequisite for the technical writer and technical illustrator. PIVADS personnel were provided the opportunity to track live targets to get the "feel" of the system from the gunner's perspective. In addition, they assisted in checking out and operating Government-furnished equipment as well as the modified system. Because procedures described in the PIVADS new-look manuals were generated from actual experience, the procedures were validated while they were being written and illustrated.

PIVADS technical writers reviewed the field manual for its predecessor, VADS, and viewed videotapes of VADS operating in a simulated combat environment. To become more familiar with the targeted audience and the system, the writers visited the facilities where maintenance operations are performed. Once liaison was established with these facilities, knowledgeable answers to operational questions were as close as the nearest telephone.

TEAMWORK

Since the center of a successful new-look program is effective communication, people who can communicate with each other and work well as a team are essential. The new-look team consists primarily of the technical writer and the technical illustrator. At times this team is augmented by members of other disciplines such as maintenance engineering, reliability, safety engineering, provisioning, and logistics. For this reason, the primary team members must be highly skilled in oral and written communication. They must have skills as information integrators and be able to disseminate information to other disciplines as required.

It is desirable for team members to have overlapping skills. For example, a technical writer with a good sense of graphics and a technical illustrator who can read and write well make an ideal new-look team. People with a commercial writing/graphics background can adapt easily to the new look since many of the new-look fundamentals are found in commercial literature.

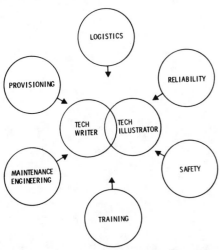

The central figure in new-look publications is the lead technical writer. The writer is responsible for researching and organizing raw data, creating manual outlines, performing task analysis, deciding in rough form what views to illustrate, and coordinating this information with other team members. As with traditional manuals, the writer provides liaison with the acquiring agency, conducts in-process reviews, and validates the completed manuals.

Because the new look relies heavily on the effective use of graphics, the illustrator is promoted from a passive supporting role to the position of a major contributor. The illustrator communicates to the reader through the selection of graphics techniques and page layouts that easily convey the required procedures. The illustrator has a responsibility to create a workable format and a style that are consistent throughout the manual, keeping in mind that the presentation must also be cost-effective.

PLANNING CONSIDERATIONS

Although organizational- and intermediate-level technical manuals can be prepared in the new-look format, it is the crew or operator manual that is most often prepared in this format. For PIVADS, only the crew manual was prepared as a new-look publication. For the XM785 Nuclear Projectile, a less complex system but one that required very detailed instructions, all manuals were prepared in the new-look format.

The preparation of new-look publications (research, artwork, and page make-up) requires more time and effort than traditional publications. Adequate research time must be allocated for the technical writer to become thoroughly familiar with the system as well as the maintenance facility and targeted audience. A new-look manual cannot be written from schematic diagrams alone. The technical writer and illustrator must also be allocated time for "brainstorming" sessions to mutually design the required illustrations and page modules. Keep in mind that a new-look procedure will require many more pages and be more detailed than the manual prepared in the traditional manner. Therefore, the time required to validate the manual will be proportionate to the increase in manual complexity and length. A complete validation of each module or set of modules is essential to achieve the essential realism that is the new look.

When preparing change pages for the new-look publications, managers must consider the level of detail and the number of pages. A relatively simple technical change that would affect a few pages in a traditional technical manual could result in a major change to a new-look manual. For example, a change could affect an illustration that appears in several modules, or possibly affect the module sequence for an entire chapter, making a new layout necessary. We experienced just such a situation with the XM785 program. Because of the delivery requirements for this program, technical manuals had to be supplied before the hardware and software designs were completed. As a result, several iterations of each manual were required. Each iteration required a *completely* new layout.

CONCLUSION

On the basis of our experience with the new look, Lockheed Electronics Company Technical Publications recommends this format primarily for crew and operator-level instructions. The new look can be successfully used for organizational- and intermediate-level manuals if the system is not excessively complex.

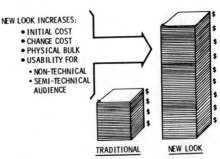

NEW LOOK IS PARTICULARLY APPLICABLE TO PROCEDURES INVOLVING MECHANICAL PARTS AND EMBEDDED DIAGNOSTICS.

New look publications are costly. Before the new look is specified, a thorough analysis of system maintenance requirements and the targeted audience is required. For example, using the new-look format for depot-level technical manuals is, in Lockheed's opinion, not cost-effective. Depot maintenance personnel are trained to perform specific, repetitive tasks. Once these specialists master these tasks, the technical manual is used only as a reference for operating parameters and tolerances.

New-look manuals at the depot level could actually prove to be a liability for experienced maintenance personnel because of the time required to locate specific information within the bulky new-look text. Therefore, higher echelons should specify the new look only after evaluating all of the tradeoffs.

The new-look manual is an extremely explicit but costly communication tool. The government is committed to it, and you will be seeing this requirement more often in Requests for Proposals. As with any tool, it must be applied judiciously to maximize its benefit. Use of the new-look format requires an experienced and creative technical publication team to achieve an accurate, realistic, and visually attractive product while keeping costs under control. Ω

Suggested Reading

DARCOM Handbook 310-1.1-81; (U.S. Army Materiel Development and Readiness Command)

MIL-HDBK-63038-1 Operator's Manual
MIL-HDBK-63038-2 Organizational and Intermediate Level Manual

BENJAMIN D. MEYER *is a Publications Engineering Specialist with Lockheed Electronics Company, Systems Division.*

The User Edit: Making Manuals Easier to Use

MARSHALL A. ATLAS

Abstract—Possibly the simplest way to make a technical manual easier to use is a "user edit"—that is, having an inexperienced user try to work with a machine, using only its manual as a guide. His errors and hesitations should tell you where the weak points are. This report describes how to set up such tests, what to be careful of, and some of the benefits you can expect.

NEARLY everyone who uses or services a complex machine—whether computer, copier, or fork-lift truck—sooner or later needs an instruction manual to figure out what to do next. Such manuals must be carefully written. It is not enough that they be technically accurate, grammatically correct, and easy to read; they must also be "usable," that is, the reader must be able to find and use the instructions quickly, easily, and without error.

One of the best tools for improving manual usability is the "user edit"; it is fast, cheap, easy, and powerful, yielding a lot of information for very little effort. Unfortunately, this technique has not been well-documented; even the U.S. Army, which uses it extensively, describes it only in very sketchy detail [1]. As a result, writers are likely to regard the user edit as either a piece of folklore or a new discovery. Indeed, when we first tried it, we had to learn most of the rules from scratch; our purpose here is to give others the headstart we lacked.

The user edit is based on a very simple idea: Find someone who knows nothing about your machine and have him work with it, using only your manual as a guide; his errors and hesitations should tell you where the weak points are. To get more detailed information, it is also a good idea to ask users to talk while they work, telling what they are trying to do, what they are looking for, what gives them trouble, and what they suggest to make the manual better.

This procedure yields so much information that you will often need to test with only one or two users. Furthermore, you can often afford to be quite casual; for example, even though your best bet is to test *typical* users of your product (people who work for other companies), you can still get good results by testing users who are much less typical (like people who work for you or your company). However, there are two things you should *not* be casual about:

1. Be sure the manual and the machine are as complete as possible. Anything you don't test is almost certain to cause you trouble later.

2. Take detailed written notes: They will be your only source of information. The best way we have found is to write on a copy of the manual, just as if you were editing it. Have at least one person do nothing but take notes (someone else can set up the machine, talk to the user, etc.). Better yet, have the writer of the manual take the notes; that way he can get a feeling for how severe each problem is (i.e., how much harm it can do, and how much work it will take to correct it) and get some practical experience with users as well.

INFORMAL TESTING

The simplest way to run a user edit is to have the user read the manual one page at a time, carrying out the instructions as he reads them. The main disadvantage to this procedure is that it does not test whether he can *find* the appropriate instructions when he needs them. Nonetheless, it can help locate the following:

Problems of Content; in particular:

Missing instructions   Unless you know exactly what you're looking for, it is extremely difficult to read a manual for missing information. The presence of something wrong is likely to be much more obvious than the absence of something necessary, simply because an incorrect item is right there for you to look at, while a missing item must be supplied from memory. A user edit, on the other hand, makes the job simple—if a vital instruction has been left out, the user will be unable to perform the task. Period.

Misleading instructions   Instructions can be accurate and detailed and still be confusing. This is most likely to happen when the procedure being described is so complicated that a diagram is necessary to make it clear.

Badly designed instructions   Some instructions are well-written but represent a poor use of the machine, producing unnecessary steps and a greater risk of error. By watching a user, you can often find out (especially if you have a human factors specialist observing the test) which procedures are unnecessary or awkward, and alter or eliminate them.

Problems of Style

Writers can often develop styles that are easy to read but still not usable. For example, one common mistake is to place warning statements only at the beginning or the end of each *set* of instructions. Such warnings are often ignored, because users expect a set of instructions—including warnings—to be presented in order. It is better to place each warning statement just before (or after) the event it is warning about.

A user edit can help uncover stylistic errors but it may not uncover all of them. With practice (and help from whoever is running the test) users can adapt to the writer's style. For this reason, the manual should be formally edited for consistency; that is, once the user edit has found a stylistic error, an editor should check the manual thoroughly for similar errors. Likewise, the editor should check for consistency of content,

Manuscript received November 7, 1980.

The author is a research psychologist. This paper was written during his association with the Information Development Dept., IBM Corp. Boulder, CO. His current address is Army Research Institute Field Unit, P. O. Box 6057, Fort Bliss, TX 79916.

Reprinted from *IEEE Trans. Prof. Comm.*, vol. PC-24, no. 1, pp. 28–29, March 1981.

especially because writers sometimes omit vital instructions (e.g., how to gain access to some part of the machine). A casual edit should also be done before the first user test—indeed, before, the manual is complete. That way, the editor can warn the writer early, before his more obvious mistakes spread through the book.

FORMAL TESTING

Testing a manual page by page can improve the usability of every section and still leave a badly flawed book. To be usable, the information in a manual must also be

• *Complete*

An informal test can find out whether information is missing from a section, but may not discover that a whole section is missing. The writer should check his work against a formal list (preferably one devised with the aid of engineers and human factors specialists) of the jobs a typical user should be able to do.

• *Accessible*

In real life, the user will seldom take time to read the manual page by page. Instead, he will need to find answers to specific questions with the aid of the table of contents, index, cross-references, and tabs. For this reason, you should give some of your users a formal test; that is, instead of asking them to work the manual page by page, give them a set of typical user problems to solve. For example, a copier user might be asked to make transparencies, add toner, and clear paper jams. This kind of test can locate

Directory problems   Entries in the table of contents and the index can often be

A. *Misleading*   For example, an entry called "Trouble-shooting" could suggest that all problems are solved in this section; an entry like "Solving Other Problems" might be better. A formal test may show that users are turning to the wrong part of the book.

B. *Cryptic*   For example, "Check Paper Path" may not be meaningful to all copier users; "Clearing Paper Jams" is more to the point. In general, avoid technical jargon and have each section heading refer to what the user will want to do, not to the part of the machine he will have to do it with. In a formal test, users will either turn to the wrong

section or give up entirely. In either case, you have a good clue to what the trouble is.

Cross-reference problems   Sometimes titles can be clear and correct, and yet users may still end up in the wrong place. For example, the user may turn to the "Control Panel Lights" page whenever a trouble light comes on. A sensible way to handle this problem is to add cross-references ("see page ——") to send him to the right location.

Getting a representative set of test items is not always easy, but it may be well worth the effort. If you can, consult a human factors specialist. If you are lucky, he may have already developed the items you need.

CLOSING REMARKS

The user edit is a simple but powerful tool for making technical manuals easier to use. Although you get the best results by being careful—testing typical users, using well-designed test problems, and having human factors consultants—you can still get a lot of useful information even if you are less thorough. Like any other kind of edit, the user edit works well as part of an edit-revision cycle. For example:

1. Run a user edit with one or two users.
2. Fix the errors you find.
3. Check for similar errors and fix those.
4. Have an editor formally edit the manual.
5. Revise to correct any errors found.
6. Run another user edit. Et cetera.

The repeated process of editing and revision refines the test procedure. The important thing to remember is that this is an *editing* tool, not a scientific one. Although the final document will not be perfect, it will be much more usable than it was in the beginning.

ACKNOWLEDGMENT

My thanks to Jeff Brand, Kathy Dowling, and Charlotte Mackey, who helped me design and run the tests on which this report is based.

REFERENCES

[1]   E. E. Miller, "Designing Printed Instructional Materials: Content and Format," Technical Report RP-WD(TX)-75-4, Human Resources Research Organization, El Paso, TX, Oct. 1975.

Preparing Instruction Manuals for Non-English Readers

Stacey Sanderlin

AS THE WORLD BECOMES MORE AND MORE A "GLOBAL VILLAGE," *companies are looking outside the United States for new markets. Many of these markets require that instruction manuals be prepared in the native language. For example, Canadian Law 101 requires that all publications used by the general public be printed in both French and English. Even if such a law does not apply, companies are discovering it is good public relations to produce documents that non-native speakers of English can easily understand. Consequently, these companies are choosing to prepare their manuals appropriately for these readers. This article summarizes the literature on the most effective techniques that writers can use in preparing such manuals.*

Much has been written about the subject of translation, including a 1982 special issue of *Technical Communication* [1] and STC's *Guide to Multilingual Publications* [2]. Most of the information in these publications concerns how to translate, how to find a translator, or how to manage a document that will be translated. In this paper, I focus instead on how writers can make it easier to produce instruction manuals for non-English readers.

Companies entering the international market can choose among three main options for preparing these manuals: using human translators, using machine translators, or using Controlled English. Whichever option a company uses, writers can smooth the process of producing manuals by following some simple guidelines governing content, prose, and visuals as they first prepare these manuals in English.

The information in this paper is probably obvious for writers working for companies that prepare manuals for non-English readers; however, other writers, especially those whose companies are considering entering the international market, should be aware of what is involved for them if their company does decide to market its products outside the United States.

BARRIERS COMPANIES FACE

The goal of writers in preparing manuals for non-English users is to produce documents that sound natural to the reader; otherwise, the writing will prevent readers from understanding the instructions. To produce natural-sounding documents, companies competing in the international market must overcome barriers that are caused by different languages, cultures, and learning styles [3].

Language

Language is the most obvious barrier to producing manuals for non-English readers. Everyone has heard examples of companies that have failed at marketing their products internationally because their marketing strategies failed. For example, one Chinese manufacturer quickly learned that automobile batteries marketed under the trade name "White Elephant" would not sell in the United States. A United States firm learned a similar lesson when it mistakenly advertised its tonic water as bathroom water in Italy [4].

This article is based on a paper prepared for the 35th International Technical Communication Conference, 10-13 May 1988.

Translating would be a simple process if every word in every language corresponded one-to-one with a word in every other language. Unfortunately, this situation is not the case. Even words in the same language can have different meanings in different countries. For example, in France, the word "glace" means "ice"; in French-speaking Canada, the same word means "ice cream." Consequently, even with the help of both general and technical dictionaries, translators often have trouble finding words with precisely the same connotative and denotative meaning as a given English word [5].

Culture

People from different countries have different values, different beliefs, different habits—in other words, different cultures [6-7]. Imagine you are in a restaurant in the United States, and a waiter places a glass of water in front of you. If you are American, you know from experience that you are supposed to drink that water. But if you come from another country, such as France, where water is not routinely served with meals, you might not know what to do with it. You might even use the glass of water as a finger bowl, as my former French professor once did when he first visited the United States.

Writers who fail to understand the differences among cultures will write unsuccessful manuals. In America, technical writers use first-person singular pronouns when it is appropriate to do so. In Japan, however, teamwork is far more important than individu-

Reprinted with permission from *Technical Communication*, published by the Society for Technical Communication. May 1988, vol. 35, no. 2, pp. 96–100.

alism, and so plural pronouns would be more appropriate [8].

Learning Style

Because people have different cultural backgrounds, they also have different learning styles. To prepare a successful manual, writers must take these differences into account. For example, Japanese people approach learning much more tentatively than Americans do. When writing for Japanese, therefore, writers need to instill a feeling of confidence in the readers by introducing new topics slowly and giving substantial background information [3, 6-8].

OPTIONS COMPANIES CAN CHOOSE FROM

Companies that need to produce manuals for non-English readers have three options to choose from in preparing them: they can use humans as translators, they can use machines as translators, or they can prepare the documents using Controlled English.

Human Translation

The traditional method of preparing manuals for non-English speakers is to hire translators, either freelance or through a translation service. Unfortunately, this method has several disadvantages. First, it can be both costly and time-consuming. In addition, translators are not always easy to find, especially for less common languages such as Bulgarian. And finally, because language is not static, translators must be up-to-date with current usage, a requirement which means companies must hire a translator who lives in the target country or who visits frequently. These disadvantages aside, human translators usually produce the most natural-sounding manuals because they can most easily understand the differences in language, culture, and learning style [9-12].

Machine Translation

Using machine translators is a newer method of obtaining translations. Although research into this area began in the early 1900s, it has proceeded slowly. Therefore, machine translation has only recently become a viable alternative for companies preparing manuals for non-English readers. And even now, machine translators are suitable only for manuals that use standard vocabulary and are being translated into common languages. If standard vocabulary is used in the manual, machine translation is about 80% accurate for translation from English into French, Spanish, and German.

Unfortunately, machine translation, like human translation, is very costly. In addition, because machine translations are not perfectly accurate, human translators usually have to check the work done by the machines. Although the machines deal with standard vocabulary very well, usual phrases or idioms can produce awkward and even humorous translations, as when one machine translated "hydraulic ram" as "water goat." However, even if human translators supplement the machine translators, translating by machines can save time [13-16].

Controlled English

One of the most popular new methods used by writers to prepare manuals for non-English speakers is the use of a limited form of English, sometimes referred to as Controlled English, Simplified English, Fundamental English, or International Service Language. This limited form of English consists of a small core vocabulary and a larger technical glossary. Each word means only one thing; for example, "right" is the opposite of "left" and "correct" is the opposite of "wrong" [17-24].

First to work with the concept of using a restricted vocabulary was C.K. Ogden in 1936 (Basic English), but his system lacked the technical terms needed. In the 1960s and 1970s, Caterpillar Tractor Company developed its own system of Fundamental English, a system that was adequate for technical manuals [18,20].

In addition to limiting the vocabulary used, Controlled English limits the sentence structures that can be used: writers must use simple structures that readers can easily understand [17-24]. Some companies use Controlled English just for service or maintenance instructions [19,23], while others use it for all types of technical manuals [18,20,22].

Both Caterpillar and NCR have published dictionaries of Fundamental English [25,26], with Caterpillar also publishing a writer's guide [27]. However, rather than devising their system just to meet the needs of non-English users, NCR developed its Fundamental English dictionary and stylistic guidelines for using Fundamental English in the early 1980s because it could not find enough experienced writers with the technical aptitude needed to produce readable manuals. McDonnell Douglas developed its limited-word dictionary for aircraft manuals to be used both by foreign technicians and by English-speaking technicians who have a limited vocabulary [18,22]. In addition to Caterpillar, NCR, and McDonnell Douglas, other companies using forms of Controlled English for manuals intended for international audiences are Sundstrand and Eastman Kodak [23,24].

To illustrate Controlled English, Kirkman et al [21] provide an example of instructions written using standard English and then rewritten in Controlled English:

Original Text When the two bi-metal springs have room temperature a gap should be provided, however, max. 0.1 mm, between the lower side of the link and contact spring 3. The distance is measured using a feeler gauge. Adjusting is executed by bending the adjusting tongues on the stiffened spring 4 using spring bender LSH 2602 or LSH 2603. After adjusting a check should be made of the simultaneousness of the twin contacts. No bending of the bi-metal springs must be made.

Rewritten Version Let the temperature of the two-metal springs become the same as room temperature. The springs will then open. The maximum acceptable distance between the lower side of the link and the contact spring 3 is 0.1 mm. Measure the distance with a feeler gauge. If you need to adjust the springs, use spring bender LSH 2602 or LSH 2603. Bend the adjusting tongues on spring 4. Do not bend the springs. Then check that the two contacts operate together.

Von Glasenapp illustrates Caterpillar Fundamental English with examples of the text for "Principles of Operation" for the SR-4 Generator, "Adjustment of Charge Pump Pressure," and "New Service Meters" [19] and Brusaw describes NCR Fundamental English with two paragraphs from an NCR customer manual for VRX (The Virtual Resource Executive) [20].

With Controlled English, non-native readers can actually read a manual in English after special training in the limited language. According to Caterpillar Tractor, which seems to have been one of the first to develop a limited form of English for use with international users, readers can learn the language in a 30- to 60-hour class [21]. Kodak has found that it normally takes about two to three months for non-English-speaking service people to become proficient in Kodak International Service Language.

If such classes are not feasible, non-English readers can use a Controlled English dictionary as they read. Or as another alternative, manuals can first be written in Controlled English and then easily translated by either human translators or machine translators because of the limited vocabulary and simple sentence structures [15]. Caterpillar translates its Fundamental English with a computer-aided translation system [23].

Controlled English owes much of its popularity to its simplicity and relative inexpensiveness. By having their writers use Controlled English, companies can avoid the errors that a translator might make. They can also keep costs down because they do not have to pay for the translation itself or for printing one manual in several languages. In addition to saving money, Kodak has found that using Kodak International Service Language allows the company to introduce products more quickly than before, while providing qualified service people [23].

In spite of having advantages, Controlled English also has disadvantages, primarily for the writers. Because writers using Controlled English have limited vocabulary and stylistic choices, they must first become familiar with the choices available to them and then become comfortable with the limits of Controlled English. Writers using it often feel restrained by the limited vocabulary. They do not feel they are producing stylistically "good" work. And often, manuals produced with Controlled English are not "good" stylistically, but they are readable [15]. With time, patience, and practice, though, writers can overcome any difficulties that they encounter.

SMOOTHING THE PROCESS

Writers can smooth the process of producing manuals for non-English readers by following certain guidelines when first preparing the manuals in English. Then translators who work with these manuals can save companies time and money because they can translate accurately without having to ask writers and technical experts to clarify obscure points [5]. With these manuals, moreover, machine translations are more accurate.

Writers whose manuals will be translated by either human or machine translators or both have more freedom, and therefore more room for error, than those writers whose manuals will remain in Controlled English when read by non-English users. Consequently, the following information applies more directly to writers preparing manuals for translation, although it may also be useful to writers using Controlled English. Writers using Controlled English must already work within certain guidelines, primarily ones limiting vocabulary and stylistic choices; some of the guidelines suggested in the following paragraphs are similar to the guidelines for Controlled English.

The most important fact that writers preparing manuals for translation must remember is simply that they are writing for translation. They should be sure that their content, prose, and visuals are suitable for instructions that will be translated.

Their *content* must translate well [9,11-12,21,28-29]. Writers should avoid using uniquely American examples: readers in New Guinea would probably not understand a reference to Muskogee, Oklahoma. Because humor usually does not translate well, writers should avoid using it in manuals intended for non-English readers. They should include just the details that users need.

To express the content clearly, writers should understand completely what they are writing about. They should organize the content logically, presenting information chronologically whenever possible; they should limit each sentence to one instruction or one idea. For example,

> *Write* "Push the on-line button. Push the form-feed button."
> *not* "Push the form feed button, after pushing the on-line button first."

In addition, writers should be sure their *prose* will translate well [9,11-12,21,28-29]. When conveying information about actions that users complete, they should limit themselves to short, direct sentences that begin with imperative verbs:

> *Write* "Fill the drilled holes with wood putty."
> *not* "The filling of the drilled holes is accomplished with wood putty."

For other kinds of information, writers should use simple, direct sentences written in subject/verb/object order, with the subject and verb kept as close together as possible:

> *Write* "All operating personnel should wear safety glasses. Safety glasses protect their eyes from steel shavings."
> *not* "Safety glasses which are required of all operating personnel protect their eyes from steel shavings."

Writers should use positive sentences, rather than negative ones; writers should tell users what to do, not what not to do. For example,

> *Write* "Leave the surge protector connected."
> *not* "Do not disconnect the surge protector."

and

> *Write* "Keep the X-74 modular unit away from water."

not "Do not let the X-74 modular unit contact water."

Dependent clauses should be expressed as separate sentences unless it is important to use a dependent clause to indicate a relationship between instructions or ideas.

Write "Push the form-feed button. Push the on-line button."

not "Before you push the form-feed button, push the on-line button."

Exception "Because high-voltage can affect the operation of the M-346, install the M-346 at least 15 feet from power lines."

but consider "High-voltage can affect the operation of the M-346. Install the M-346 at least 15 feet from power lines."

Writers should use clear, descriptive English. They should use nomenclature consistently, always referring to items by the same name. In addition, they should be careful to use words or phrases that will translate well, particularly avoiding words considered jargon, slang, or idioms.

Incorporating extensive *visuals* can also smooth the task of translating manuals for the international market [5,9,11-12,23,29]. According to Kathy Strong at Eastman Kodak, "We feel that visual communication is universally understood and is superior to verbal communication whenever there is a choice" [23]. Writers at Kodak try to express actions in visuals, rather than to express them in prose suitable for translation. Visuals decrease the amount of text needed, thus usually increasing the accuracy of the instructions. Visuals also can improve users' access to information. In addition to helping the final readers understand the text, visuals help translators understand the information they are translating.

Table 1 summarizes the preceding information about content, prose, and visuals in manuals for non-English users.

Those writing instructions for audiences whose native language is English should note that many of their users have reading levels similar to those of non-English readers. Such users would benefit from writers pre-

Table 1. Guidelines for Writers Preparing Manuals for Non-English Users

Content
- Avoid uniquely American examples.
- Avoid humor.
- Understand completely what you are writing about.
- Organize logically, using chronological order if possible.
- Limit sentences to one instruction.
- Include only essential details.

Prose
- Use short, direct sentences.
- Use imperative verbs.
  Begin sentences with verbs.
  or
  Use simple sentences in subject/verb/object order.
  Keep subject and verb close together.
- Use positive, not negative sentences.
- Change dependent clauses to separate sentences unless needed to indicate relationships between instructions.
- Use clear, descriptive English.
- Use consistent nomenclature.
- Avoid jargon, slang, and idioms.

Visuals
- Use visuals extensively.
- Use visuals instead of text.

paring instructions as if they were preparing them for non-English readers. Even though they are native speakers, these audiences have reading levels below those of sixth and seventh graders (some technicians in the military). Or even if they are fluent readers, they may not have fluency in the subject area (novice users of computer documentation). Although the guidelines given in this article are for writers who are producing manuals to be used by non-English readers, many of these guidelines might increase the effectiveness of instructions for users whose native language is English. Ω

REFERENCES

1. Fred Klein, ed. *Technical Translation Today*, special issue of *Technical Communication* 29, no. 4 (Fourth Quarter 1982):4-32.
2. John Archibald and Alan Darisse, *A Guide to Multilingual Publishing* (Washington, DC: STC, 1981).
3. Hiraku Amemiya, "Defining a Good Japanese User Manual," *Intercom* 33, no. 3 (October 1987):7.
4. Karl-Wilhelm Weber, "International Commerce with Verbal Savoir Faire," *Business Marketing* 32 (1984):85-92.
5. Carol MacKay, translator for O'Sullivan Co., Lamar, MO, personal interview, November 27, 1987.
6. "Writing User Manuals for Japanese Readers," *Simply Stated* 7 (July 1980).
7. Izumi Aizu and Hiraku Amemiya, "The Cultural Implications of Manual Writing and Design," in *Proceedings*, 32nd International Technical Communication Conference (Washington, DC: STC, 1985), WE33-WE35.
8. John C. Condon, *With Respect to the Japanese: A Guide for Americans* (Yarmouth, Maine: Intercultural Press, 1984).
9. Gretchen H. Schoff and Patrician A. Robinson, "Manuals for International Markets," in *Writing and Designing Operator Manuals* (Belmont, CA: Lifetime Learning, 1984), 134-148.
10. Kurt Gingold, "The In-House Translator in U.S. Industry," *Technical Communication* 29, no. 4 (Fourth Quarter 1982):8-10.
11. Deanna Tiefenthal, "English is Not Enough," in *Proceedings* 27th International Technical Communication Conference (Washington, DC: STC, 1980), W39-W42.
12. Grace Tillinghast, "Getting the Word to the World," in *Proceedings* 27th International Technical Communication Conference (Washington, DC: STC, 1980), W43-W45.
13. Karl-Heinz Brinkmann, "The TEAM Multi-lingual Terminology Bank," *Technical Communication* 29, no. 4 (Fourth Quarter 1982):6-7.
14. Fred Klein, "Automatic Translation of Natural Languages: A Status Report," in *Proceedings*, 32nd International Technical Communication Conference (Washington, DC: STC, 1985), ATA65-ATA68.
15. Judith Ramey, "Lincoln's Dog, 'Readable' English and Machine Translation," in *Proceedings*, 29th International Technical Communication Conference (Washington, DC: STC, 1982), E 90-E 93.
16. Barbara N. Stafford and Sharon A. Clark, "Machine Translation: History and Prospects," in *Proceedings*, [see #7] (Washington, DC: STC, 1985), ATA 61-ATA 63.
17. Timothy Dykstal, "Living with a Limited Vocabulary," in *Proceedings*, 31st International Technical Communication Conference, Vol. 1 (Washington, DC: STC, 1984), WE 41-WE 43.
18. Becky Gingras, "Simplified English in Maintenance Manuals," *Technical Communication* 34, no. 1 (First Quarter 1987):24-28.
19. Bernt W. von Glasenapp, "Caterpillar Fundamental English," in *Proceedings*, 19th International Technical Communication Conference (Washington, DC: STC, 1972), 81-85.
20. Charles T. Brusaw, "Dismantling the Tower of Babel in Computer Documentation," *Journal of Technical Writing and Communication* 10, no. 2 (1980):133-139.
21. John Kirkman, Christine Snow, and Ian Watson, "Controlled English as an Alternative to Multiple Translations," *IEEE Transactions on Professional Communication* PC-21 (1978): 159-161.

22. Joseph M. Kleinman, "A Limited-word Dictionary for Technical Manuals," *Technical Communication* 29, no. 1 (First Quarter 1982):16-19.

23. Kathy L. Strong, "Kodak International Service Language," *Technical Communication* 30, no. 2 (Second quarter 1983):20-22.

24. E. N. White, "Using Controlled Languages for Effective Communication," in *Proceedings* 27th International Technical Communication Conference (Washington, DC: STC, 1980), E 110-E 113.

25. *Dictionary for Caterpillar Fundamental English* (Peoria, IL: Caterpillar Tractor Company, 1972).

26. *NCR Fundamental English Dictionary* (Dayton, OH: NCR Corporation, 1978).

27. *Writer's Guide for Caterpillar Fundamental English* (Peoria, IL: Caterpillar Tractor Company, 1972).

28. Lynn Hedwig Downs, "So You're Multi-national: How to Prepare Your Manuscripts for Translation," in *Proceedings*, 27th International Technical Communication Conference (Washington, DC: STC, 1980:, W 47-W 49.

29. John A. Conrads, "'Illustrations': Increasing Safety and Reducing Liability Exposure," *IEEE Transactions on Professional Communication* PC-30 (1987):133-135.

STACEY SANDERLIN *is a technical writer for DST Systems in Kansas City, Missouri.*

FEW TECHNICAL documents need more careful preparation than a proposal. You can annoy a lot of customers if you sell equipment accompanied by unclear instructions, but a poorly written proposal will guarantee that you never get those customers in the first place. The articles in this section, with their emphasis on meeting the requirements of the customer-audience, offer you a great deal of help on how to write winning proposals.

THE OVERALL PICTURE

T. M. Georges's one-page synopsis gets to the point right away: your proposal will succeed only to the extent that it meets your customers' needs. Georges's article could just as well end this section as begin it since the author's 15 questions can serve either as a preliminary outline or final checklist for your proposal. His closing comment on packaging introduces a theme further developed by other authors in this section.

WHEN BREVITY COUNTS

The next two articles look at short proposals whose formats have not been mandated in a Request for Proposal (RFP). These articles will be useful if you have an idea for an unsolicited proposal and decide to write it up. Although Sherry Sweetnam concentrates on short proposals for training programs, you can adapt her methodical approach to each segment of the document—from purpose statement to follow-up—to any brief, informal proposal. Following Sweetnam, Bernard Budish and Richard Sandhusen illustrate a short proposal in the form of a one-page letter that even omits headings. Their article analyzes why long proposals often fail and encourages you to use short proposals early in the negotiation process when feasible.

LONGER PROPOSALS

Not all proposals are one page long, of course, and the next two articles focus on writing a substantial document in response to an RFP. First, Robert Greenly presents ten strategies for putting together such a proposal, from preliminary organization to final packaging. He suggests preparing proposal drafts even *before* you receive an RFP. Since large proposals are usually not read in their entirety by all evaluators, section summaries assume great importance, as do carefully presented data and graphics. This article also introduces you to the mechanics of setting up a proposal room and using techniques such as storyboarding (presenting sketches of material on panels) to generate and review your proposals.

TELLING THE STORY AS A TEAM

Storyboarding is investigated in depth by Robert Barakat in the next article. Long, formal proposals require detailed planning and efficient teamwork if they are to succeed—that is, win the bid. Barakat defines and describes storyboarding, showing how it enables you and other members of a team to concentrate on individual tasks and yet work together to produce a unified document. The process is broken down into five steps: outlining, creating storyboards, supporting the message, illustrating storyboards, and reviewing. Everyone on the team is involved in each step, so no matter how lengthy your final document might be, it will be highly organized and coherent.

ASKING THE CUSTOMER QUESTIONS

The final two articles by Annette Reilly and Clark Beck investigate different facets of producing a winning proposal: procuring information not fully provided in an RFP, and putting yourself in the place of a proposal evaluator. How do you get more specific information on the customer's needs after an RFP has been issued? And how do you ask clarification questions (which may be made public) without divulging your own approach to your competitors? Reilly's article looks at the intricacies of this problem and considers the rhetorical subtleties involved in asking open questions in a highly competitive environment. Her ten guidelines for writing such questions will show you how to obtain information from your potential customer, while presenting further evidence that you have the best answer to that customer's needs.

BEING EVALUATED

Beck's closing article emphasizes in-depth reading of an RFP, particularly the ''technical exhibit'' section that outlines what you as the bid winner will be required to do. Your proposal will be judged on how you respond to this section and on whether you convince the proposal evaluators that you are the best person or company for the job. In a research proposal you may have some leeway in your approaches to the problem, but only a meticulous study of the RFP will tell you just how much freedom you have. Beck's five suggestions for final format, and his comments on what will be looked for during the technical evaluation, perceptively summarize the stages of a proposal's life just prior to the decision on whether it becomes the winner.

Fifteen Questions to Help You Write Winning Proposals

T. M. GEORGES

Abstract—Providing the answers to these 15 questions can help ensure that your proposal is complete and oriented to your customer. The questions can also be used as a guide for reviewers and as a checklist for completed proposals.

WHEN you write a proposal, you are usually offering to do some work or provide some product at a specified price for a specific customer. Your proposal has the best chance of succeeding if your product closely matches your customer's needs, if your price is one he can afford, and if you can deliver your product on time.

Here are a few questions to ask yourself while you are writing your proposal. The questions force you to think about the most important things your prospective customers will look for as they read and evaluate your proposal. Make sure your proposal answers all these questions clearly.

Keep in mind that many of your readers are busy executives who are buried in paper and will be able to take only a few minutes to digest your message. Be sure they can find what you want them to find in those few minutes.

1. Who, specifically, are you writing your proposal to?

2. Suppose you were that person and had to review your proposal and decide whether to buy your product or service. What information would you look for first to help you make that decision? Is that information up front and easy to find?

3. What specific need or problem does your product or service address?

4. How will your product or service make your customer's life easier?

5. What best qualifies you to supply your product or service over any of your competitors?

6. Have you convinced your customer that you can actually supply the proposed product or service?

7. Are the costs of your product or service clearly spelled out?

8. How might you break down costs so your customer can select parts of your product or service that best suit his needs?

9. What extra-cost items, not included in your proposal, are likely to be needed?

10. What products and services that some readers might expect are *not* included in your proposal? Should you list them explicitly?

11. If you are offering a service, how will your customer know when the job is done?

12. Do you clearly state how long it will take to deliver your product or service? If your product is a large or complex one, should you provide a delivery timetable with milestones?

13. If you are proposing a research program, do you clearly state what you are going to do and how you will know the job is done? Do you avoid using vague words like "investigate" and "develop" unless you couple them with specific processes and goals?

14. If you are selling a product, what after-sale arrangements do you offer for training, maintenance, parts, and service?

15. Have you relegated most of the technical details and specifications to clearly labeled appendices, where interested readers can easily find them?

You can use these questions in three ways:

- You can answer them before you begin work on your proposal, to form its core outline. The more detailed and specific your answers, the easier it will be to flesh out your outline.

- You can use them as a checklist for your completed proposals.

- You can attach them to copies you submit for approval to speed the review process.

Your careful answers to these questions will help you package the information your customers are looking for and will help them make their decision quickly. The way you present your product tells your customers a lot about the kind of work you do. It pays to show them right at the start that you know how to take their needs into account.

FOR FURTHER READING

Georges, T. M. *Business and Technical Writing Cookbook—How to Write Coherently on the Job*. Boulder, CO: Syntax Publications; 1983.

Received September 28, 1982.
The author is a communication consultant, 340 Norton St., Boulder, CO 80303, (303) 530-2692.

Reprinted from *IEEE Trans. Prof. Comm.*, vol. PC-26, no. 2, p. 84, June 1983.

Training Proposals that Sell Themselves

Here are some practical guidelines to help you make sure your hard work doesn't go unnoticed and unused.

By SHERRY SWEETNAM

How many times have you put your heart and soul—not to mention endless hours—into writing a training proposal and nothing happens? To keep this from occurring again, I'd like to suggest three stages to proposal writing that will help increase your success rate, and help you to write proposals faster.

■ *Stage #1: Interview your reader/buyer.* First and foremost, you must get to know your reader, preferably by talking to him directly. Set up a short but formal interview with him, the purpose of which is to come to a mutual agreement about the content of the proposal. To ensure a successful meeting, give your buyer an agenda. Some possible topics are

■ the training problem
■ the targeted audience
■ the audiences' training experience and interest level in training
■ the buyer's training solutions and concerns
■ your ideas for training solutions
■ what kind of support is available from upper management
■ time constraints on training time
■ the scope and potential of future programs
■ scheduling needs
■ cost constraints

The benefits of interviewing your reader are twofold. First, you can brainstorm and come to a mutual agreement about training problems and solutions. Next, you can learn a second language—your reader's language—which includes what your reader's favored ideas and words are. This

Sherry Sweetnam is a communications consultant and president of Sweetnam Communications, New York City. This article is excerpted from her book, The Executive Memo: A Guide to Persuasive Business Communications, to be published in September by John Wiley & Sons, Inc.

knowledge will guide you in deciding which ideas to emphasize and which words to include or avoid.

Reader-knowledge gives you new communication power by enabling you to write from your reader's point of view, not just from your limited perspective. Your reader will feel subconsciously, "we speak the same language," because you will be using *his* language and emphasizing what *he* thinks is important.

■ *Stage #2: Write a "yellow-pad" proposal.* This is a pre-proposal or zero-draft proposal. It takes only a few minutes to write and serves as a safety valve to make sure there are no surprises in the final proposal.

To begin a yellow-pad proposal, jot down notes on a legal yellow pad either during or after the meeting. *As soon as possible* after the meeting, call or meet with your buyer and summarize the points you discussed. Once you've agreed on the content, then you're ready to write the proposal.

■ *Stage #3: Use a guide for proposal writing.* You can write either by starting with a plan, or by writing randomly and seeing what emerges. It doesn't matter which method you choose, so long as the end product answers the reader's questions and is well organized. Six categories of information that you should include in your proposal are

■ Purpose
■ Problem
■ Solution
   Benefit
   Feature
■ Implementation
■ Follow-up

Figure 1 is an example of a training proposal that uses this model. It was written by the head of a training department to the director of human resources, who had shown interest in establishing training programs for secretaries.

Organizing your proposal

A proposal comprises five different paragraphs.

■ *Paragraph #1: The Purpose Statement.* This statement establishes trust, and tells your reader exactly why you are writing to him and what you are proposing. Be sure that the word "proposal" appears in the purpose statement toward the beginning. Otherwise, the reader is forced to go on a wild goose chase around the page to figure out the writer's intentions.

The purpose statement also helps you get started. Whenever you are stuck, you can begin with phrases such as "The purpose of writing is to propose that...," "The purpose of this proposal is to...," "This is a proposal...," and "I propose...".

Additionally, you can insert an agenda statement within the first paragraph. This information tells the reader what will follow, puts him at ease, and piques his interest.

■ *Paragraph #2: The Problem Statement.* This paragraph focuses on the training needs of the audience. Here is where you tell your reader about the audience's present performance and how it needs to be changed to increase productivity and effectiveness.

■ *Paragraph #3: The Solution Statement.* This paragraph is devoted to training solutions. It describes the new training program, idea, product, or service that will fix the problem. There are two sections to the statement: *benefits* and *features*.

The benefit statement is the most neglected message in training proposals. This statement tells your reader what's in it for him personally, for the participants, and for the company. It explains how the audience's work will be made easier, faster, and/or more productive and can also mention how to save the company money or time.

The feature statement describes the physical features of a training program: the materials, the subject matter, or the stages of the design. For example, Figure 1 discusses needs assessment, interviews with managers, and customized training design. If the writer of the letter in Figure 1 had decided to include subject matter, he might have added telephone training skills, time management, and word processing skills.

■ *Paragraph #4. The Implementation Statement.* This paragraph gives the steps and time frames involved in designing and conducting the training program. In Figure 1, the writer incorporated the features of the program design and their sequence into this paragraph.

■ *Paragraph #5: The Follow-up Statement.* This last paragraph answers the question "Now what?" Here you explicitly spell out who is responsible for the next move. Too many proposals fail at this point because it is not clear who is responsible for what.

If you are highly committed to creating action, I recommend that *you* take responsibility for suggesting the next step. Use an action statement which includes a verb and specific date. For example: "*I will call you* (verb) *April 12* (date) to discuss how you want to proceed." Vague statements like "Let's get together" or "Please advise" create no action. Without a date, there is no commitment to action.

Final proposal writing tips

Here are just a few proposal writing things to keep in mind when you prepare your proposal.

■ *Organize your ideas according to your reader's interests.* If your reader is concerned about cost, don't bury it. If he is concerned about benefits, highlight them.

■ *Use single-theme paragraphs.* This technique will increase your clarity by leaps and bounds.

■ *Don't become a slave to this model.* Each communication is unique. Use this model as a guide and starting point, not as a written-in-stone formula.

■ *Strive for one-page proposals.* No one has time to read long proposals, so keep it short. With each additional page you add to a proposal, you diminish your chances of being read and remembered.

Figure 1–Model of a Proposal

Subject: Proposal for Secretarial Skills Training

Purpose

The purpose of this proposal is to recommend the Successful Office Skills Program which will increase the skills of our secretarial staff. I will outline the problems,, and the benefits and steps we need to take to implement the program.

Problem

Our problems with the secretarial staff are numerous. We have a constant problem hiring and retaining qualified secretaries. What's more, we are paying an enormous amount of money to personnel agencies just to find the secretaries we do have.

Solution/ Benefits

As I see it, the benefits to the firm are numerous. They include:
■ dollar savings in agency fees
■ increased efficiency and productivity
■ decreased turnover
■ improved morale by providing workers with a career path
■ preparation for office automated systems

Implementation/ Features

These are the steps that need to occur. First, Mr. Smith wants to prepare a needs analysis questionnaire and then he would like to speak to three or four key managers. After the data has been accumulated and analyzed, Mr. Smith will offer a customized program for us. He estimates that he will be prepared to begin needs analysis by the middle of May and then begin training in September.

Follow-up

Mr. Smith needs to receive a go-ahead from us. I will call you on Monday to set up a meeting so we can discuss your reactions to the proposal and to confirm start-up dates.

Yours truly,

Roxanne Johnson
Director of Training

The Short Proposal: Versatile Tool for Communicating Corporate Culture in Competitive Climates

BERNARD ELLIOTT BUDISH

RICHARD L. SANDHUSEN

Abstract—Proposals, applicable in a broad range of corporate communication situations, are typically overwritten, poorly researched documents that force recipients to search for ideas of pertinent concern and rarely accomplish their objectives. This paper suggests short proposals, introduced early in the negotiation process, as replacements for formal proposals, or as interim documents leading to more productive, persuasive formal proposals.

SITUATIONS in which corporations communicate to, through, and among their diverse internal and external publics can differ appreciably. Some are subtle and complex; others simple and straightforward. Some are welcome, as when the firm apprises customers, suppliers, or financial intermediaries of a new product or project venture. Others are sought, as when the firm solicits favorable publicity for a newsworthy venture. Still other situations are unwelcome, as when the firm confronts hostile stockholders, employees, or citizen action groups.

The degree of risk associated with communication failure also differs appreciably from situation to situation. In some situations, the risk is negligible; in others, terminal—as such diverse groups as air traffic controllers, professional football players, and corporate raiders continue to discover.

Regardless of context, content, degree of complexity or risk, however, most corporate communication situations exhibit common characteristics that help define their essential nature and suggest strategies for success. Invariably, each involves an exchange of information, a negotiation, implying varying degrees of accommodation among the participants and, if successful, culminating in a "win-win" outcome in which all parties perceive gain. The seller gains the order, the buyer the product benefit; union members get their pay increase, the company an increase in productivity; the company gets its funds, its stockholders a share of the profits. In addition to perceiving gain, each party to a successful negotiation also accepts a commitment to maintain the conditions that produced this gain. As cases in point, recently negotiated agreements to reduce atomic stockpiles or sell the New York Post would be little more than scraps of paper without such commitments.

Another characteristic common to negotiation situations is the early appearance of a proposal, offered by one party to be accepted, rejected, or countered by the others. Typically, this proposal, written or oral, presents one party's interpretation of the situation which precipitated the negotiation—a problem, an opportunity, a grievance—and makes an offer designed to address this situation. Frequently, this proposal, and the response it engenders, is the best motivator and measure of a successful outcome to the negotiation: if the proposal is read and clearly understood by all parties, and written and sequenced to perform its persuasive purpose, it can be the single most important element in generating a mutually satisfactory outcome. Alternatively, if it is rejected, or diverts the negotiations into unproductive bickering over messages and meanings, it can be the key element in ensuring failure.

Unfortunately, most proposals don't succeed in achieving their persuasive purpose. Hillman, for example, sampled a diverse population of multipurpose proposals and concluded that "approximately 90 out of 100 proposals are rejected." [1] This ratio is documented in other studies in the generally sparse literature on proposals. For example, surveys of government procurement sources by Wexler and Carmel indicate that 75 percent of all proposals submitted are either "inadequate or nonresponsive," with another 15 percent judged as only "barely adequate." [2] Another study by Department of Defense procurement officials found that, of a representative sample of 1103 proposals surveyed, fewer than 15 percent were deemed worthy of further consideration. [3] An informal survey undertaken by the authors among partners at four "Big 8" accounting firms further reinforces this conclusion. Each partner was queried as to the success ratio of proposal submissions designed to generate new business. The consensus: even when representatives of the accounting firm had consulted with prospective client personnel prior to proposal submission, and the proposal itself had been prepared with the

Bernard Elliott Budish is Professor of Management at Fairleigh Dickinson University.

Richard L. Sandhusen is a member of the adjunct faculty in the Marketing Department at Fairleigh Dickinson University and is completing a dissertation in the Doctorate of Professional Studies program at Pace University.

Reprinted from *IEEE Trans. Prof. Comm.*, vol. 32, no. 2, pp. 81–85, June 1989.

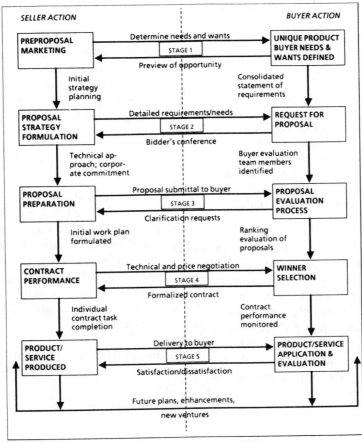

Figure 1. Proposal Preparation Process

assistance of communication specialists—the usual scenario—the rate of success was never above 20 percent.

Aggravating the impact of this high failure rate are high direct and indirect costs involved in the preparation of a typical formal proposal. Figure 1, for example, depicts a five-stage model of buyer and seller actions identified by Horowitz and Jolson as the typical sequence followed by industrial firms and government procurement agencies in the proposal preparation process. [4] At each stage, direct costs of the process itself, plus indirect costs of human resources that could be productively employed elsewhere, can accumulate dramatically.

Is this high cost of preparing proposals a necessary cost? This paper suggests not. Specifically, we suggest an alternative proposal preparation process that employs interim short proposals to address expensive problems endemic in the traditional process of preparing formal proposals. Typically, these short proposals obviate the need for long formal proposals, or ensure that they are used much more productively and profitably.

WHY LONG PROPOSALS FAIL

To understand the value of short proposals in the negotiation process, as well as characteristics and components of these proposals, we first examine problems implicit in the

preparation of long proposals that are effectively addressed by short proposals.

One obvious problem is the sheer length and complexity of the process, as illustrated in figure 1. Given competitive pressures and time constraints, this frequently means that insufficient time is devoted to important information-gathering activities required to ensure a document that is responsive to receiver needs. For example, regardless of proposal context or content, the following information is invariably required in preparing a proposal calculated to lead, persuasively, to a mutually beneficial outcome:

- The roles of people who will read and receive the proposal. Some, for example, will influence decisions leading to an agreement; others will actually make the decisions; still others will have to live with these decisions. The proposal must speak, persuasively, to each role.
- The goals and needs of recipients assuming these diverse roles in the decision process. Again, proposal content and structure should address this diversity in presenting a cogent, compelling rationale for accepting the proposal offering.

Much more often than not, however, roles, goals, and needs of proposal recipients are addressed, if at all, in a scattershot manner, with great amounts of data and information piled high, often almost at random, in the apparent assumption that individual readers will select information of interest. This self-selection assumption is implicit, for example, in this excerpt from a proposal preparation training course offered to managers and partners at a large accounting firm:

> A written proposal should be a document that covers in complete detail all the prospect needs to know about the firm and its approach to the prospect's problems. It is available to be studied at the reader's convenience, to be picked up, put down, and possibly picked up again later. If a part of the written proposal is not immediately clear at first, the prospect can reread it several times. Also, the written proposal can be read at various times by different readers and then discussed later.

Also reflective of this perception of the reader as willing to leaf leisurely through the proposal, selecting items of interest from a sort of literary smorgasbord, are the weighty, elaborate formats for formal proposals suggested in the proposal preparation literature. Herewith, three examples, from these three sources: *What Makes a Good Proposal* (F. Lee and Barbara I. Jacquette) [5], *Guidelines for Preparing Proposals* (Meador) [3], and a content analysis by the authors of characteristics of formal proposals submitted by accounting firms:

- Sellers are not responsive to buyer's needs. (Reflect little knowledge of customer's business, fail to do technical and marketing homework, do not follow Request for Proposal requirements etc.)
- The message is poorly stated. (Wordy and repetitious, lacks innovation and creativity, jumbled and crowded, writing style dwarfs substantive content, conclusions unsupported by data, inherently dull, too much boilerplate or technical overkill, etc.)

HOW SHORT PROPOSALS SUCCEED

The short proposal that addresses communication problems implicit in the traditional proposal preparation process is a versatile, multifaceted document with a useful life that begins well before it is actually written. Specifically, it comes into being at the point during the negotiation when the first substantive point of agreement is arrived at. Typically, this first point of agreement implies a cooperative effort to arrive at a solution, and can include an agreement to resolve a dispute.

Examples of these early points of agreement that trigger the short proposal preparation process include the following:

Jacquette	Meador	Accounting Firms
Summary	Cover letter	Transmittal letter
Defense	Title page	Table of contents
Biographies	Table of contents	Abstract
Budgets	Proposal summary or	Engagement services
Organizational	abstract	(tax, audit,
arrangements	Introduction	management services)
Competencies	Statement of the research	Engagement benefits
Feasibility	problem or program	Engagement team
Study importance,	Objectives and expected	resumes
utility	benefits of the project	Firm's profile
Venture creativity,	Description of the project	Practice clients
originality	Timetable for the project	in area
Appropriateness to	Key project participants	Fee estimate
client focus	Project budget	Appendixes
Leverage prospects	Administrative provisions	
Appendixes	and organizational chart	
	Alternative funding	
	Postproject planning	
	Appendixes and support materials	
	Bibliography and references	

More often than not, a formal proposal is a ponderous, lengthy, uninteresting document, often pieced together with inventoried boilerplate in the hope that individual recipients will plow through it in search of items of personal interest. Realistically, this is almost invariably an invalid assumption, as suggested, for example, by the response to queries among proposal recipients pertaining to "major areas of weakness" in currently submitted proposals: [4]

- From an employer to employee representatives: Let's agree to clarify issues separating us in these negotiations.
- From a corporate raider to stockholders of a target company: Let's agree to the possibility, at least, that this company can be made much more productive and profitable.
- From a company representative to a prospective cli-

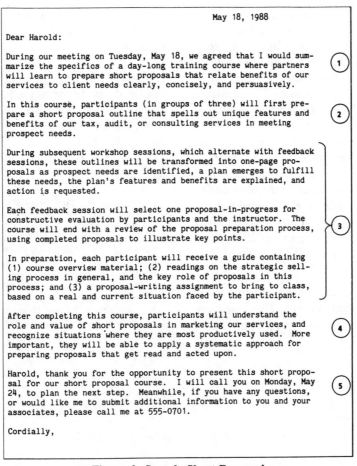

Figure 2. Sample Short Proposal

ent: Let's agree that you can use our services to solve a pressing problem or meet a pressing need.

Note that these agreements are arrived at early in the negotiating process, they are generalized, and they are couched in a spirit of further inquiry, not finality. Invariably, they are points much easier to agree on than agreements arrived at during later, more irrevocable stages of the process. For example, a prospective customer can agree on the need for a certain type of product, without agreeing that the product should be purchased from the seller company.

This early agreement aspect of the short proposal highlights a significant difference between it and its long proposal counterpart; thus, while the long proposal is largely perceived as a document for summarizing data relating seller services to buyer needs, the short proposal is perceived as, among other things, a device for accumulating the data in the first place. Typically, for example, the proposer will suggest the short proposal as a device for gathering and pulling together data required to solve a problem.

The short proposal might end with a preliminary presentation of the accumulated data, and a clear exposition of information that will be required to formulate a final proposal. So used, the short proposal serves as an interim document between the preliminary agreement and the final settlement, generally ensuring that the final proposal will be a more focused, pertinent document than the typical scattershot version.

Alternatively, the short proposal, with minor modifications, might also serve as the final agreement, thereby streamlining and simplifying the negotiation process, increasing the likelihood of a win-win outcome, and dramatically reducing associated direct and indirect costs.

That this short proposal, although viewed mainly as a preliminary, interim document, will indeed double as the final proposal is usually a strong probability, given inherent qualities that make it a much more efficient, interesting, persuasive document than its long, boring counterpart.

Figure 2, a one-page short proposal for a short proposal course, illustrates these short proposal qualities:

- *Structured and sequenced logically and economically:* The main reason the short proposal is short is that it dispenses with most of the headings (see above) included in the typical long proposal—often gratuitously, inflexibly, and with little concern for reader needs. The short proposal reduces this laundry list of topics to the four or five of real reader concern, and sequences them logically in terms of reader interests. In figure 2, for example, the sequence includes (1) needs agreement to put the ensuing discussion into context; (2) a concise proposal of what will be done to facilitate agreement—the plan; (3) a brief discussion of how the plan works; (4) a summary of benefits, and (5) a description of action to be taken by the sender and receiver of the proposal.

- *Easier to read:* The fact that the short proposal has many fewer topic categories than the long proposal is usually sufficient, in itself, to ensure that it will be read in its entirety by all involved in the final decision; that it will be more effectively targeted to audience goals, needs, and roles; and that it will be more effectively comprehended. Aiding this efficient comprehension is the simple, direct writing style of short proposals.

The short proposal serves, initially, as the stimulus for further inquiry into areas of mutual concern. Then, data emerging from this inquiry are organized and presented in the short proposal format designed to facilitate quick, efficient comprehension by all parties to a mutually satisfactory agreement.

Frequently, the short proposal, with minor additions and modifications, becomes the long proposal; alternatively, it can help provide a base of understanding which makes the long proposal, if necessary, a much more efficient, productive document.

REFERENCES

1. Hillman, H., *The Art of Writing Business Reports & Proposals,* New York: The Vanguard Press, 1981.
2. Wexler, J. A., and Carmel, C. A., *How to Create a Winning Proposal,* Santa Cruz, CA: Mercury Communications, 1977.
3. Meador, R., *Guidelines for Preparing Proposals,* Chelsea, MI: Lewis Publishers, 1985.
4. Horowitz, H. M., and Jolson, M. A., "The Industrial Proposal as a Promotional Tool," *Industrial Marketing Management 9,* 1980, pp. 101–109.
5. Jacquette, F. L., and Jacquette, B. I., *What Makes a Good Proposal?,* Washington, DC: The Foundation Center, 1973.

Technical Writing and Illustrating Strategies for Winning Government Contracts

ROBERT B. GREENLY

Abstract—Procurement of government systems, equipment, and services is a complex process. It normally begins with the government advertising its intention to solicit bids for the supplies or services sought. The government issues a Request for Proposal, or RFP, which is a detailed statement of the requirements, the specifications, the elements required in the contractor's bid, and the relative criteria against which bids will be evaluated. The technical writer's task is to organize, structure, and package the proposal into a coherent, readable, and salable whole. Here are ten writing and illustrating strategies that can help create winning proposals.

THIS ARTICLE gives guidance to technical writers, engineers, proposals specialists, and managers who compete for government contracts. If you follow the recommendations of this article you can

- Improve your company's win/lose ratio,
- Reduce the long hours you and your staff typically invest in the preparation of proposals, and
- Make proposal assignments a welcome challenge.

Contracts, including those from the government and those from industrial enterprises, are awarded on the basis of written proposals and supporting marketing activities. Contracts are won by offering the best proposed solution to the "problem" posed by the Request for Proposal (RFP). The name of the game is to score the highest number of points with the proposal evaluation committee, often called the SSEB, or Source Selection Evaluation Board, in U.S. Government procurements.

When you have developed your strong team and excellent solution to the "problem," then you must use modern and appropriate writing and illustrating techniques—"on-the-walls visualization," storyboards, the graphics-driven writing process, what have you, that best *communicate* that message to the evaluators. Modern writing and illustrating techniques are what this article is all about.

STRATEGY 1—START EARLY

A key to winning contracts is to get started *before* the RFP is released. The official release is signaled when the procurement is announced in the *Commerce Business Daily*, a U.S. Department of Commerce publication.

Savvy proposal specialists know, however, that *Commerce Business Daily announcements are too late.* Initial drafts should be prepared *before* the formal CBD advertisement. The usual excuse is, "We can't do anything until we have the RFP," which is not true. You can make your own "strawman" RFP, one that anticipates what will appear in the real thing. You know for example, that it will ask about your experience, facilities, organizational structure, and your cost and schedule control system. So get started on these first drafts. Resumes are another area where the real RFP probably will shed no additional light. Set up interviews with key personnel and "customize" their resumes to the requirement.

Experience shows an unmistakable correlation between starting early and winning. Starting from scratch at RFP release is not an impossible position to be in, but it is dangerously close to it. If you start at RFP release your chances probably hinge on the favorite stumbling or falling out-of-favor before the contract is awarded.

STRATEGY 2—STRUCTURE YOUR PROPOSAL IN THE CORRECT SEQUENCE

Treat the RFP as a series of questions. It asks about your technical approach, management plan, experience, facilities, etc. Answer the questions in the same sequence in which they are asked. The RFP will state what needs to be included in the proposal and the order in which things are asked for is the *required* order of a high-scoring proposal response. In the absence of RFP directions, order your proposal headings in the sequence of the statement of work (SOW) headings or in the sequence of the specification topics. Resist the temptation to restructure the order of things, no matter how elegant your restructuring might seem. It takes less time to evaluate a proposal whose organization is familiar than one which is not. A predictable organization takes the drudgery out of the evaluator's job by putting things in the expected order, the order in which things are spelled out by the RFP.

In the management section the traditional approach is to

Robert B. Greenly is a Program Acquisition Leader at the Lockheed Missiles & Spaces Company, Inc.

Reprinted from *IEEE Trans. Prof. Comm.*, vol. PC-28, no. 2, pp. 7–12, June 1985.

describe the parent corporation, then the company or division within the corporation, then the program or project. The last part, the project structure, is the real heart of the matter. I contend that the usual narrative sequence of corporation-company-project is wrong. Customers are most interested in who is going to do *their* work. The order of discussion should place the project, (the evaluator's real interest) *first*, company organization *second*, corporate structure *last*. Corporate organization may be an important and necessary part of the overall writeup, but it does not really deserve the lead-off position.

STRATEGY 3—BEGIN EVERY VOLUME AND EVERY MAJOR SECTION WITH A SUMMARY

The higher the proposal evaluator is in the customer's organization, the less of the proposal he reads. The older, more experienced evaluators are likely to skim rather than read with great care. For these reasons, summaries take on magnified significance. Every volume, every major section and, indeed, the total proposal should have a summary. Summaries also serve to verify in the writer's mind that all the pertinent questions: Who? What? Why? When? Where? and How? have been answered.

For example, a writer can develop a summary by asking: Who will perform the work? What are the expected results? When will significant milestones be reached? Where will the work be done? Why are we best qualified among all of the potential bidders? How will we manage the work? Answering questions such as these, first in isolated sentences, then in a smoothly connected narrative is one journalistic formula for creating effective summaries.

STRATEGY 4—WRITE YOUR PROPOSAL USING THE LANGUAGE OF THE RFP

You should thoroughly study and *understand* the RFP in

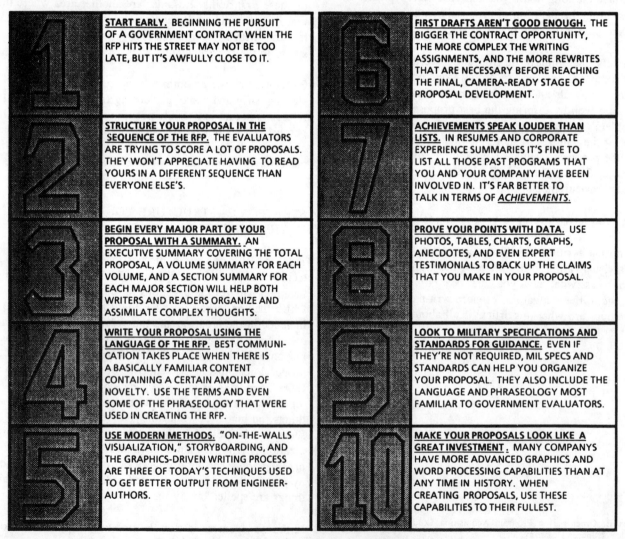

1 START EARLY. BEGINNING THE PURSUIT OF A GOVERNMENT CONTRACT WHEN THE RFP HITS THE STREET MAY NOT BE TOO LATE, BUT IT'S AWFULLY CLOSE TO IT.

2 STRUCTURE YOUR PROPOSAL IN THE SEQUENCE OF THE RFP. THE EVALUATORS ARE TRYING TO SCORE A LOT OF PROPOSALS. THEY WON'T APPRECIATE HAVING TO READ YOURS IN A DIFFERENT SEQUENCE THAN EVERYONE ELSE'S.

3 BEGIN EVERY MAJOR PART OF YOUR PROPOSAL WITH A SUMMARY. AN EXECUTIVE SUMMARY COVERING THE TOTAL PROPOSAL, A VOLUME SUMMARY FOR EACH VOLUME, AND A SECTION SUMMARY FOR EACH MAJOR SECTION WILL HELP BOTH WRITERS AND READERS ORGANIZE AND ASSIMILATE COMPLEX THOUGHTS.

4 WRITE YOUR PROPOSAL USING THE LANGUAGE OF THE RFP. BEST COMMUNICATION TAKES PLACE WHEN THERE IS A BASICALLY FAMILIAR CONTENT CONTAINING A CERTAIN AMOUNT OF NOVELTY. USE THE TERMS AND EVEN SOME OF THE PHRASEOLOGY THAT WERE USED IN CREATING THE RFP.

5 USE MODERN METHODS. "ON-THE-WALLS VISUALIZATION," STORYBOARDING, AND THE GRAPHICS-DRIVEN WRITING PROCESS ARE THREE OF TODAY'S TECHNIQUES USED TO GET BETTER OUTPUT FROM ENGINEER-AUTHORS.

6 FIRST DRAFTS AREN'T GOOD ENOUGH. THE BIGGER THE CONTRACT OPPORTUNITY, THE MORE COMPLEX THE WRITING ASSIGNMENTS, AND THE MORE REWRITES THAT ARE NECESSARY BEFORE REACHING THE FINAL, CAMERA-READY STAGE OF PROPOSAL DEVELOPMENT.

7 ACHIEVEMENTS SPEAK LOUDER THAN LISTS. IN RESUMES AND CORPORATE EXPERIENCE SUMMARIES IT'S FINE TO LIST ALL THOSE PAST PROGRAMS THAT YOU AND YOUR COMPANY HAVE BEEN INVOLVED IN. IT'S FAR BETTER TO TALK IN TERMS OF *ACHIEVEMENTS.*

8 PROVE YOUR POINTS WITH DATA. USE PHOTOS, TABLES, CHARTS, GRAPHS, ANECDOTES, AND EVEN EXPERT TESTIMONIALS TO BACK UP THE CLAIMS THAT YOU MAKE IN YOUR PROPOSAL.

9 LOOK TO MILITARY SPECIFICATIONS AND STANDARDS FOR GUIDANCE. EVEN IF THEY'RE NOT REQUIRED, MIL SPECS AND STANDARDS CAN HELP YOU ORGANIZE YOUR PROPOSAL. THEY ALSO INCLUDE THE LANGUAGE AND PHRASEOLOGY MOST FAMILIAR TO GOVERNMENT EVALUATORS.

10 MAKE YOUR PROPOSALS LOOK LIKE A GREAT INVESTMENT. MANY COMPANYS HAVE MORE ADVANCED GRAPHICS AND WORD PROCESSING CAPABILITIES THAN AT ANY TIME IN HISTORY. WHEN CREATING PROPOSALS, USE THESE CAPABILITIES TO THEIR FULLEST.

Technical writing and illustrating strategies for winning government contracts.

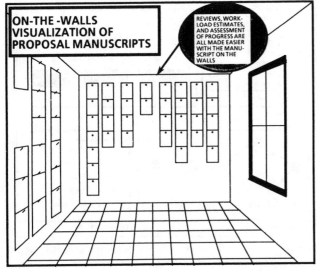

Fig. 1. On-the-walls visualization of proposal manuscripts.

space is the place to do it. Work progress and the magnitude of the workload are made more apparent, permitting you to better prepare for and schedule additional help, temporary typists and outside illustrators, for example.

Storyboarding, the communication technique originated in Hollywood, is another way to create proposals that are superior textual/visual entities. Five elements make up each storyboard module: title, theme sentence, narrative, graphic, and caption. The sequential steps of the process are illustrated in Fig. 2.

Storyboarding begins with the creation of theme sentences, compelling statements that summarize each subsection of the proposal. Narratives and captioned illustrations are then developed which support the themes. A storyboard control form like the one shown in Fig. 3, facilitates this development. The form is designed for quick, hand-lettered and hand-drawn entries. For the finished proposal, the storyboarded material is formatted into a paired-page spread, as illustrated in Fig. 4.

order to see clearly what is required and how the various requirements are connected. You should have a mind's eye view of how your capabilities match up with the requirements, especially in ways that show that you are *uniquely* capable of providing a solution. You should then write your response, answering the requirements using language that is familiar to the evaluator. Use many of the same words and some of the phraseology used in the RFP itself. Add supporting rationale and examples in the form of line drawings, photos, charts, tables, graphs, anecdotes, and even testimonials.

STRATEGY 5—USE MODERN METHODS

On-the-walls-visualization, the taping or pinning of the proposal (or any writing project) to the walls is the way to get the most out of manager, editor, and engineer-author reviews. A proposal room, like Fig. 1, with lots of wall

Storyboarding makes the writing dramatically easier. It can be an effective process for improved communication, but be forewarned that successful storyboarding can involve a Herculean *planning* effort in the early stages of the proposal's preparation. This is because storyboarding is a method that attempts to describe, by use of themes, highly technical and complex matters, often before a clear idea of these matters exists. Writing the themes can be a very tricky and difficult assignment. "Off-the-mark" themes, a hazard of the storyboard method, can be disastrous to the proposal's chances for success. Hundreds of person-hours can add up to nothing if the central point is missed. But theme sentences done well—even though they are more work in the beginning—can save you much work in the later stages of the proposal's preparation.

A variation of the storyboard method called the graphics-driven writing process places initial emphasis on *first* doing the illustrations and data exhibits (summaries of all the relevant parametric data), and doing the writing *last*.

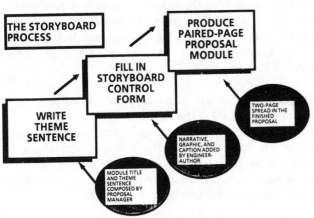

Fig. 2. The storyboard process.

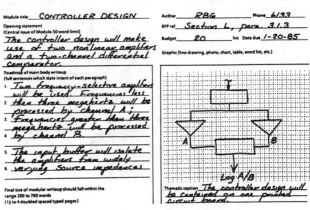

Fig. 3. Storyboard control form.

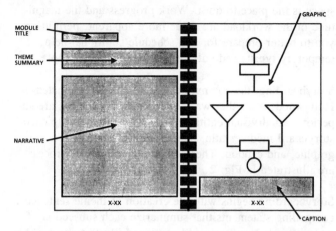

Fig. 4. Paired-page storyboard format as it might appear in the finished proposal.

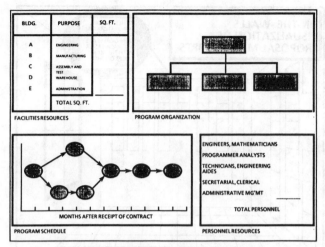

Fig. 6. An example of the graphics-driven approach to proposal preparation.

While storyboarding might be summed up by "Do all the theme sentences first," the graphics-driven writing process is oppositely polarized; it says "Do all the illustrations first." Simply put, the graphics-driven approach, illustrated in Fig. 5, consists of four steps.

1) Create graphic data exhibits that support your proposal: tables, drawings, photographs, flow diagrams, schematics, equations, etc. Consolidate these graphics onto generously-sized, profusely-illustrated proposal pages, even foldouts. Figure 6 is an example.
2) Sequence and format these parametric graphics such that they answer the requirements of the RFP.
3) Review the illustrated pages for responsiveness, completeness, accuracy, and clarity. Rework and reposition the graphic elements if required.
4) Write accompanying proposal text that talks to the graphics.

If you have an aversion toward foldouts, as some do, the graphics-driven writing process still has merit. It can be an effective tool for developing better first and intermediate drafts. If you want a conventional-appearing final proposal, you can cut the graphics apart and place them in the

text, "river raft"* fashion as was typically done before. You'll probably have a better proposal having followed the "do the graphics first" discipline.

The number of hours spent writing will decrease when using the graphics-driven writing process. First drafts have been shown to be of markedly higher quality because well-thought-out graphics give better focus and purpose to the writers. But illustrator hours may increase as much as two-fold because of the greater emphasis on graphics. After a bit of practice and with one or two "pacesetters" on the proposal team who are adept at thinking graphically, the graphics-driven process can be very effective.
Figure 7 compares the relative levels of planning, writing, graphics design, and production efforts for the graphics-driven writing process, the storyboard method, and the conventional, i.e., river raft way of doing things.

STRATEGY 6—FIRST DRAFTS AREN'T GOOD ENOUGH

General Douglas MacArthur wrote complex letters and reports, setting down his thoughts clearly and accurately on his first attempt, and only rarely making corrections. Most of us are not that gifted. Most good writers will rewrite everything that they create several times and few would even dare to send out the first draft of an important proposal, letter, or memo. A proposal whose parts come together, cover-to-cover, for the first time on the day before delivery is destined to contain many errors and will likely receive a low score.

Rewriting is a vital part of the proposal writing process. How many proposal drafts should you plan for? Experience shows that elaborate proposals, those involving larger contract amounts, require the most rewrites. Small proposals for contracts less than one million dollars and containing only a few thousand words can usually be finalized in two rewrite cycles following the initial draft. (A "re-

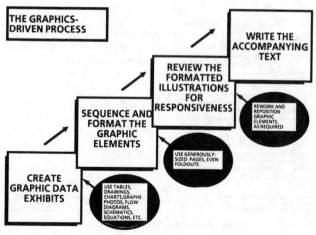

Fig. 5. The graphics-driven process.

---

* "River raft" format is effectively a scroll with unpredictable figure placement. It is the traditional format of most textbooks.

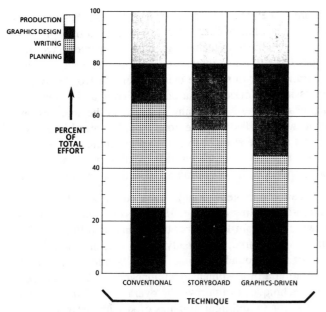

Fig. 7. Levels of effort in three methods of proposal preparation.

write'' is defined here as a complete pass through a review, edit, and correction process.) Large, complex proposals for contracts of 100 million dollars or more with many volumes and books containing tens of thousands of words may have to be rewritten as many as seven times before reaching the camera-ready, ready-to-print stage.

Figure 8 summarizes the recommended number of proposal drafts that you should plan for. In cases where an earlier proposal, study report, or research and development report can be used as a point-of-departure document, your proposal effort can hit the ground running and the height of the bars in Fig. 8 might be reduced by one or two draft cycles.

At the risk of stating the obvious, the reverse interpretation of Fig. 8 is not necessarily true. A proposal rewritten seven times is not automatically worthy of a billion dollar contract.

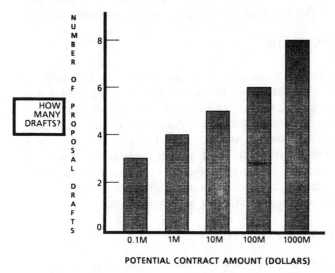

Fig. 8. The number of drafts you should plan for is largely determined by the dollar value of the potential contract award.

STRATEGY 7—ACHIEVEMENTS SPEAK LOUDER THAN LISTS

Evidence of recognized superior prior work from customers, e.g., testimonial letters, awards, early completion of high quality work within budget, etc., can be especially effective towards achieving a top score for the experience section of your proposal. Don't just list what you have done, state the *significance* of what you have done. Also, company experience should not be restricted to the management volume. It should be infused throughout the proposal to substantiate and explain claims. For example: ''......using a similar technique on the ABC Project, we were able to reduce the time needed to prepare a failure modes analysis from 56 hours to 1/2 hour.''

Likewise, in resumes, do not just list where you worked and what you did. Tell *how well you did* each of your assignments. And, in doing so, lean heavily on action verbs: ''designed,'' ''directed,'' ''created,'' ''caused,'' etc.

STRATEGY 8—PROVE YOUR POINTS WITH DATA

Visual impact and style, while important, are qualities that can only count as secondary (but they can be decisive). What really matters, is relevant parametric data, absolute proof that your proposal is quantitatively superior to all others.

Graphically-oriented proposal specialists can often visualize a proposal graphic filled with parametric data before hard data exist. They sometimes know what point the data should make even before the actual data have been gathered. A case-in-point is to summarize the company's facilities and resources in a way that not only states floor area, but also helps show diversified functions, capabilities, and resources. Instead of merely stating that one million square feet of floor space is available, create an interesting table, like the one illustrated below, that shows the allocation of space and the various purposes that are served.

Building	Purpose	Square feet
A	Administration	------
B	Electronic Data Processing	200,000
C	Engineering	350,000
D	Manufacturing	------
E	Word and Graphics Processing	200,000
	Total Sq. Ft.	1,000,000

Few people will have all of the needed detailed data entries at their fingertips. But, somebody does. The point is, tables can be constructed in the proposal draft without having the actual data at hand. By inserting blanks and fake data a busy, highly-paid engineer need not ponder over the numbers. Someone else can be given the data fill-in assignment: ''Find out what the right entries should be,'' or simply, ''Fill in the blanks.''

161

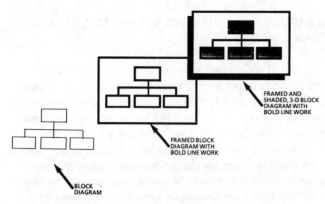

Fig. 9. Progressively greater highlighting of a block diagram.

STRATEGY 9—LOOK TO MILITARY SPECIFICATIONS AND STANDARDS FOR GUIDANCE

Mention Mil specs and standards to engineers and their thoughts turn to higher costs. But consider the work that has gone into these documents, some having been revised many, many times. For the most part, they are well-organized and comprehensive and provide valuable guidance on management, engineering, manufacturing, logistics support, and other topics. They can be useful and valuable references *even when they are not requirements of the RFP*. MIL-STD-881, for example, is all about work breakdown structure. If you want to organize a proposal, report, program, or even a company, applying the WBS technique can bring order out of chaos.

STRATEGY 10—MAKE YOUR PROPOSALS LOOK LIKE A GREAT INVESTMENT

Let's face it, even good proposals are not the most stimulating reading. Proposals are plagued by a sort of flatness of style that typifies books-by-committee, and most are under-illustrated. More and better illustrations are what proposals need. Albert Einstein rarely thought in words. Ideas came to him in images, and only later did he try to express these in words.

Following Einstein's lead, instead of verbalizing, use illus-

trations to heighten reader interest and break the monotony of otherwise dull reading. Because of a flood of rapid advances in publishing technology most companies have the proper tools to make interesting graphics. (I designed the illustrations for this article using a Xerox 8010 "STAR" Professional Workstation, a powerful computer-based graphics and word processing system.) In the space of only about two decades we have gone from hot-metal line casters that set type at eight to ten lines per minute to laser photocomposition equipment that handles intermixed text and graphics while operating at 1000 lines per minute.

How can you make a graphic stand out? Figure 9 is an attention-demanding example. Here an organization chart is drawn using plain line work, then using bold lines and a surrounding frame, and finally, using bold line work, shading, a surrounding frame, and a "shadow," giving the illustration a three-dimensional appearance. You can use these techniques to improve the overall visual impact of a proposal or to highlight graphics of key importance.

Sometimes, three dimensions are more interesting than two. Many scientists and engineers rarely venture beyond two-dimensional plots in their technical illustrations. Yet, an additional dimension can be helpful in the study and understanding of complex relationships. Three-dimensional illustration is often the "natural way" when data have X, Y, and Z components: height, width, and depth; x-position, y-position, and altitude; range, bearing, and elevation; speed, weight, and amplitude (as in vibration analyses); volume, weight, and cost; wear rate versus load versus velocity.

Three-dimensional graphics should be exploited. For example, returning to the facilities and personnel resources depicted in the upper left corner of Fig. 6, we might try to visualize how to illustrate a company's facilities and personnel resources, showing how these parameters have changed over the years. With a little research and using the graphics capability of our in-house computerized system, we can show these relationships in one three-dimensional illustration, Fig. 10, even projecting into the future. I call this technique "illustrating in three-attribute space."

CONCLUSION

Writing and illustrating strategies will not, on their own, win government contracts. Much more is required than just the proposal. Qualifying past performance is often an overriding credential. But keep in mind that the most qualified contractor does not always get the job. Just as the race is not necessarily to the swift, nor the battle to the strong, government contracts are often won, not by the most qualified, but by the one who presents himself most persuasively, in person and in his written proposal.

REFERENCES

Greenly, R.B., *How to Win Government Contracts*, Van Nostrand Reinhold Company, Inc., 1983.

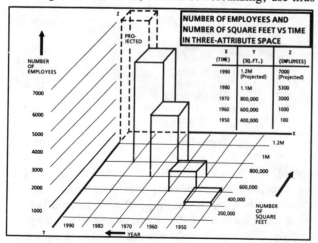

Fig. 10. Number of employees and number of square feet vs time in three-attribute space.

Storyboarding Can Help Your Proposal

ROBERT A. BARAKAT

Abstract—Storyboarding is an efficient technique that provides proposal managers and writers a disciplined, yet flexible, framework for planning, developing, and reviewing proposal text incrementally and sequentially. The technique facilitates intragroup communication so that all team members are aware of the "message" being developed to sell an approach to a customer's problem. A major benefit of storyboarding to proposal teams is that it encourages consistency which, in turn, ensures that the proposal complies with the customer's requirements and the bidder's win strategy.

PREPARING PROPOSALS can be an interesting, challenging, and satisfying experience, especially when the proposals turn out to be "winners." Usually, winning proposals are the result of careful and thoughtful planning by proposal managers who know what they want from proposal teams and can communicate ideas clearly to them. Everyone is working to the same objectives, plan, and schedule; all goes smoothly.

Unfortunately, not all proposal development efforts are well planned or well executed. Many a proposal manager has burned the midnight oil trying to undo the confusion resulting from misdirections, ineffective communication, or foggy thinking. For example, writers will write their sections with only a hazy idea of what they need to accomplish, groping for direction and guidance at a time in the proposal development effort when they should be operating from a firm base. In addition, proposal managers often devise outlines that are inadequate for their intended purposes. Often they simply mimic the Request for Proposal (RFP) in a fashion that satisfies the RFP requirements but does little to sell the substantive content of the deliverables in the proposal.

Storyboarding is a technique that can make the process more effective and efficient because it gives proposal managers more control over the development process.

WHAT IS STORYBOARDING?

Storyboarding is an efficient and effective technique for controlling and coordinating the *incremental* and *sequential* development of technical and management information

for proposals into a coherent story. It involves writing ideas and phrases on sheets of paper or cards and placing them, in order, on a wall in a dedicated proposal control room. This technique enables a team of writers to work together in laying out a blueprint for the message they want to convey to the evaluators. It ensures that all topics to be covered in the final proposal are included and treated in the most effective manner possible, and that each topic relates to the total proposal theme and win strategy (figure 1).

Storyboarding is the focal point of proposal planning. From it stem the pieces of the overall proposal message which, when properly assembled, constitute the final version of the proposal. The storyboarding phase is a critical transition that links the planning with the writing. Inadequate planning at this transition period affects all later stages of the proposal development effort.

As a tool for developing mammoth proposals, storyboarding can be indispensable because it helps the proposal manager and team to devise complete outlines of items to be covered in the actual proposal. Storyboarding allows writers to focus their attention on individual writing tasks instead of the potentially overwhelming whole proposal project. Further, the technique allows proposal managers and writers to plan, visualize, review, adjust, and revise the proposal's organization and contents well before any draft writing begins. Additionally, storyboarding encourages, through reviews, intragroup communication and facilitates, among writers, a common understanding of what should be considered and included in the final version of the proposal. It is this last point that makes the process efficient.

To create effective storyboards, writers still have to outline, select topics, write thematic sentences, develop key thoughts, and conceive illustrations to support the thesis or central message. Each of these elements, including rough sketches of relevant illustrations, is placed on the storyboard. Once completed, the storyboards can be reviewed quickly, revised or discarded, as necessary, even before writers generate rough-draft text.

Working in this fashion, writers will discover that rewrite times are cut considerably and the entire proposal development cycle is also shortened, because reviews and revisions take place earlier than on conventionally developed proposals. Figure 2 depicts the several steps in the storyboarding process.

Robert A. Barakat is Director of Training and Staff Development for Arthur D. Little, Inc.

Reprinted from *IEEE Trans. Prof. Comm.*, vol. 32, no. 1, pp. 20–25, March 1989.

PROPOSAL STORYBOARD

SECTION NO. & TITLE: _TRAINING 5.16 (SOW 3.11)_
TOPIC: _TRAINING DESIGN/DEVELOPMENT/IMPLEMENTATION_

RFP REQUIREMENTS: _DETAILED DESCRIPTION OF OUR APPROACH TO TECHNICAL TRAINING FOR DATA PROCESSING PERSONNEL_

PROPOSAL: _XYZ PROGRAM_
VOLUME: _1_
NO. PAGES: _5_
AUTHOR: _RAB_
EXT: _2250_
DATE: _6/6/87_

STRATEGY: _EMPHASIZE THE BENEFITS TO THE AIR FORCE OF USING THE ISD MODEL TO DEVELOP TRAINING. ALSO, STRESS THE ADVANTAGES OF PERFORMANCE-BASED TRAINING._

THEMATIC SENTENCE: _WE USE AN EIGHT-STEP ISD MODEL BASED ON THE CLOSED-LOOP MODEL USED SO SUCCESSFULLY BY THE AIR FORCE TO DEVELOP PERFORMANCE-BASED TRAINING FOR DATA PROCESSING PERSONNEL._

POINTS TO BE MADE

- SYSTEMS APPROACH
- PERFORMANCE-BASED
- CONSISTENT WITH AIR FORCE'S TRAINING SYSTEMS TECHNOLOGY
- BENEFITS — TIME, COST, COMPETENCE
- EXAMPLES OF SIMILAR WORK DONE USING ISD MODEL
- DISCUSS HOW/WHY OUR 8-STEP MODEL DIFFERS FROM AIR FORCE'S 5-STEP MODEL

ILLUSTRATIONS

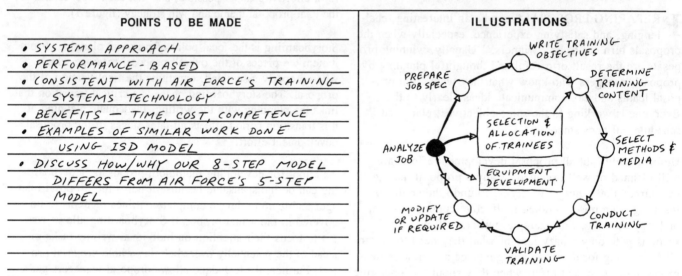

Figure 1. This completed storyboard illustrates a representative form as well as examples of information placed in each block.

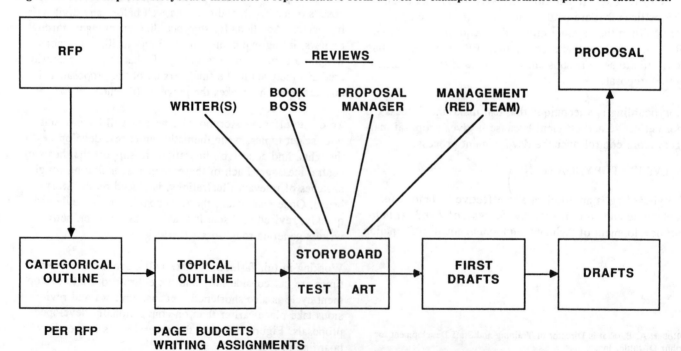

Figure 2. Storyboarding is an integral part of the proposal development process that helps managers plan and control the group writing process.

STEP ONE—OUTLINING THE PROPOSAL

As the preliminary step to creating a storyboard, the proposal manager and selected team members must devise two types of outlines:

- Categorical outline (of the RFP requirements)
- Topical outline (for each proposal section)

Categorical Outline

Typically, this outline specifies who is to write each section and subsection, as well as the appropriate RFP references to which they respond. The categorical outline, therefore, should include only highest-level categories. Figure 3 illustrates portions of a categorical outline subdivided to third-level headings. It is critical that a proposal manager or a designated individual check and recheck the categories against the RFP requirements before distributing the outline to individuals responsible for each volume who, in turn, distribute it to writers.

Since categorical outlines divide thoughts and ideas into discrete categories subordinated with letters or numbers, the proposal manager can rank these categories by their relative importance. In addition, categorical outlines help ensure that all topics are covered. Completeness is what proposal managers must strive for when devising this outline.

However, the categorical outline does not specify *how* writers are to treat the headings. This is the primary function of the topic or topical outline.

Topic or Topical Outline

Transforming a categorical outline into a more detailed topical outline makes the headings more specific and helps the writers flesh out the contents of their sections. In the process of performing this transformation, writers will realize they can formulate a message that they can weave into their narratives to emphasize the win strategy. Care must be taken not to create an outline that is too long or exceeds reasonable page allocations, if these are considered at this time in the proposal development effort.

Topic headings are not difficult to create. Once formulated, they serve as aids for writing thematic sentences and as guides for ensuring that the narrative conveys the exact message intended. Writers simply ask themselves one or both of these questions: "What about this topic?" or "What do I really want to tell the evaluators of this section or subsection?" The following examples illustrate the process:

- *Example:* Operator Training
  What about it?
  We will use the U.S. Air Force-approved Instructional System Development model to design, develop, evaluate, validate, and implement XYZ System operator training.

```
1. INTRODUCTION AND SUMMARY
2. HARDWARE DEVELOPMENT AND INTEGRATION
   2.1  Hardware Development
        2.1.1  End Item Design and Development (SOW 3.4.1)
        2.1.2  Fabrication (SOW 3.4.2)
        2.1.3  Factory Support Equipment
        2.1.4  Maintenance Support Equipment (SOW 3.5.1)
   2.2  Hardware Integration
        2.2.1  Planning and Control
        2.2.2  Integration (SOW 3.4.1.2)
   2.3  Interface Development
        2.3.1  ICDs
        2.3.2  Joint Operating Procedures
                 •
                 •
4. TEST AND EVALUATION
   4.1  Test Planning and Conduct (SOW 3.4.1.2; 3.7)
        4.1.1  Test Schedule
   4.2  Development Test and Evaluation (SOW 3.4.1.2)
   4.3  Development Integration Testing (SOW 3.4.1.2.2.2)
   4.4  Test Hardware Requirements
   4.5  Test Documentation
   4.6  Test Facilities
5. SUPPORTING ACTIVITIES
   5.1  Reliability (SOW 3.9.2)
   5.2  Maintainability (SOW 3.9.3)
   5.3  Standardization (SOW 3.9.14)
                 •
                 •
   5.16  Training (SOW 3.11)
   5.17  Technical Publications (SOW 3.12)
   5.18  Quality Assurance Program (SOW 3.8.5)
```

Figure 3. A bare-bones categorical outline responds exactly to the RFP requirements but says nothing about how each item is to be treated.

- *Example:* Material Control
  What about it?
  Material control is accomplished through a totally integrated data system that monitors and controls the flow of material.

It is important that writers create headings that are specific, informative, and descriptive. Once created, the authors' messages become clear and coherent. It is important to keep in mind the bidder's position on the topic as well as the topic itself. For example, "Reduction of data base errors" is better than "Data base errors," and "How to simplify operator training and make it more effective" is better than "Operator training."

The topical outline is not an expanded version of the categorical outline. It specifies precisely the manner in which writers will approach their topics. A thorough and complete outline allows writers to achieve consistency and coherency throughout their narratives. Figure 4 illustrates a carefully conceived and executed topical outline.

STEP TWO—CREATING STORYBOARDS

The storyboard process begins in earnest when writers formulate thematic sentences that state the central message of their sections. These sentences make a challenging, refutable statement that stimulates the reader to say "Prove it!", and then read on for the details used to support the writer's claim. Thus, thematic sentences contain a good deal of "punch" and specific evidence in support of their

5.18   QUALITY ASSURANCE PROGRAM

　5.18.1   Organizing for Control of Quality

　　5.18.1.1   The Importance of the QA Organization
　　　• QA assures quality, reliability, and safety of product
　　　• Safety is designed in; QA assures it is built in (i.e., not degraded during build)
　　　• Our QA was in place for the build of over 500 units for the XYZ Qual Program

　　5.18.1.2   The QA Engineering Function
　　　• Establishes and reviews criteria
　　　• Assures personnel awareness of quality requirements
　　　• Assures timely and effective corrective action

　　5.18.1.3   The Quality Control Function
　　　• Analyzes material
　　　• Inspects and tests parts, material and assemblies

　5.18.2   Planning for Control of Quality

　　5.18.2.1   Our Policies and Procedures
　　　• Policy 20.1 provides access to vice president and general manager for resolution of quality issues
　　　• Operating procedures in place and approved by local DCAS

　　5.18.2.2   QA Engineering Quality Planning
　　　• All required inspections and tests are identified. Instructions can be used as is or with slight modification
　　　• No new inspection equipment is necessary
　　　• Results of XYZ Qual Program have been used to firm up criteria

　　5.18.2.3   The Quality Assurance Program Plan
　　　• Defines quality role to other departments to assure timely inter-department coordination of tasks

　5.18.3   Implementing the Quality Program

　　5.18.3.1   Design Review and Approval
　　　• Assures quality requirements are incorporated in the design
　　　• Assures manufacturability, inspectability of item

　　5.18.3.2   Control of Purchases
　　　• Assures selection of qualified suppliers
　　　• Assures quality requirements are imposed on suppliers

Figure 4. A topical outline that answers the question "So what?" is likely to stimulate authors to write interesting and sales-oriented narratives.

assertions. Working together at different levels, the proposal manager, book bosses (persons responsible for individual volumes), section leaders (persons responsible for individual volume sections), and writers can produce effective thematic sentences that guide both the writer and the reader.

The following are some suggestions for formulating effective thematic sentences:

• Ask yourself a series of questions about your topic.
Example: Suppose the topic is "transportability of hazardous waste." Take some time to think about the topic, then ask:
"What about the transportability of this hazardous waste?"
"Who needs it?"
"Is it required by the RFP?"
"Does the waste present any dangers during transit?"
Write down all the questions and answers as to *how and why* the company can achieve this requirement.

Some of these points may have come up in your thinking:

— We are certified by the DOT to transport hazardous waste over public roads to our EPA-approved hazardous waste treatment facilities.
— We use double and triple bagging procedures and sealed barrels that minimize risks, yet remain cost effective.

— Our staff is well trained to handle any emergency that might arise, including spills.

Now formulate a thematic sentence based on these points and rework the sentence until you have a persuasive or refutable statement. The final sentence might read

"Our company is fully certified by the DOT to transport carefully packaged hazardous waste over public roads to our EPA-approved waste treatment facilities, where our trained staff can handle any emergencies that might arise."

• Keep the sentences short, preferably 25 words or less. The impact on the reader will be greater. If you need to write a longer sentence, do so. Occasionally, two sentences might be necessary.
• Use words that demonstrate a clear line of reasoning, like these:

because	moreover
since	consequently
therefore	otherwise
however	furthermore

• Take a position by using qualitative words, like these:

better	optimal
benefits	fastest
advantage	most
limitations	smaller

• Make the sentences an argument:
Irrefutable, weak: "The widget has been designed to meet the requirements."

Refutable, strong: "We have adopted Model A of the widget design because it contains components capable of withstanding extreme conditions in all environments in which it will be deployed."

• State the purpose of the unit:

"The widget must be capable of withstanding EMI and still transmit signals, without degradation, to satellites."

In summary, the thematic sentence is derived from the win strategy, expresses the key topic of the section, addresses the RFP requirement, and makes a challenging statement that stimulates the reader to say: "Prove it!" When writing the narrative, writers should keep this sentence in mind at all times because it is their guide—and the readers—to the section. It may be used as a standalone headline at the beginning of the section or as the opening sentence of the section's first paragraph.

STEP THREE—SUPPORTING THE MESSAGE

The thematic sentence represents the point of reference for developing the key thoughts of the narrative. Once this sentence is finalized and written on the storyboard, writers must begin to jot down convincing facts to support the statement. The facts selected must relate directly to the statement. Writers must think carefully about how they formulate and organize the supporting facts. By asking and answering a series of questions about the statement, writers will come up with a random list of facts which then can be logically sequenced for maximum impact on the reader.

For example, "Our company uses data from the System Test of the XYZ Program units because these units are similar to those proposed for the Widget System Program." The points to support the thematic sentence (above) could include the following:

- We can provide early estimates.
- We can conduct tests to verify our analysis.
- We can generate inputs for Electric Power Analysis CDRL requirements in conjunction with similar data regarding the Widget System Program requirements.

Although it is not essential, writers should compose their points in complete sentences rather than in short phrases or single words. By doing so, they will more clearly define their points and also help the storyboard reviewers.

Usually at this point, writers should begin to think about the graphics they will need to illustrate their points. Placing rough sketches of illustrations on the storyboard forms is an effective means of developing text and illustrations simultaneously.

STEP FOUR—ILLUSTRATING STORYBOARDS

The next major step in developing storyboards involves the creation of illustrations such as flow charts and tables. It is important for writers to understand that graphics are powerful vehicles for conveying complex ideas, concepts, information, and messages to evaluators. For maximum impact, graphics have to be designed with the reader in mind. Getting the evaluator's attention is one function that graphics serve; there are others:

- Reinforcing a message or parts of a message conveyed in the text
- Conveying a complex idea or concept that would otherwise require hundreds of words to describe

In addition, illustration captions offer writers another opportunity to sell their messages. Therefore, captions should relate to the entire concept being sold, identify the particular feature that forms part of that concept, and have punch as a selling point. For example:

- *Instead of* "Figure 2.1—Subsystem Block Diagram" *Try* "Figure 2.1—The key feature of this subsystem is the microprocessor."

- *Instead of* "Figure 2.4—Master Program Flow Diagram." *Try* "Figure 2.4—A highlight of our program flow is our ability to identify and minimize risk areas."

The figure captions used with this paper are further illustrations of the above points.

STEP FIVE—"RED TEAM" REVIEWS

The storyboarding technique provides proposal managers and writers a means of developing proposal text incrementally within a well-defined review process called the *Storyboard* or *Red Team Review*. This review helps control and channel the efforts of writers and simultaneously enhances the effects of the win strategy. This review also encourages intragroup communication and ensures that the win strategy is addressed throughout the proposal.

Storyboards should be kept in a dedicated proposal control room where the current version of the proposal plan is instantly visible to all team members. Posting the storyboards on the control room wall also allows individual writers to comment on other parts of the proposal. Reviewers can best perform their task because they can visualize the organization and flow of the proposal. If the storyboards need to be reorganized or if the proposal manager wants to experiment with different organizations, then the storyboards can be rearranged easily and quickly until the most effective organization is obtained.

The first formal team review should be conducted within a week or so after storyboards are assigned and before draft writing begins. Reviewers should check each storyboard and group of storyboards for the following:

- Compliance with the RFP requirements
- Presentation and placement of the win strategy
- Coherence and relevance
- Consistency among thematic sentences, illustrations, and strategy
- Proper sales message
- Effective use of illustrations
- Logical ordering of storyboard modules
- Oversights, misinterpretations, and inconsistencies among storyboards

Storyboard reviews are group reviews that take place in two stages:

- Continuous reviews by the proposal manager, book bosses, and writers
- Management Red Team reviews by representatives not associated with the proposal development effort

This latter team puts their final stamp of approval on the storyboards or recommends revisions before draft writing takes place.

Typically, each review team has an appointed chairperson who is responsible for setting review dates, compiling agendas, maintaining order, recording action items, stimulating discussions, and resolving conflicts. However, these reviewers should remember that storyboards are simply plans for the writing to follow. Reviewers' comments should be limited to content, viewpoint, and organization. Nitpicking of grammatical and stylistic faults is to be avoided since these can be fixed easily after draft writing is completed. Some writers might find it difficult to compose relevant, challenging thematic sentences, especially if they have never written them before, so reviewers should offer suggestions for improving the sentences. Also, suggestions for improving the illustrations should be offered.

However, vague comments like "needs work" or "rewrite" are not helpful and should be avoided. Clear, concise, and incisive suggestions will help writers improve their storyboards and, ultimately, their narratives.

IN SUMMARY

Storyboarding is a technique and a process which proposal managers and writers can use to their advantage. In fact, when storyboarding is used to its full potential, proposal teams are able to telescope an otherwise tedious, inflexible proposal development effort into a neat, coherent, and winning proposal. Since storyboarding exploits the creative and management skills of the entire proposal team, it provides control over the planning and writing without sacrificing flexibility and early reviews. Each proposal team member has a common set of guidelines for developing text and illustrations.

Writers benefit because storyboards allow them to focus their knowledge and creativity on developing the appropriate message for their respective sections. Proposal managers have a baseline for measuring performance of the writers, as well as a means for ensuring cross-fertilization among team members. The Storyboard or Red Team Review, in particular, is valuable to the proposal manager because it is a means for quickly pinpointing any text that is weak, unsupported, or redundant. Similarly, the reviewers no longer have to read through completed rough drafts to check the writers' message; this can be accomplished before or during the review cycle. Editors benefit because they receive text that is well thought out and complete.

Clarification Questions That Work

ANNETTE D. REILLY

Abstract—In technical marketing to the government, the purpose for writing clarification questions is to improve the firm's chances of preparing a winning proposal. Written clarification questions are a valuable means of obtaining information from the potential customer once the RFP has been released. Well-worded clarification questions can also persuade the government to accept one approach and to suspect competitors' approaches. Since the government releases all written clarification questions and answers, they provide valuable competitive information to all bidders. Thus, clarification questions must simultaneously present a position to the government, and conceal it from competitors.

WITHIN THE GOVERNMENT PROCUREMENT cycle, clarification questions are an important step in developing a proposal (figure 1). Potential bidders submit clarification questions to the government as soon as they have reviewed the government's Request for Proposal (RFP). They may continue to submit additional questions to the government until their proposals are submitted, or as specified in the RFP.

The government may also ask clarification questions, but at a later stage in the procurement process. After proposals are submitted, the government issues clarification requests and deficiency reports (CRs and DRs) to obtain clarifications and corrections in the vendors' proposals. However, this article considers clarification questions only as a technical marketing technique used by potential contractors.

The potential contractor's purpose for submitting clarification questions is to improve its chances of preparing a winning proposal. Since contact between the government and vendors is limited after the RFP is released (FAR 15.612 (e)(2)), written questions are a valuable means of obtaining information from the potential customer, either in the form of amendments to the RFP, or as direct answers to the questions.

The opportunity to ask clarification questions in a negotiated procurement produces a complex rhetorical situation. Understanding the rhetorical situation and applying strategic writing and editing guidelines will produce better clarification questions. This article briefly discusses the rhetorical factors influencing clarification questions. It then

Annette Reilly works for Martin Marietta Data Systems in Greenbelt, MD, as a documentation and proposal manager for systems integration projects.

presents editorial guidelines to improve them. These guidelines are also useful in preparing questions for interviews, marketing presentations, legal proceedings, and other situations.

RHETORICAL SITUATION

The immediate audience for clarification questions is the government agency that issued the RFP. Reading the RFP and comparing it to any previous drafts or other RFPs the agency has issued adds to a firm's understanding of its audience. The firm needs to consider the audience's strengths and biases in technical areas, its efficiency, and its relative interest in the technical, management, and cost factors of the procurement.

With this audience profile in mind, the immediate purpose for clarification questions is to obtain information that was not in the RFP. Clarification questions usually seek information in the following areas:

- Does the RFP lack information needed to prepare a good proposal?
- Are there discrepancies between sections of the RFP that hinder preparing a consistent response?
- Does the RFP limit the solution to one vendor's product?
- Does the RFP require outmoded technology or technology beyond the state of the art at fixed prices?

Thus, the purpose of asking clarification questions is not only to obtain information, but also to persuade the government to favor one approach and to draw away from competitors' approaches.

However, the government is not the sole audience for clarification questions. In the interest of fairness, the government contracting officer distributes all clarification questions and their answers to all bidders. Thus, the clarification questions will be read by competitors, and a prudent company will weigh carefully what sensitive information its questions disclose to them. Conversely, although a firm is unlikely to see its competitors' proposals, it will see their questions. The proposal team can learn more about the competition by studying the complete question and answer packet:

- How many serious competitors are there?
- Do the questions suggest that the competitors misunder-

Reprinted from *IEEE Trans. Prof. Comm.,* vol. 31, no. 2, pp. 93–95, June 1988.

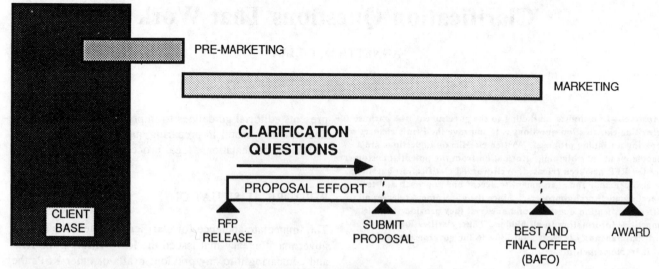

Figure 1. The Business Development Process

stand the requirements, or are they sophisticated analysts?

- Do the revisions the competitors want in the RFP suggest weaknesses in their approach?
- Did no other bidder question a requirement that is difficult for the proposal team to meet, indicating a weakness in its approach?

In short, clarification questions must simultaneously present a firm's position to the government, and conceal it from its competitors. Skillful editing can improve the chances that the questions will achieve their purposes.

EDITING GUIDELINES FOR CLARIFICATION QUESTIONS

In many respects, the task of preparing a set of clarification questions is similar to other editorial projects. The editor or proposal manager assigns members of the proposal team to review all sections of the RFP, collates and edits their questions, and arranges for internal reviews of the package. For competitive reasons, questions are usually submitted on plain paper as an attachment to a cover letter. The questions are arranged in order of RFP reference, or grouped by topic if they are voluminous. Each question needs references to an RFP paragraph and page number; the government is prone to ignore questions with incorrect or missing references.

Beyond this general procedure, here are some specific guidelines for editing clarification questions.

Ask Direct, Concise Questions

Edit to pare each question to the essentials; omit lengthy explanations. Circumlocutory indirect requests are easily overlooked, or condensed by the government so that the real question is evaded. Reword comments into questions. For example, the technical staff may submit items like "Section C.6.4.2.5 requests an unreasonable manpower

requirement." The editor needs to find out what staffing level the firm considers reasonable, why the firm wants that level, and what question will persuade the government to amend the requirement. The question might be worded, "May the offeror determine the level of staffing required to meet the 4-hour response requirement?"

Don't Repeat the Requirements

Restating the RFP wastes space. Summarizing the RFP runs the risks of misinterpreting it or of drawing competitors' attention to an advantageous interpretation. Avoid preaching to the government about what the requirements mean.

Ask One Question at a Time

Burying multiple questions in a paragraph makes it unlikely that each question will be answered. Moreover, a rapid-fire series of short questions in one paragraph, in an adversarial style, may not endear a firm to the customer.

Use the Indicative, Not the Subjunctive

Suppose the question is asked, "Would the government consider the use of offsite test facilities?" Even if the answer is yes, the firm has no assurance that the government will actually accept the offsite approach after hypothetically considering it. An improved wording is, "Are offsite test facilities acceptable to the government?" Or, more subtly, "Will travel costs to offsite test facilities be reimbursed by the government?"

Don't Disclose the Solution

Watch out for questions that explain a technical approach. For example, the question, "Can the requirement for 24-hour coverage be met by equipping an Acme Corporation technician with a beeper?" lets competitors know that the firm doesn't plan to bid three technicians on 8-hour shifts. Also, never mention the company or product name in a question; the government is sometimes remiss in editing them out of the response package.

Don't Disclose Weaknesses

The comment, "Our operating system does not provide the automatic time-out feature," is unlikely to win any change in the requirement, and pinpoints a vulnerability in the solution. Requests for softening requirements are more effective if they point out that the requested feature is either commercially unavailable or vendor-specific.

Justify the Request from the Customer's Point of View

The changes and clarifications you request should be justifiable on the grounds of reducing cost, simplifying evaluation, allowing full and open competition, and the like. Good questions show the government that the firm thoroughly understands the requirements and intends to provide the best solution.

Ask for a Specific Answer

A general question may produce a great deal of unwanted information. Asking, "What does the government mean by 'dynamic reconfiguration' capabilities?" may produce a wish list that far exceeds what a firm has to offer. Instead, the question can be worded, "Which of the following dynamic reconfiguration capabilities does the government require?" taking care to cite only those that the firm is prepared to provide—and which its competitors may not have.

Include the Desired Answer in the Question

Avoid either/or questions. "Should configuration drawings be provided for all 654 sites, or in generic form for each of the five standard configurations?" can be asked only if either answer is acceptable. Otherwise, ask, "Should configuration drawings be submitted for each of the five standard configurations?" Rather than asking, "How many copies of the documentation would the Navy like?" the question should ask, "Should the offeror submit two copies of all referenced documentation?"

Don't Ask Needless Questions

Weigh each question carefully. Is it worth the risk of alerting others to pitfalls in the RFP by asking about them? Can the firm justifiably proceed on its best judgment of what the government wants? Or will a mistaken assumption cause more trouble later? When in doubt, remember that the only dumb question is an unasked question.

Applying these guidelines asks more of an editor than a cursory check of page references and pronoun antecedents; it means substantive revisions of draft clarification questions. The guidelines are more readily applied when a firm has a well-developed technical approach and a clearly understood competitive position, and when the technical editor is familiar with the firm's approach and strategy.

Although this ideal situation is rarely achieved, careful scrutiny of clarification questions and responses is even more valuable when the firm's marketing position is not as strong as might be desired. An astute editor can alert management to possible problems by asking questions about draft clarification questions. Anticipating the answers to clarification questions before submitting them allows a firm to assess alternative approaches, perceive its comparative weaknesses, and determine how to exploit its strengths. As a byproduct, the process of reviewing the draft questions can itself contribute to building a productive proposal team early in the marketing cycle. Thus, clarification questions are an important step in technical marketing, for they contribute to

- Developing a technical approach
- Understanding the customer and the competition
- Building the proposal team

Good clarification questions can lead to a winning proposal—the ultimate step in the new business cycle.

Proposals: Write to Win

CLARK E. BECK

Abstract—Knowing ahead of time what a proposal evaluator looks for will help increase your chance of writing a winning proposal. Winning proposals contain (1) a proposed line of investigation, (2) a method of approach to the problem, (3) recommended changes (if appropriate), (4) logical work units, and (5) estimated completion time. Other tips include writing directly and specifically and making the proposal complete and easy to read.

ARE you involved in any facet of technical proposal writing? If so, knowing how it will be evaluated might suggest how to write to enhance its chance of success. This article is written from the perspective of one who requests and evaluates technical proposals.

Winning proposals require the best of all four forms of writing: exposition, description, narration, and persuasion. As an instrument of persuasion, it must be aimed directly at the interests and objectives of the customer. Company interests, prejudices, and conflicting goals must be avoided in the proposal. The writer must achieve a customer-oriented tone and resist a natural tendency to offer personal views.

Only a small percentage of proposals are winners in technical circles—as in social circles. But proposal preparation represents no small investment of time, money, and facilities. Some degree of success must be achieved to justify continued investments in proposal writing.

Research-and-development proposals are among the most difficult to write, and often they are the most difficult to evaluate. The reason? The limited knowledge available—for both the writer and the evaluator. In early research work there is usually no universally agreed and predetermined best way or approach. Both the preparation and the evaluation are based on predictions and estimates, usually with a host of unknown complicated factors. Writers and evaluators understandably feel that theirs is a difficult task.

PROPOSAL REQUEST

The request for proposal (RFP) includes a section that outlines the work for which the proposal is solicited. This section is most important to the proposal writer. Often called the "technical exhibit," it offers the *only* specific guidance and criteria about the work the winner will do. Your proposal *must* respond to this section better than all other competitors. Keep in mind that (1) the work described in the technical exhibit *will* be done (there is no need for your proposal to convince anyone that it should be done)

Manuscript received August 2, 1982.
The author is a technical manager in the Flight Dynamics Laboratory, Wright-Patterson Air Force Base, OH 45433; (513) 255-2274.

and (2) *someone* will do the work. Your proposal must persuade the evaluators that *you* should be the one to do it. Sometimes a problem develops at this point. The possible problem is a combination of communication and understanding. The technical exhibit was written by one who presumably knows what is wanted (the result) but may not know for sure the best way to achieve it. If it had been known, the request would not be for a research proposal but for hardware. A list of specifications would have been provided with the request for you to design, assemble, fabricate, or build the item to satisfy the given specifications.

If the technical exhibit is not well written or if you, as a proposal writer, misinterpret some part of that technical exhibit, the result could be disastrous. You may write an excellent proposal—offering the best possible solution—but for the wrong problem!

Don't read the technical exhibit just once and charge into the proposal preparation. Study the technical exhibit to see if there could be something more (or different) really asked for. The sender and the receiver may be on different wavelengths.

PROPOSAL CONTENTS

A technical exhibit might have more than one possible interpretation; your proposal does not have that luxury! A technical proposal must be specific and complete. A competitor with a more specific proposal will make you look bad. If your proposal is incomplete or vague, the evaluators are likely to consider you incapable.

Your technical proposal has to stand alone as advocate and defender during the evaluation process. Only information contained in the proposal is going to be considered during the evaluation. If your company is recognized as the leader in the specific technology needed to solve the problem, state that fact, with justification in the proposal to increase your chance of being favorably reviewed.

Several acceptable proposal formats exist; however, any good research proposal includes

1. *A proposed line of investigation*   Brief but specific.
   The customer is entitled to know exactly what steps you plan to take. Restate the specific problem and define the specific solution you expect to offer.
2. *A method of approach to the problem*   An especially important element in research. Often the probability of

Reprinted from *IEEE Trans. Prof. Comm.*, vol. PC-26, no. 2, pp. 56–57, June 1983.

success depends on how you approach the problem. Convince the evaluators that you plan to spend the money you will receive in a manner that offers the highest probability of success.

3. *Recommended changes* Perhaps not an obvious proposal inclusion. In research work, however, your company experts may know (or infer) that a part of the work requested in the technical exhibit has already been done and will contribute nothing to the solution of the problem. Also, some work that needs to be done may have been omitted from the technical exhibit. In these cases, the evaluators will appreciate your recommendations to improve the effort or decrease the cost. After all, one evaluator will probably be assigned to monitor the program; its degree of success will affect his or her future.

4. *Logical work units* A list of logically sequenced phases with reasons why they optimize chances of success.

5. *Estimated completion time* A timetable to indicate when results will be available and to define the proposed expenditure of funds and labor.

Your proposal should not merely offer to conduct an investigation in accordance with the technical exhibit. Even unqualified companies can say "yes, yes" and still not be able to produce. Clearly state in the proposal what, how, why, and when things will be done.

TECHNICAL EVALUATION

Once submitted, your proposal is on its own. In the evaluation arena each proposal is examined thoroughly and a winner selected. Rarely is there a prize for "place" or "show."

The criteria for evaluation certainly vary, but two broad categories cover much of the technical information of interest to evaluators: (1) qualifications of the organization and (2) the scientific/engineering approach.

Qualifications

1. *Specific experience* List the specific, related, and pertinent experience of the company and the personnel who will perform the proposed work.

2. *Technical organization* Consider three aspects of organization:
   (a) How the proposal is organized. Often the proposal format and appearance indicate how reports will be presented.
   (b) How the personnel, resources, and facilities working on the program will be organized.
   (c) How the company is organized. Do the researchers on this program have access to top management in the company?

3. *Special equipment and facilities* Indicate that all necessary equipment and facilities needed to complete the program are available.

4. *Analytical capacity* Ensure that adequate computational and analytical skills necessary for this program are available.

5. *Level of effort and support* Be sure that enough person-hours will be dedicated to this program. Is the mix of the time and skills proper to complete the program? Will the company adequately support its personnel working on this program? How much of the total program effort must be obtained from other divisions of the company—or subcontracted from other companies?

Scientific/Engineering Approach

1. *Understanding of the problem* The proposal must reflect a good understanding of the problem. The first step in the solution of any problem is a complete and in-depth understanding of the real problem.

2. *Soundness of approach* There may be more than one way to approach the problem. The proposal must show justification and sound technical reasons for the one offered.

3. *Compliance with requirements* The technical exhibit often calls for specific requirements such as reporting schedule and format. The proposal should indicate clearly that all such requirements will be satisfied.

4. *Special technical factors* If there are any special benefits that favor your company's doing the proposed work, mention them in the proposal. When two proposals are essentially equal, consideration of items in this category could make a difference.

SUMMARY

Advice for writing a winning proposal includes

1. Be sure you understand the problem.
2. Address the problem directly; be specific.
3. Write descriptively.
4. Make the proposal easy to read.
5. Make the proposal complete and self-supporting.
6. Include all pertinent information; leave nothing for the evaluators to assume (they aren't supposed to, anyway!).

Finally, recognize that the evaluators will infer from the quality of writing in your proposal the quality to expect in a final report.

Part 6
Revising and Editing
Refining Your Document

FEW OF US joyfully approach revising a paper, but, as H. F. Lippincott points out in the second article in this section, if your report is worth writing, it's worth editing. Only a rare writer can produce a flawless document the first time around, yet even if you never expect to be a great writer you can still train yourself to become an effective rewriter. These selections provide numerous pointers on how to edit, improve, and even perfect your written work.

CHECKING AND RECHECKING

Effective revision is often difficult due to our tendency to think that if a document is completed, it must be all right. Gary Blake's opening article is a checklist of 20 aspects of a finished document where improvement might still be needed, however. These range from organization and brevity to grammar and word choice, and although Blake's approach might be limited (as the author admits), his "report card" enables you to evaluate your own written work and that of your organization.

Lippincott also emphasizes that you should never be satisfied with a first draft and gives seven pointers on what to check for in a completed report. You should use technical jargon and acronyms with care, for example, and reexamine your document through the eyes of its intended audience. Although Lippincott's list of helpful publications might seem dated, it is still valid, and the first book mentioned, H. J. Tichy's *Effective Writing for Engineers, Managers, Scientists*, was recently published in a new edition.

EDITING IN LAYERS

The next article provides a different approach to the mundane task of editing. Roger Masse shows how you can methodically improve and perfect your editing by dividing it into separate "levels." A table outlines these levels and suggests questions to ask at each one. Helena Chytil also takes a systematic approach to making a document as effective as possible. She divides the task into two parts: fundamental editing and creating a master checklist applicable to specific documents. Finding a "proof buddy" is advised, as is allowing yourself a cooling-off period before revising your work.

TEAM EDITING

Chytil's suggestion of a colleague helping you proofread is expanded in Charles Stratton's discussion of collaborative writing. Although collaboration might constitute plagiarism in a freshman English class, it is often desirable and necessary in industry, where documents can be hundreds of pages long. However, the traditional division of labor can be improved on, as Stratton shows, by stratifying tasks rather than by assigning them individually. Such stratification enables individuals to work where their strengths are while their efforts are coordinated by a project manager.

THE OUTLINE AS EDITOR

Like Stratton, Dietrich Rathjens focuses on perfecting longer documents. You might have thought of an outline as something to write first and then use as a blueprint for the rest of the paper, but it can also be a useful tool if you apply it to a document *after* completing a draft. Rathjens gives five steps to follow if you wish to use this kind of "reverse engineering" to analyze and restructure your document to improve its final coherence and impact.

START EARLY—AVOID THE RUSH

You can avoid a lot of last-minute revision, Herbert Michaelson points out in the conclusion of this section, if you begin writing a project report in the early stages of the project and develop your report as an integral part of the project. Writing a report concurrently with ongoing research and development will create both a better report and a better project. Moreover, if you start the writing process early, much of your document will be completed as the project is, and you will have more time for editing and perfecting a final report you can be proud of.

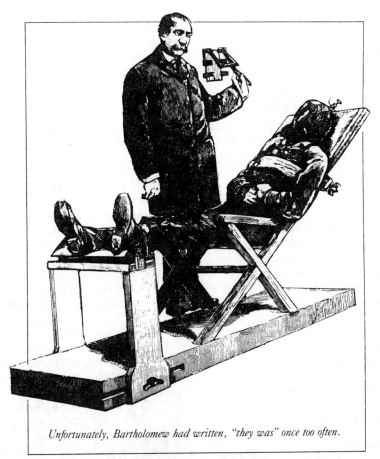

Unfortunately, Bartholomew had written, "they was" once too often.

A Writing Report Card

Don't let poor writing skills speak badly of your company. Pinpoint them precisely with this practical tool.

By GARY BLAKE

At a New York brokerage firm, more than 100 managers in a recent survey were shown two memos; same message, different styles. One of the memos used wordy, stuffy phrases such as "at this point in time" and "initiate a modification." The other memo was concise. More than 66 percent of the managers judged the wordy memo to be "more professional," yet 95 percent confessed that they'd rather receive the concise one.

Discovering who needs writing skill improvement, and measuring a writing course's effectiveness are good reasons for having a systematic method to catergorize particular writing problems. This

Gary Blake is director of The Communication Workshop, New York, N.Y.

evaluation—a "writing report card"—gives you a chance to rate, as objectively as possible, a person's skill level in each of 20 critical writing areas. It may be used by a training director, outside consultant, or supervisor who is unhappy with the quality of writing but doesn't know why. And because the report card has been designed to evaluate the writing skills of everyone from clerk to company president, it may also be used to perform a company-wide assessment.

By rating each writer on a 1 to 5 scale in each of 20 categories, we can arrive at a number to show the relative skills among individuals, among departments—or even among companies.

Admittedly, any such writing evaluation is fraught with dangers. The evaluation certainly will be prejudiced by the whims of individual graders. Some people will

want to substitute categories. Some may feel that a few categories aren't important to the type of writing that is done in their organization. But, even with these caveats, the writing report card can be a helpful tool for evaluating the level of writing within an organization, determining who should attend a writing course, and measuring how well they do after the course.

Look for quality

To pin down the overall quality of a particular writer, make a 1 to 5 evaluation of the writer's level of accomplishment in the following areas:

Organization. Score 5 for a high sense of organization, and 1 for a complete lack of it. Organized writers quickly get to the point. Well-organized writers often use subheads to help keep the reader on track and to make the document easy to scan.

While poor writers usually wander from point to point, organized writers recognize that writing is a three-step process: organizing, writing, and editing. They give the reader a road map.

Conciseness. A "5" makes his or her point with the fewest words possible (e.g., "usually" instead of "in the majority of instances"); a "1" uses three words where one will do. Concise writers make each word count. Most letters, for example, are kept to one page. It's almost impossible to prune a word from a "5."

Getting To The Point. Most writers take too long to get to the point. They start with a lengthy review of background information. Score a 5 for writers who usually get to the point in the first paragraph. And score progressively lower for people who bury their main ideas somewhere in the middle or end of a document. In persuasive writing, however, good writers say what the reader needs before discussing a way to meet those needs. The point—what the *writer* needs or wants to do—may appear later.

Sentence Length. Most people err on the side of length. Generally, if you can't complete a sentence without taking a breath, it's probably too long. Knock off points, too, if the writer includes too many choppy sentences (12 words or less) that don't provide links necessary for understanding related thoughts. Typical sentences should run from 15 to 25 words. Writers should lose points if they customarily clog introductory clauses with too many thoughts.

Paragraph Length. Is the writer sensitive to keeping his or her paragraphs short? Or does the writer overwhelm the reader with too many lengthy (more than 12 lines) paragraphs? To score 5, a writer has to show sensitivity to paragraph length, making sure that paragraphs of more than 12 lines are not placed at the beginning of a document, where the reader needs to be courted into reading on.

Redundancy. Does the writer repeat ideas instead of elaborating on them? Does the writer fall into using redundancies such as "consensus of opinion," "plan ahead," "very unique," or "end result"? Even excellent writers can fall prey to subtle redundancies such as "IRA account," "safe haven," "chemotherapy treatment," or "tuition fees." But a 5 avoids these, using only words that add new meaning.

Phrasing. When you read the words, does phrasing sound convoluted or smooth? Does it sound eloquent or does it fall flat instead? Deduct points if the writer slips into such antiquated phrases as "enclosed please find," "under separate cover," "pursuant to your request," "please do not hesitate to call," "kindly," "herewith," or "I remain . . ."

Word Length. The simpler the words, the higher the score. Good writers want to express, not impress. They don't use big words when small ones will do. Does the author write "use" instead of "utilize"? "Total" instead of "aggregate"? "Best" instead of "optimum"? Reward simplicity.

Hedging. To get a 5 in this category, a writer must write authoritatively. A writer who hedges—often using words such as "perhaps," "maybe," and "seems"—is afraid to state things in a definite way. Another example: "It has been recommended." Who recommended it, and why doesn't the writer say so?

Whenever her sister began, "It has come to my attention that . . . ," Gwendolyn wondered where the tattler hid.

Passive Language. Hedging and passive language are close friends. The writer relies on passive constructions such as "it is believed" (instead of "I believe") or "a decision was made" (instead of "I decided . . ."). The more tentative the writing, the lower the score.

Persuasiveness. Writers who earn 5s in this category make compelling arguments for their beliefs, appealing to the interests of the reader before making a case for themselves. They close their messages with a strong call for action. No "If you concur, please let me know" for them. Instead, they write, "I'll call you next week to discuss this further." Weak writers try to compel their readers to do what they want with weak, annoying phrases such as "thanking you in advance for your cooperation."

Tone. A writer's tone should fit the situation, the attitude of the reader, and the subject matter. Good writers dwell on the positive, looking for ways to change a bad situation instead of clobbering people for what they've done in the past. Most good writers avoid a stuffy, front-office tone, opting instead for a conversational style that reflects the brightest, most colorful aspects of the way we speak. Score a 1 for any writer who hides his or her personality under a heap of formalities such as "pursuant to your request," "thanking you in advance," "yours of the 26th," "I trust," or "Please note."

Spelling. Andrew Jackson once said, "It's a damn poor mind that can think of only one way to spell a word." Even good spellers miss the difficult ones—"privilege," "accommodate," or "supersede," for instance. Midlevel spellers are stumped by words like "embarrass," "disappoint," or "benefited." Poor spellers have trouble with the likes of "written," "describe," and "personnel." Score a 5 if the writing sample you're reviewing is free of spelling errors. Score 1 if there are more than two spelling errors per page.

Capitalization. Poor writers overcapitalize and are unsure as to which words truly deserve capital letters. Take a point off each time you see a capitalization mistake. Good writers don't capitalize for emphasis; take points off if the writing sample uses capitals when underlining is appropriate.

Punctuation. Even excellent writers will make an occasional mistake with commas, hyphens, or apostrophes. Midlevel writers are often confused about when to use semicolons. Poor writers use commas arbitrarily, often forgetting to put them at the end of introductory clauses. Deduct points for each gross error such as confusing "it's" and "its" or linking two independent clauses with a comma.

Grammar. Give a 5 to the writer who correctly uses modifiers, maintains subject-verb agreement, and selects proper tenses. Give a 1 to anyone who uses "myself" when it should be "me," and who is unsure of which tense is correct.

Format. Score a 5 if the writing sample is centered on the page, has ample white space (wide margins) and if its format is appealing to the eye. When necessary, good writers use subheads and underlining, and know how the inside address of a letter should look. Give a 1 to the memo that lacks a "Subject" line.

Specificity. Reward colorful, concrete language with a 5. Deduct points for murky or general words or phrases like "as soon as possible," "nice," "interesting," "at your earliest convenience," or "I look forward to hearing from you in the near future."

Word Choice. Does the person mean to write "uninterested" instead of "disinterested"? "Invariably" instead of "inevitably"? "Farther" instead of "further?" A 5 knows the difference between "eager" and "anxious," "affect" and "effect," "continual" and "continuous." A 1 chooses the word "hopefully" instead of "I hope," and might write, "He makes $20,000 a year" instead of "He earns $20,000 a year." Deduct points for the use of such ugly words as "prioritize," "strategize," and "finalize," and for anyone who uses "irregardless."

Overall Care. Score a 5 for the writer who keeps things parallel (e.g., clustering items in a list, starting porcedures with verbs, staying with a chosen format such as numbers or bullets) and proofreads carefully.

Once you've assessed the overall quality of your company's writing, and decided what to do with the people who need help, you might use the writing report card in recruitment, to make sure that the people your company hires are up to the high standard you've already set.

Rating the Writing

90-100—Superb
80-89—Excellent
70-79—Very Good
60-69—Good
50-59—Fair

Below 50—probably in immediate need of a writing class. Writers at this level are hurting the image of your organization, and the writing probably results in errors that cost both time and money.

Some Tips for Clear Writing

H. F. LIPPINCOTT

Abstract—Making sense is the hallmark of purposeful writing, yet too few people take the time necessary to revise what they have written. No one should be satisfied with a first draft. Writing is a process and, as such, requires taking whatever time is necessary to produce worthwhile writing. Included are seven tips to consider in the revision process: (1) keep subjects, verbs, objects, and complements close together; (2) maintain an average sentence length of less than 20 words; (3) prefer the active voice; (4) eliminate the indefinite "this"; (5) simplify verb tenses; (6) ensure that all paragraphs have clear topic sentences; and (7) avoid "this is" and "there are" constructions.

WRITING is a highly complex process that most people unfortunately do not know enough about. The act of writing has an effect on the way we think, because only in putting our thoughts down on paper can we see them clearly. The writing process transforms ideas into sense. What we end up with is not necessarily what we started with. The person who laments "How do I know what I want to say until I write it down?" is probably right. Moreover, many people find it hard to write because they do not have enough ideas to start with. They attempt the writing process before they are prepared. If we do not know our subject, writing makes us vulnerable. The fact that we are not prepared becomes evident immediately.

Too often we do not allow the writing process to work for us completely. Sometimes we are satisfied with a first draft that may not express all we think or feel about a matter. The first-draft syndrome is also the excuse of busy people who think they have no time to revise. But if the report or the article or the letter is worth writing, it is always worth revising until it makes sense.

Making sense is what writing is all about because only sense communicates. Often the difference between a good writer and a poor one is not native ability but the fact that a good writer takes the time to be clear. Hard writing makes easy reading, as the saying goes.

TRANSLATING TECHNICAL JARGON

Another problem with writing today is that our world has become so complex that most of us must be content with fairly narrow specialties—subject areas we really feel comfortable with. When we are writing for others in our specialty, we can afford to take shortcuts and use initials or acronyms (words like radar or sonar, formed from initials) and the special vocabulary of our field that we call jargon. But any time we write for a wider audience—and that is most of the time—we have to keep in mind that our jargon does not make sense or communicate to others. The following sentence might be appropriate when writing for diehard computer buffs: "Implementing DDP will make maximum use of the mainframe computer." However, even in a specialized journal, "DDP" should be spelled out as "distributed data processing," then defined; and some in the audience might even need "mainframe" computer explained.

The real expert in a field is one who can take the complex jargon dealt with daily and translate it into generally understandable terms. To attempt to communicate using just the jargon is elitist and insulting to the general reader. More important, jargon simply fails to communicate clearly, even within a field.

In this process of translating technical material, two things are obvious: The first is that translation takes time and effort; it's much easier to get the jargon down on the page with no regard for who will read it or for its lack of clarity than it is to translate it. The second is that in the process of translation some subtleties may be lost.

These tasks, revising drafts and translating jargon, can make reports and articles clear. Here are some specific tips for improving your writing:

- **Keep subject, verb, and object or complement close together.** The subject and verb acting on an object (or complement) are the essential parts of a sentence that help express a complete thought. Every other sentence element is a modifier or a "hedger." Consider this sentence:

  A comprehensive *evaluation* of the operational retraining program and of the progress and performance of the 10,000 to 15,000 airmen who change specialties annually *is underway*.

Technically, we call this a suspended sentence, for the completion of the thought is suspended until we get to the verb and its complement at the end. Recast the sentence as "A comprehensive evaluation is underway . . ." or, even better, tell us who is evaluating. Revising to put subject, verb, and complement as close together as possible tremendously enhances readability.

- **Write sentences averaging less than 20 words.** Rudolf Flesch's studies of readability in the forties show that sentence length is a most important factor. If you know that you have a tendency to be verbose, consciously write shorter sentences. Do not be afraid to be curt or choppy at the start. You can always combine or, better still, revise some by using subordinate clauses or phrases.

Reprinted with permission from *Logistics Spectrum*, vol. 16, no. 1 pp. 43–45, Spring 1982; copyright 1982 by the Society of Logistics Engineers, Huntsville, AL.

The author is retired from the U.S. Air Force, P. O. Box 11281, Montgomery, AL 36111, (205) 263-2375.

Reprinted from *IEEE Trans. Prof. Comm.*, vol. PC-26, no. 1, p. 11–12, March 1983.

- **Prefer the active voice to the passive.** The natural form of the English sentence, refined through the centuries, predominantly uses the active voice (John saw the dog), which has the subject acting through the verb on the object. The frequent use of passive voice (The dog was seen by John) is a relatively recent phenomenon—perhaps influenced by German scientific writing of the 19th century. Its use is usually artificial, because the actor in the sentence is either missing or buried in a prepositional phrase. The argument that you must throw in a passive sentence from time to time for variety is hokum. If you use more than one or two passive sentences in each page of writing, you are not consciously thinking in active voice. Find the actor and make him or her the subject of your sentences wherever practical. A great writing teacher at Michigan, Sheridan Baker, says, "Avoid the passive. Avoid the passive. Avoid the passive."

- **Eliminate the indefinite "this" at the beginning of a sentence.** Consider this passage:

   The UDB will make logistics and HR data available to the designers, logisticians, and engineers at an earlier stage of the WS design process. *This* will allow logistics and HR factors to impact design to a much larger degree.

Quite apart from the meaning of the two sentences (they are far from clear), the floating "this" at the start of the second has no identifiable antecedent. It refers to an abstraction—the whole idea expressed in the previous sentence. Often the fix is to insert a noun after "this," such as "this change" or "this innovation." Sometimes the fix is more difficult because "this" hides a shift in thought and the second sentence does not actually follow from the first. Careful writers always make antecedents clear and avoid a floating "this."

- **Simplify verb tenses.** Whenever possible, keep to the simple present (he goes), simple past (he went), or simple future tense (he will go). Avoid compound tenses (he is going, he would have gone). "Today we feel" is clearer than, "Certain conditions facing the nation today are being felt severely." Remember that when you cite from a book, the fact that someone has written the passage should usually be expressed in the present tense (as Rosenstein *says* in his preface, or Aristotle *reminds* us continually in his *Poetics*). The fact that in one case the author is living and in the other he is dead has nothing to do with how we cite him. But when specifying a point past in time use past tense: "As early as 1934, Sarnoff expressed interest in ...."

- **Avoid "there is" and "there are."** These expressions are lifeless because they substitute for a finite verb and lead only to simple enumeration. They also lead to sentences like "*There are* several universities around the country *which* are striving to set up a logistics academic specialty ...." Notice that the italicized words can be deleted entirely, making the sentence more concise.

- **Write a clear topic sentence for every paragraph (or group of paragraphs).** Be continually aware of the "so what?" of writing. If you force yourself to establish paragraph direction in the topic sentence and extend from it logically to a conclusion, the reader will follow your argument more easily. The same goes for forecasting. Avoid empty statements like "The purpose of my paper is ...." Instead tell the reader what to expect in a solid declarative sentence with an action verb such as, "Diminishing manufacturing sources (DMS) for older defense systems will plague the DOD logistician the next decade." With a sentence like this readers know exactly where they are headed, and if they do not want to read further, they do not have to. Psychologists tell us that adults, especially, like clear direction both in the classroom and in dealing with written material. There can be no more reprehensible criticism of a piece of writing than to have its readers continually ask, "What's this article about? What's the point?"

WHERE TO GO FOR HELP

Here are some books that may be helpful: A classic is H. J. Tichy's *Effective Writing for Engineers, Managers, Scientists* (New York: John Wiley and Sons; 1966). Buzz words and jargon to avoid are found in Rudolf Flesch's, *The ABC of Style: A Guide to Plain English* (New York: Harper and Row; 1964), recently republished in a paperback (Perennial Library P83). Although not specifically concerned with technical writing, a book of the "underground grammarian" Richard Mitchell, *Less Than Words Can Say* (Boston: Little, Brown; 1979), is an eye opener on how language can be misused.

Although books on writing can help us, in the long run nothing substitutes for thinking clearly and expressing thoughts as simply as possible. A person who speaks pompously probably writes pompously, too, and may not think clearly. Writing is a transparent process. Poor writing fails to communicate and also casts doubt on our basic competence. Sir Ernest Gowers, a distinguished British civil servant, reminds us: "To be clear is professional; not to be clear is unprofessional."

Theory and Practice of Editing Processes in Technical Communication

ROGER E. MASSE

Abstract—A conscious and organized study of editing processes can introduce basic editorial values to technical communication students. Through research and practice, students can learn systematic methods of editing with levels of edit (or types of edit) for written manuscripts and with editorial dialogue for conferring with writers. This work can lead to development of individual theories and practices of editing.

TECHNIQUES for editing in technical communication can be learned in practice on the job but, without guidance, the self-taught editor could be severely limited. To be really good, to meet the challenges of editing well, to effectively help someone else communicate technical information clearly and consciously, the editor needs to develop a broad understanding of the theory and practice of editing. This foundation for developing basic editorial values can be learned from courses or workshops that are organized to teach editors how to edit efficiently and effectively. These courses can be structured to help students examine research on editing, practice editing, and develop their own theories and practices of editing.

For the graduate students who take my Advanced Workshop in Technical and Professional Communication, I have developed such a course to provide the foundation for understanding and using editing processes. The advanced workshop is limited to ten graduate students from any discipline, including technical communication. While the students work on the writing they need to do in their particular fields and on writing processes for their manuscripts, they also learn much about editing for peer evaluation of each other's work. The editing helps students in all fields learn how editors work and how their own writing can be critiqued. And as they learn to edit the writing of other students, they also learn what to examine in their own writing.

To learn the theory and practice of editing, my students and I concentrate on two main activities:

1. Developing and using levels of edit and editorial dialogue in editing workshops
2. Developing theories of editing process through research and experience.

The author is an Associate Professor of English, New Mexico State University, Box 3443 University Park, Las Cruces, NM 88003; (505) 646-3931.

The first activity is accomplished within the editing workshop by all the students; the second activity is accomplished by the graduate students who are majoring in technical communication.

DEVELOPMENT OF EDITING TECHNIQUES IN THE GROUP

In the group workshop, my students and I develop and use levels of edit and examine and use editorial dialogue to edit manuscripts written by the students.

Levels of Edit and Types of Edit

To develop techniques for editing papers in our workshop, my graduate students and I examine several articles on editing. Some of those articles are nicely collected in a 1981 issue of *Technical Communication*, a special issue on technical editing [1]. The issue contains useful articles by Lola Zook, Don Bush, Alberta Cox, Harold Osborn, Eva Dukes, Lee Shimberg, and Mary Fran Buehler. We concentrate on Buehler's essay [2] and on one that Buehler wrote in 1977 [3].

Buehler's essays provide us with an organized approach to editing through the concept of "levels of edit." The concept, which Buehler developed for editing at the Jet Propulsion Laboratory (JPL), involves defining types of editing and then combining them into different levels according to the amount of editing needed in a manuscript. The types of editing activity include coordination, policy, integrity, screening, copy clarification, format, mechanical style, language, and substantive. These nine types are combined into five levels, which are used according to what is needed for a publication at JPL. A level 1 edit, for instance, would involve all nine types of editing while a level 5 edit would mean only two types of editing would be needed on a manuscript.

Most of Buehler's "tag" words for the types of editing are easily understood though overlapping seems to exist. For a "coordination edit" an editor coordinates or schedules and monitors production processes for a manuscript. For a "policy edit" the editor ensures that the manuscript reflects the company's policy on parts of a report, references, and units of measurement. For an "integrity edit" an editor checks that the parts of a publication match. A "screening edit" is a minimal editing of language and graphics. "Copy clarification" refers to legible copy. For

Reprinted from *IEEE Trans. Prof. Comm.*, vol. PC-28, no. 1, pp. 34–42, March 1985.

a "format edit" the editor marks a manuscript for correct format. In a "mechanical style edit" the editor checks for correct and consistent usage. For a "language edit" the editor considers all aspects of language. For a "substantive edit" the editor reviews the manuscript for content, coherence, emphasis, subordination, and parallelism.

The concept of levels of edit provides an approach for almost any editing situation. Buehler demonstrates its use for manuscripts at the Jet Propulsion Laboratory; I modify the concept to give students a tool for evaluating manuscripts written by other students in the workshop and for later use in their professional lives. I also modify the concept to structure my graduate workshop. Editing experiences in workshops have shown me that editors need organized approaches to editing, especially when ten people are doing the editing with one writer. Because a "shot gun" approach to editing only confuses a writer, I have my students determine main areas that need emphasis in editing a manuscript. If the editing is not organized, beginning editors cover various areas in a manuscript in a hit-or-miss fashion as they jump from a content weakness to a subject-

TABLE I

LEVELS OF EDIT AND QUESTIONS FOR EACH TYPE

Content—Knowledge of subject matter and transfer of that knowledge, information, or message

1. Is information or a message being transferred to the reader?
2. What is the message?
3. Are specific details provided to explain or prove generalizations?
4. Have the best materials been selected to explain the message?
5. Are the ideas fully explained?
6. Are unnecessary materials included?

Structure—Organization of whole piece of writing, of each section, and of each paragraph with clear beginnings, middle parts, and endings

1. Can a definite structure be seen?
2. Is that structure logical?
3. Is another structure better for the material?
4. Does the introduction set up all the parts?
5. Do the middle sections fulfill the promises of the introduction?
6. Is there logical coherence between the parts?
7. Does the conclusion summarize all the parts?

Style—Pattern of sentences and use of words

1. Is the writing clear?
2. Is the writing concise?
3. Is the writing strong?
4. Are the style of sentences and use of words appropriate to the subject?
5. Does the style interfere with the intended message?
6. Is the language appropriate for the intended audience?
7. Does the writer use effective parallelism, subordination, and coordination?
8. Is there an absence of wordiness, compound phrases, and redundancy?
9. Is there good use of agents for action to avoid passive constructions?
10. Is there good use of specific verbs, adverbs, and adjectives to suggest action?
11. Is the diction clear, concise, and connotative?

Format—Specialized physical arrangement and appearance

1. What format is used?
2. Is the format appropriate for the material?
3. Is that format the correct format for a report, a memo, an article, a proposal, a thesis, or a class paper?
4. Are the graphic aids effectively prepared and placed?
5. Are the headings used correctly and spaced correctly?
6. Is the material referenced correctly?
7. Are footnotes or citations used correctly?
8. Is the bibliography or list of references set up correctly?
9. How can the general appearance of the writing be improved?

Mechanics—Use of language according to established rules of grammar

1. Is the grammar of all the sentences correct?
2. Are the sentences structured correctly?
3. Are all the words spelled correctly?
4. Are punctuation marks used correctly?
5. Are typical errors avoided such as subject-verb disagreement, dangling modifiers, incorrect pronoun reference, pronoun-antecedent disagreement, incorrect parallelism, and poor subordination?

Tone—Voice or persona of the writer

1. Is the tone appropriate for the subject and the audience?
2. Is the writer present in the writing?
3. Should personal pronouns be used?
4. Is a persona created for the writer in the writing?

Policy—Conventions that the writer should follow for a journal, company, or organization

1. What policy conventions should the writer be following for the publishing agent or for the intended audience?
2. Are those conventions followed?
3. Is the writing non-sexist?
4. Is the writing free of other prejudices, biases, and imbalances?

verb disagreement to a format inconsistency to a graphics problem to a parallelism error to a paragraph misplaced to a word misspelled. If ten people are using the hit-or-miss approach at once, the writer understandably feels overwhelmed and totally confused. Therefore, to organize editing sessions early in a semester, I have my students develop guidelines for editing. After they have read Buehler's essays, we decide main areas to examine for editing a manuscript.

Invariably, we come up with seven areas or types of editing: content, structure, style, format, mechanics, tone, and policy. For each type, we then define the type and develop questions such as those in Table 1.

Loosely using Buehler's concept of levels of edit but reversing it, we decide that a level 7 edit includes all types of edit, a level 6 edit includes any six types, a level 5 edit includes any five types, and so on. Then, for an editing workshop, students read a manuscript written by another student, determine which types of edit and how many types need to be worked on, comment in writing in the text of the manuscript, and provide an overall reaction to the writing with suggestions on what needs the most work.

In the workshops, we then agree on which types of edit in a manuscript need the most work and which level of edit we will be involved with. Usually as a group we concur that a level 3 or 4 edit is needed. Very rarely do we decide that a level 7 edit is needed because by the time that we are editing papers in the course, students who are writing manuscripts are quite aware of what editors will be concentrating on and work hard in their writing on the areas that we will be examining.

Then, if a manuscript needs a level 2 or 3 or 4 edit, we decide which area needs the most work and concentrate on that area until it has been fully discussed with the writer by all the editors. We also examine other areas of editing that need more work. Near the end of the editing session, we may make comments on minor revisions needed in the remaining types of edit. In this way, we organize the editing session so that the most important types of editing are concentrated on for that particular manuscript and so that the writer is not overwhelmed with disorganized comments.

Editorial Dialogue

Organizing the editing sessions is not enough. To produce worthwhile editing sessions in the workshops, I have my students learn to use editorial dialogue to confer with a writer. Before any editing is done, the students read articles published by Mary Sigurdson Hageman, Louise Merck Vest, and Patrick M. Kelley [4].

The articles on editorial dialogue provide techniques that allow the workshop editors to work on talking *with*, not *at*, the writers. Editorial dialogue helps the editors work at

making an editing session not a hatchet job but a team effort as editors and writer discuss together the editing that is needed in a piece of writing. Editorial dialogue helps the editors and writers work as partners as the editors suggest changes to improve the writer's manuscripts. As explained by Hageman, Vest, and Kelley, editorial dialogue emphasizes empathy and mutual respect in the editor-author relationship. Editorial dialogue makes use of Richard L. Johannesen's components of dialogue, which include the following [4; pp. W 39–40, W 64–65, W 107–108]:

- *Genuineness*—Being yourself and expressing what you think and feel, not what you think you ought to express
- *Accurate empathic understanding*—Comprehending and understanding the other person in a relationship
- *Unconditional positive regard*—Affirming the other person as a partner in dialogue
- *Presentness*—Being consciously and actively present in a dialogue and concentrating on the other person
- *Spirit of mutual equality*—Seeing the other person as an equal
- *Supportive psychological climate*—Communicating without preconceptions.

These components of dialogue can be used in an editing session through the following techniques [4; pp. W 107–108]:

1. *Receptive listening*—The editor actively listens to the writer and asks questions and the writer actively listens to suggestions.
2. *Guide for analysis*—The editor uses a guide [such as the questions for levels of edit] to check what has been done and the writer uses a guide to see what needs work.
3. *Notetaking*—The editor takes notes as the writer talks so that the writer's words can be used in a revision.
4. *Role mirroring*—Both the editor and the writer take the other's role to understand problems.

To use editorial dialogue, the student editors and I work to build a positive relationship with a writer to help the writer successfully transfer information to readers. The editorial dialogue allows us to make our editing sessions a team effort that is also a humane effort, and the concept of levels of edit allows us to organize our criticism and suggestions.

DEVELOPMENT OF INDIVIDUAL THEORIES OF EDITING PROCESSES

In addition to working on team editing, the technical communication students research editing processes to lead to a description of their own theory of the editing process. They are given the following assignment:

> Study editing processes in technical and professional communication to develop a knowledge of theory on editing process and to develop your own theory on editing process.

Complete the following reading and writing assignments:

1. Read articles and books on editing.
2. Write an essay for technical communication professors, writers, and students (or for publication) on your literature search and on your theory of editing process in technical and professional communication. Include a literature review of the reading materials with the explanation of your own theory of editing process. Either make the literature review a separate section of your essay or integrate it into the explanation of your theory.

The word "theory" is used in this assignment to mean a proposed explanation, perhaps conjectural, to explain the operation of certain acts or behaviors. It is used in the sense of a looking at, a contemplating, a speculating of how something operates or how something is done. It is a view, a perception of reality, an attempt to describe reality—the reality being the editing process.

For their research on editing processes, I give students a bibliography for beginning their literature searches. The bibliography, which is presented in the appendix, contains titles of articles and books on editing. I also urge students to use the lists of references at the end of the articles and books for titles of other sources. I urge them to check technical communication journals, such as *Technical Communication, The Technical Writing Teacher, Journal of Technical Writing and Communication, IEEE Transactions on Professional Communication,* and *Proceedings of ITCC.* I tell them to check the Society for Technical Communication publication *Technical Editing: Principles and Practices,* edited by Lola Zook [5]. I suggest that they have computer searches done. In addition I suggest that they consider their editing practices in past and present editing jobs and their editing practices in their present work in our workshops. I suggest that they examine other people's editing processes.

The students then combine their research with a description of their editing processes. Often students are able to use their own experiences at jobs as technical writers and editors to explain their theories of editing processes. For instance, one student combined research with her own editing experience at the Atmospheric Sciences Laboratory to explain her theory of editing process. Her process includes four stages: (1) reviewing—determining the areas to be edited; (2) repairing—making necessary changes during each type of editing; (3) conferring—discussing the manuscript with the author; and (4) evaluating—reviewing the edited product. Another student described her editing process in three stages: (1) getting in shape, which includes developing editorial attitudes, editorial competency, and editorial thinking; (2) finding the problems, which includes reading through a manuscript twice, first for understanding of content and second for locating possible problems; and (3) solving the problems, which includes suggestions to the writer to ensure the transfer of information to an audience.

A third student explained that her editing process consists

of understanding the nature of editing, finding and correcting barriers in a manuscript through three readings, and convincing authors of needed revisions. A fourth student developed a theory based on the concept that an editor must discover facts on content, structure, and style in a manuscript to be able to explain what is wrong with a manuscript and how it can be improved. She developed an editing process that has five stages: (1) Read the manuscript to obtain a general idea of content and to underline grammatical errors to discover if a pattern of grammatical errors exists; (2) outline the manuscript by finding out what the structure is and by commenting on it; (3) edit the manuscript for content by indicating what needs more support and for structure by indicating where logic could be strengthened; (4) edit the manuscript for composition to suggest objectively revisions in style and emphasis in sentences and paragraphs; and (5) comment to the author on major and minor problems in the manuscript and suggest possible revisions. A fifth student looked for ways to achieve professional attitudes and to develop workable editing techniques. He discovered that editors are not mere proofreaders, that an editor has a responsibility to ensure clear communication, and that an editor needs to become personally involved in editing decisions. He developed a systematic method for editing that involved understanding content, discovering organization, and analyzing and solving the writer's problems with content, structure, and style [6].

The research and development of individual theories of editing thus go beyond simply teaching editing skills. The students develop a broad understanding of the scope of editing and are thereby prepared for the editing challenges they will meet in their future jobs.

SUMMARY

The work in editing ensures that graduate students become aware of editing processes through experience and research, through practice and theory. By developing and using levels of edit and editorial dialogue in editing workshops and by researching and examining their own and other people's techniques, the students develop their own theories of editing processes. Thus, through their group editing and their individual editing experiences and research, the graduate students learn basic editorial values and editing techniques that they can use to help others communicate well.

REFERENCES

1. *Technical Communication.* Ed. Frank Smith. Fourth Quarter 1981; 28(4).
2. Buehler, M. F. "Defining the Terms in Technical Editing: The Levels of Edit as a Model." *Technical Communication.* Fourth Quarter 1981; 28(4):10–14.
3. Buehler, M. F. "Controlled Flexibility on Technical Editing: The Levels-of-Edit Concept at JPL." *Technical Communication.* First Quarter 1977; 24(1):1–4.
4. Hageman, M. S.; Vest, L. M.; Kelley, P. M. "Editorial Dialogue: An Alternative Writer-Editor Relationship." *Proceedings of the*

28th International Technical Communication Conference. Washington, DC: Society for Technical Communication; 1981: pp. W 38–40. Kelley, P. M. "Charting a New Course for Technical Writing and Editing: Technical Writing and Editing as Dialogue." *Proceedings of the 29th International Technical Communication Conference.* Washington, DC: Society for Technical Communication; 1982: pp. W 62–65. Vest, L. M.; Hageman, M. S. "Developing an Alternative Writer-Editor Relationship: A Workshop in Editorial Dialogue." *Proceedings of the 29th International Technical Communication Conference.* Washington, DC: Society for Technical Communication; 1982: pp. W 106–109.

5. *Technical Editing: Principles and Practices.* Ed. Lola M. Zook. Washington, DC: Society for Technical Communication; 1975.
6. The students cited include Marie Richardson, Mary Lou Vocale, Martha Delamater, Susan Bagby, and Robert Toland.

APPENDIX: SELECTED RESOURCES ON EDITING PROCESSES

A

Abshire, Gary M.; Culberson, Dan. **The Art of Technical Writing and Editing.** Report 05.225. Boeblingen, Germany: IBM Corporation; 1978.

Adams, Tom. **Be Your Own Editor.** *Industrial Supervisor.* 1980; 44: 6.

Amsden, Dorothy Corner. **Exercise Your Visual Thinking.** In *Proceedings of the 29th International Technical Communication Conference.* Washington, DC: Society for Technical Communication; 1982: pp. W 12–15.

——. **Get in the Habit of Editing Illustrations.** In *Proceedings of the 27th International Technical Communication Conference.* Washington, DC: Society for Technical Communication; 1980: pp. W 147–154.

Applewhite, Lottie. **An In-House Editorial-Tutorial Program for Developing Communication Skills.** *Technical Communication.* First Quarter 1983; 30: 5–7.

——. **An Individual Development Program.** In *Proceedings of the 29th International Technical Communication Conference.* Washington, DC: Society for Technical Communication; 1982: pp. E 14–16.

Atkins, Eldred E. **At the Outset of Technical Editing.** *Electronic Engineer.* November 1969; 28: 29–30.

Atlas, Marshall, A. **The User Edit: Making Manuals Easier to Use.** *IEEE Transactions on Professional Communication.* 1981; PC-24(1): 28–29.

B

Bagby, Susan A. **Editing as a Matter of Fact.** In *The 577 Papers: Writing Processes, Editing Processes, Written Products,* Volume II. Eds. Roger E. Masse and Martha Delamater. Las Cruces, NM: Technical Communication Programs, New Mexico State University; 1983: pp. 125–137.

Barnow, Renee K. **Setting and Cleaning a Table: How an Editor Can Get Out of the Kitchen.** In *Proceedings of the 29th International Technical Communication Conference.* Washington, DC: Society for Technical Communication; 1982: pp. W 18–21.

Batchelder, Susan K. **Friends or Foes? The Relationship Between Writer and Editor.** In *Proceedings of the 30th International Technical Communication Conference.* Washington, DC: Society for Technical Communication; 1983: pp. W&E 73–74.

Behnke, Lynn. **Stranger in a Strange Land: My First Year as an Editing Manager.** In *Proceedings of the 29th International Technical Communication Conference.* Washington, DC: Society for Technical Communication; 1982: pp. C 11–14.

Bennett, John Barnard. **Editing for Engineers.** New York: Wiley-Interscience; 1970.

Boomhower, E. F. **Producing Good Technical Communications Requires Two Types of Editing.** *Journal of Technical Writing and Communication.* Fourth Quarter 1975; 5: 277–281.

Brett, Carlton E. **Editor's Bootstraps.** *Journal of Technical Writing and Communication.* Fourth Quarter 1971; 1: 307–316.

Briggs, Nelson. **Editing by Dialogue.** In *Technical Editing: Principles and Practices.* Ed. Lola M. Zook. Washington, DC: Society for Technical Communication; 1975: pp. 56–61.

Briles, Susan M. **Designing a Training Program for a Technical Editing Department.** In *Proceedings of the 29th International Technical Communication Conference.* Washington, DC: Society for Technical Communication; 1982: pp. C 15–18.

Brogan, John A. **A Pitfall for Professionals.** In *Technical Editing: Principles and Practices.* Ed. Lola M. Zook. Washington, DC: Society for Technical Communication; 1975: pp. 87–91.

Bronson, Judith Gunn. **Prevention of Donkeyism: The Role of the Medical Author's Editor.** In *Proceedings of the 28th International Technical Communication Conference.* Washington, DC: Society for Technical Communication; 1981: pp. W 10–13.

Buehler, Mary Fran. **Defining Terms in Technical Editing: The Levels of Edit as a Model.** *Technical Communication.* Fourth Quarter 1981; 28: 10–15.

——. **Situational Editing: A Rhetorical Approach for the Technical Editor.** *Technical Communication.* Third Quarter 1980; 27: 18–22.

——. **Controlled Flexibility in Technical Editing: The Levels of Edit Concept at JPL.** *Technical Communication.* First Quarter 1977; 24: 1–4.

——. **Patterns for Making Editorial Changes.** In *Technical Editing: Principles and Practices.* Ed. Lola M. Zook. Washington, DC: Society for Technical Communication; 1975: pp. 1–6.

Burr, William. **Why Technical Editors Act That Way.** Report 64-825-1195. Oswego, NY: IBM Corporation; 1964.

Bush, Don. **The Trouble with Definitions.** In *Proceedings of the 30th International Technical Communication Conference.* Washington, DC: Society for Technical Communication; 1983: pp. W&E 28–31.

——. **Content Editing, an Opportunity for Growth.** *Technical Communication.* Fourth Quarter 1981; 28: 15–19.

——. **Strategies for a Technical Editor.** *Technical Writing Teacher.* 1979; 7: 19–23.

——. **Semantics (Words are Chameleons).** In *Technical Editing: Principles and Practices.* Ed. Lola M. Zook. Washington, DC: Society for Technical Communication; 1975:

pp. 33–37.

Butcher, Judith. **Copy-Editing.** Cambridge, England: Cambridge University Press; 1975.

C

Carmichael, Edna. **Management's Responsibility in Training the Technical Communicator.** In *Technical Editing: Principles and Practices.* Ed. Lola M. Zook. Washington, DC: Society for Technical Communication; 1975: pp. 82–86.

Cathcart, Margaret E. **Training the Technical Editor.** In *Proceedings of the 30th International Technical Communication Conference.* Washington, DC: Society for Technical Communication; 1983: pp. RET 13–15.

Cederborg, Gibson A. **The Role and Rationale of Technical Editors.** *Journal of Technical Writing and Communication.* Fourth Quarter 1975; 5: 283–286.

Clements, Wallace. **Jargon and the Technical Writer.** In *Technical Editing: Principles and Practices.* Ed. Lola M. Zook. Washington, DC: Society for Technical Communication; 1975: pp. 38–41.

——; Waite, Robert G. **A Guide for Beginning Technical Editors.** In *Proceedings of the 26th International Technical Communication Conference.* Washington, DC: Society for Technical Communication; 1979: pp. W 32–36.

Coggshall, Gordon. **Using the Core Sentence to Edit Poorly Written Technical Manuscripts.** *Technical Communication.* First Quarter 1980; 27: 19–23.

Colby, John. **Paragraphing in Technical Writing.** In *Technical Editing: Principles and Practices.* Ed. Lola Zook. Washington, DC: Society for Technical Communication; 1975: pp. 42–46.

Corrigan, Anne M. **The Technical Editor as Teacher: How to Explain What's Wrong and Why.** In *Proceedings of the 27th International Technical Communication Conference.* Washington, DC: Society for Technical Communication; 1980: pp. R 167–170.

Cox, Alberta. **Copy Editing—The Final Word.** *Technical Communication.* Fourth Quarter 1981; 28: 18–20.

——. **The Editor as Generalist as Well as Specialist.** In *Technical Editing: Principles and Practices.* Ed. Lola M. Zook. Washington, DC: Society for Technical Communication; 1975: pp. 7–10.

D

Dalla Santa, Terry M. **Managing the Editing Function on Large Publication Tasks with Short Flow Times.** In *Proceedings of the 30th International Technical Communication Conference.* Washington, DC: Society for Technical Communication; 1983: pp. W&E 21–24.

Delamater, Martha. **Editors: Trolls or Fairy Godpersons?** In *The 577 Papers: Writing Processes, Editing Processes, Written Products,* Volume II. Eds. Roger E. Masse and Martha Delamater. Las Cruces, NM: Technical Communication Programs, NMSU; 1983: pp. 111–123.

De Quattro, James. **Getting It Right with the Author.** In *Proceedings of the 26th International Technical Communication Conference.* Washington, DC: Society for Technical Communication; 1979: pp. W 46–48.

Dukes, Eva P. **Some Authors I Have Known.** *Technical Communication.* Fourth Quarter 1981; 28: 27–30.

——. **The Art of Editing.** In *Technical Editing: Principles and Practices.* Ed. Lola M. Zook. Washington, DC: Society for Technical Communication; 1975: pp. 62–66.

——. **The Simple Joys of Editing.** *Technical Communication.* Third Quarter 1972; 19: 7–8.

F

Farkas, David K.; Farkas, Nettie. **Manuscript Surprises: A Problem in Copy Editing.** *Technical Communication.* Second Quarter 1981; 28: 16–18.

Fearing, Bertie E. **The Education of an Academic Journal Editor.** In *Proceedings of the 29th International Technical Communication Conference.* Washington, DC: Society for Technical Communication; 1982: pp. E 41–43.

Fourdrinier, Sylvia. **The Editor as a Teacher.** In *Technical Editing: Principles and Practices.* Ed. Lola M. Zook. Washington, DC: Society for Technical Communication; 1975: pp. 67–70.

G

Gamer, Roy W. **Improving the Effectiveness of Technical Documentation through Analysis.** In *Proceedings of the 30th International Technical Communication Conference.* Washington, DC: Society for Technical Communication; 1983: pp. MPD 13–15.

Garber, Reeta. **Terminal Oversight: The Editor and the Word Processor.** In *Proceedings of the 31st International Technical Communication Conference.* Washington, DC: Society for Technical Communication; 1984: pp. ATA 50–52.

Genin, Michael S. **Editing Report Art Differs from Editing Presentation Art.** In *Proceedings of the 31st International Technical Communication Conference.* Washington, DC: Society for Technical Communication; 1984: pp. VC 53–55.

——. **Turning Adversaries into Allies—Avoiding Tension in an Author-Editor Relationship.** In *Proceedings of the 26th International Technical Communication Conference.* Washington, DC: Society for Technical Communication; 1979: pp. W 59–60.

Gibson, Martin L. **Editing in the Electronic Era.** Ames, IA: Iowa State University Press; 1979.

Griffin, C. W. **Theory of Responding to Student Writing: The State of the Art.** *College Composition and Communication.* October 1982; 36: 296–301.

H

Hageman, Mary Sigurdson. **High Touch in the Workplace: Integrating Scientific Research and Scientific Writing (and Editing).** In *Proceedings of the 31st International Technical Communication Conference.* Washington, DC: Society for Technical Communication; 1984: pp. WE 112–113.

——; Vest, Louise M.; Kelley, Patrick M. **Editorial Dialogue: An Alternative Writer-Editor Relationship.** In *Proceedings of the 28th International Technical Communication Conference.* Washington, DC: Society for Technical Communication; 1981: pp. W 38–40.

Hallinan, Edward J. **Practical Writing and Editing Techniques.** In *Proceedings of the 31st International Technical Communication Conference.* Washington, DC: Society for Technical Communication; 1984: pp. WE 34–36.

Harrington, J. Y. **Editing Computer Manuals.** *Technical Communication.* Fourth Quarter 1980; 27: 14–17.

Hartley, James. **The Role of Colleagues and Text-Editing Programs in Improving Text.** *IEEE Transactions on Professional Communication.* March 1984: PC-27(1): 42–44.

Hasch, Jean; Chepeleff, Val. **Wearing the Production Editor's Hat.** In *Proceedings of the 29th International Technical Communication Conference.* Washington, DC: Society for Technical Communication; 1982: pp. G 23–26.

Haughness, Norman. **The Technical Editor as Tact-ician.** *Technical Communication.* Third Quarter 1968; 15: 18–19.

Heffner, Maxine. **Stalking the Troublesome Hyphen.** In *Technical Editing: Principles and Practices.* Ed. Lola M. Zook. Washington, DC: Society for Technical Communication; 1975: pp. 53–55.

Henderson, Arnold C. **Editing for the First Half-Second: The Perceptual Process and the Technical Editor.** In *Proceedings of the 28th International Technical Communication Conference.* Washington, DC: Society for Technical Communication; 1981: pp. W 49–52.

Heiken, Jody H.; Norton, Diane D. **Mechanized Editing: We Won't, We Can't, We Did!** In *Proceedings of the 31st International Technical Communication Conference.* Washington, DC: Society for Technical Communication; 1984: pp. WE 128–131.

J

Jack, Judith. **Teaching Analytical Editing.** *Technical Communication.* First Quarter 1984; 31: 9–11.

——. **Teaching Technical Editing: A Structured Approach.** In *Proceedings of the 30th International Technical Communication Conference.* Washington, DC: Society for Technical Communication; 1983: pp. RET 11–12.

Jackson, Purvis M.; Dunkle, Susan B. **A Systematic Approach to Editing.** In *Proceedings of the 31st International Technical Communication Conference.* Washington, DC: Society for Technical Communication; 1984: pp. WE 121–124.

Jarmon, Brian. **Coping with Crash Editing.** In *Proceedings of the 27th International Technical Communication Conference.* Washington, DC: Society for Technical Communication; 1980: pp. W 9–12.

K

Kantrowitz, Bruce M. **Recipe for a Cooperative Technical Editing Program.** In *Proceedings 1978 of the Council for Programs in Technical and Scientific Communication.* Ed. David L. Carson. Troy, NY: Council for Programs in Technical and Scientific Communication; 1979: pp. 62–68.

Keedy, Hugh. F. **Musings of an Engineering Professor on Leave as a Technical Editor.** In *Proceedings of the 28th International Technical Communication Conference.* Washington, DC: Society for Technical Communication; 1981: pp. W 64–67.

Kelley, Patrick M. **High Tech/High Touch: A Trend in Technical Writing and Editing.** In *Proceedings of the 31st International Technical Communication Conference.* Washington, DC: Society for Technical Communication; 1984: pp. WE 106–108.

——. **Charting a New Course for Technical Writing and Editing: Technical Writing and Editing as Dialogue.** In *Proceedings of the 29th International Technical Communication Conference.* Washington, DC: Society for Technical Communication; 1982: pp. W 62–65.

Kellner, Robert Scott. **A Necessary and Natural Sequel: Technical Editing.** *Journal of Technical Writing and Communication.* First Quarter 1982; 12: 25–33.

Koski, Raymond; Mann, Gerald A. **The Editor's Role in Reducing Future Shock.** *Technical Communication.* Second Quarter 1974; 21: 2–5.

L

Layton, Edward. **Editor-Author Relationships: Both Can Win.** *IEEE Transactions on Professional Communication.* September 1973; PC-16(3): 57–59, 172.

Leavitt, William D. **The Proof of Your Editorial Pudding Is in Its Tasters.** In *Proceedings of the 30th International Technical Communication Conference.* Washington, DC: Society for Technical Communication; 1983: pp. MPD 16–18.

Lehr, Dolores. **Three Roles of a Technical Editor.** In *Proceedings of the 31st International Technical Communication Conference.* Washington, DC: Society for Technical Communication; 1984: pp. WE 65–66.

Lien, Patricia L. **Text Editing on the LLL Octopus.** In *Proceedings of the 27th International Technical Communication Conference.* Washington, DC: Society for Technical Communication; 1980: pp. C 209–212.

Lindberg, Helen. **Keeping a Sense of Humor.** In *Proceedings of the 26th International Technical Communication Conference.* Washington, DC: Society for Technical Communication; 1979: pp. W 87–91.

Love, Earl A. **An Abrupt Awakening: Or, an Editor Comes of Age.** In *Proceedings of the 26th International Technical Communication Conference.* Washington, DC: Society for Technical Communication; 1979: pp. W 92–94.

Lynch, Denise. **Easing the Process: A Strategy for Evaluating Compositions.** *College Composition and Communication.* October 1982; 33: 310–314.

M

Mann, Gerald A. **Minimal Editing: How Much Is Too Much?** In *Proceedings of the 27th International Technical Communication Conference.* Washington, DC: Society for Technical Communication; 1980: pp. W 5–7.

Masse, Roger E. **Editing in Technical Communication: Theory and Practice in Editing Processes at the Graduate Level.** Las Cruces, NM: NMSU; 1984. ERIC document ED 229 790.

—— Ed. **The 577 Papers: Writing Processes, Editing Processes, Written Products, Volume I.** Las Cruces, NM: Technical Communication Programs, NMSU; 1982.

——; Delamater, Martha. Eds. **The 577 Papers: Writing Processes, Editing Processes, Written Products, Volume II.** Las Cruces, NM: Technical Communication Programs, NMSU; 1983.

Mazzatenta, Ernest. **GM Research Laboratories Improve Chemistry Between Science Writers and Editors.** In *Teaching Technical Writing and Editing—In-House Programs That Work.* Ed. James G. Shaw. Washington, DC: Society for Technical Communication; 1976; pp. 55–61.

McCarron, William E. **Confessions of a Working Technical Editor.** *The Technical Writing Teacher.* Fall 1978; 6: 5–8.

McCormick, Barbara S. **How to Function as a Schizoid Editor.** Washington, DC: Society for Technical Communication; 1977.

McDonald, John W. **Taking the Noise Out of Technical Writing.** In *Technical Editing: Principles and Practices.* Ed. Lola M. Zook. Washington, DC: Society for Technical Communication; 1975: pp. 28–32.

McGough, David L. **Production Editor: Key to New Effectiveness.** In *Technical Editing: Principles and Practices.* Ed. Lola M. Zook. Washington, DC: Society for Technical Communication; 1975: pp. 71–76.

Meckel, Susan R.; Sauer, Kathleen H. **Five Concepts for Effective Interaction Between an Artist and Editor/Writer.** In *Proceedings of the 29th International Technical Communication Conference.* Washington, DC: Society for Technical Communication; 1982: pp. G 38–40.

Mott, Wesley T. **Editing Business and Institutional Publications: A Course for the Working Editor.** *The ABCA Bulletin.* 1981; No. 1: 3–7.

Myers, Barbara Y. **A Classification of Author-Editor Relationships: Toward Team-Centered Relationships.** In *Proceedings of the 31st International Technical Communication Conference.* Washington, DC: Society for Technical Communication; 1984: pp. WE 116–119.

Mullins, Carolyn J. **The Computer as Nitpicking Copy Editor.** In *Proceedings of the 27th International Technical Communication Conference.* Washington, DC: Society for Technical Communication; 1980: pp. C 197–199.

O

Osborne, Harold F. **Intuition, Integrity, and the Decline of Editing.** *Technical Communication.* Fourth Quarter 1981; 28: 21–26.

——. **Criticism and Creativity.** In *Technical Editing: Principles and Practices.* Ed. Lola M. Zook. Washington, DC: Society for Technical Communication; 1975: pp. 17–19.

P

Peterson, Dart G. **Developing the Editor-Author Relationship.** In *Proceedings of the 23rd International Technical Communication Conference.* Washington, DC: Society for Technical Communication; 1976: pp. 85–86.

Power, Ruth M. **Who Needs a Technical Editor.** *IEEE Transactions on Professional Communication.* September 1981; PC-24(3): 139–140.

R

Rathbone, Robert R. **Communicating Technical Information.** Reading, MA: Addison-Wesley; 1966.

Richardson, Marie. **The Editor Won't Let You.** In *The 577 Papers: Writing Processes, Editing Processes, Written Products,* Volume I. Ed. Roger E. Masse. Las Cruces, NM: Technical Communication Programs, NMSU; 1982: pp. 75–91.

Rohne, Carl F. **Editing the Small Study Proposal.** In *Technical Editing: Principles and Practices.* Ed. Lola M. Zook. Washington, DC: Society for Technical Communication; 1975: pp. 77–81.

Rosner, Mary. **Sentence-Combining in Technical Writing: An Editing Tool.** *The Technical Writing Teacher.* Winter 1982; 9: 100–107.

Ross, Peter Burton. **Slash for Quick Editing.** *Technical Communication.* Third Quarter 1977; 24: 11–14.

Rutter, Russell. **Starting to Write by Re-Writing: A Unit on Teaching Editing and Revision.** *The Technical Writing Teacher.* 1980: 8: 22–26.

S

Sealine, Barbara A. **Using Interpersonal Communication Skills as a Managing Writer/Editor.** In *Proceedings of the 27th International Technical Communication Conference.* Washington, DC: Society for Technical Communication; 1980: pp. M 133–140.

Shear, Marie. **Fixing Rotten Writing: A Cameo Case History.** *The Journal of Business Communication.* 1981; No. 2: 5–14.

Shimberg, H. Lee. **Editing Authors' Style—A Few Guidelines.** *Technical Communication.* Fourth Quarter 1981; 28: 31–35.

Shipley, L. J; Gentry, J. K. **How Electronic Editing Equipment Affects Editing Performance.** *Journalism Quarterly.* Fall 1981; 58: 371–374, 387.

Sideras, George. **Creativity in Technical Editing.** In *Proceedings of the 18th International Technical Communication Conference.* Washington, DC: Society for Technical Communication; 1971: pp. 6–10.

Simons, John L. **The Technical Editor as a Decision-Maker.** In *Proceedings of the 27th International Technical Communication Conference.* Washington, DC: Society for Technical Communication; 1980: pp. W 27–30.

Sims, Robert L. **Advantage of Dialogue from a Management Perspective.** In *Proceedings of the 31st International Technical Communication Conference.* Washington, DC: Society for Technical Communication; 1984: pp. WE 114–115.

——. **Dialogue: The Key to Professionalism in Technical Communication.** In *Proceedings of the 30th International Technical Communication Conference.* Washington, DC: Society for Technical Communication; 1983: pp. W&E 35–37.

Smith, Frank R. **The Education of a Society Journal Editor.** In *Proceedings of the 29th International Technical Communication Conference.* Washington, DC: Society for Technical Communication; 1982: pp. E 110–111.

Smith, Herbert J. **Training the Technical Editing Student in Interpersonal Skills.** In *Proceedings of the 30th International Technical Communication Conference.* Washington, DC: Society for Technical Communication; 1983: pp. RET 52–54.

Smith, Patricia N. **Here, Edit This!** In *Proceedings of the 28th International Technical Communication Conference.* Washington, DC: Society for Technical Communication; 1981: pp. W 92–94.

Sommers, Nancy. **Responding to Student Writing.** *College Composition and Communication.* May 1982; 33: 148–156.

Stocker, Deborah J. **Managing a Successful Editorial Group in a Multidisciplinary Consulting Firm.** In *Proceedings of the 31st International Technical Communication Con-*

ference. Washington, DC: Society for Technical Communication; 1984: pp. MPD 87–89.

Stohrer, Freda F. **Training Apprentice Editors.** In *Proceedings 1978 of the Council for Programs in Technical and Scientific Communication.* Ed. David L. Carson. Troy, NY: CPTSC; 1979: pp. 62–68.

Stratton, Charles R. **Ambiguity: An Exercise in Practical Semantics.** In *Technical Editing: Principles and Practices.* Ed. Lola M. Zook. Washington, DC: Society for Technical Communication; 1975: pp. 47–52.

Swain, Deborah E. **Dynamic Online Editing: A Proposal.** In *Proceedings of the 31st International Technical Communication Conference.* Washington, DC: Society for Technical Communication; 1984: pp. WE 132–135.

Swaney, J. H.; *et al.* **Editing for Comprehension: Improving the Process Through Protocols.** Washington, DC: American Institutes for Research; 1981. ERIC document ED 209 642.

T

Taylor, P. **How to Get Along with Authors.** *The Editorial Eye.* January 1981; 53: 6–7.

Thralls, Charlotte. **Editing of Professional Reports: A Rhetorical Modes Approach.** In *Proceedings of the 27th International Technical Communication Conference.* Washington, DC: Society for Technical Communication; 1980: pp. 199–203.

Toland, Robert. **Attitudes and Method: Complements of an Editing Process.** In *The 577 Papers: Writing Processes, Editing Processes, Written Products,* Volume II. Eds. Roger E. Masse and Martha Delamater. Las Cruces, NM: Technical Communication Programs, NMSU; 1983: pp. 139–152.

V

Van Buren, Robert; Buehler, Mary Fran. **The Levels of Edit,** 2nd edition. JPL 80-1. Pasadena, CA: Jet Propulsion Laboratory, California Institute of Technology; 1980.

Van Eps, Barbara J. **Editing Computer-Based Education Lessons.** In *Proceedings of the 28th International Technical Communication Conference.* Washington, DC: Society for Technical Communication; 1981: pp. W 114–116.

Vaughn, David E. **A Logical Approach to Editing Proposals, Reports, and Manuals.** In *Technical Editing: Principles and Practices.* Ed. Lola M. Zook. Washington, DC: Society for Technical Communication; 1975: pp. 20–27.

Vest, Louise M. **Toward Human-Centered Technology: '...A Sense of Obligation.'** In *Proceedings of the 31st International Technical Communication Conference.* Washington, DC: Society for Technical Communication; 1984: pp. WE 109–111.

——; Hageman, Mary S. **Developing an Alternative Writer-Editor Relationship: A Workshop in Editorial Dialogue.** In *Proceedings of the 29th International Technical Communication Conference.* Washington, DC: Society for Technical Communication; 1982: pp. W 106–111.

Vocale, Mary Lou. **Overcoming the Manuscript: An Editing Process.** In *The 577 Papers: Writing Processes, Editing Processes, Written Products,* Volume I. Ed. Roger E. Masse, Las Cruces, NM: Technical Communication Programs, NMSU; 1982: pp. 57–73.

W

Wagner, Carl B. **The Technical Side of Technical Editing.** In *Proceedings of the 29th International Technical Communication Conference.* Washington, DC: Society for Technical Communication; 1982: pp. W 112–115.

Wales, Ruth W. **A Taxonomy of Editing Tasks.** In *Proceedings of the 29th International Technical Communication Conference.* Washington, DC: Society for Technical Communication; 1982: pp. W 116–117.

Wall, Florence E. **Requirements and Responsibilities of a Technical Editor.** *Journal of Chemical Education.* October 1953; 31: 516–521.

Weil, Benjamin. **Technical Editing.** Westport, CT: Greenwood Press; 1975.

Whittaker, Della. **Editor? Teacher?** In *Proceedings of the 22nd International Technical Communication Conference.* Washington, DC: Society for Technical Communication; 1975: pp. 407–408.

Wood, M. **The Sharp Pencil: Editing Corporate Jargon.** *Editors Workshop.* July/August 1981; 2: 5.

Z

Zimmerman, Muriel. **Reducing by Design: A Checklist for Editors.** In *Proceedings of the 30th International Technical Communication Conference.* Washington, DC: Society for Technical Communication; 1983: pp. W&E 18–20.

Zook, Lola M. **Technical Editors Look at Technical Editing.** *Technical Communication.* Third Quarter 1983; 30: 21–26.

——. **Editing and the Editor: Views and Values.** *Technical Communication.* Fourth Quarter 1981; 28: 5–9.

——. **Even an Editor Can't Have Everything.** *The Editorial Eye.* January 1981; 53: 2–3.

——. **Lessons Learned—Not Always By Choice.** In *Proceedings of the 27th International Technical Communication Conference.* Washington, DC: Society for Technical Communication; 1980: pp. W 31–36.

——. Ed. **Technical Editing: Principles and Practices.** Washington, DC: Society for Technical Communication; 1975.

——. **Training the Editor: Skills Are Not Enough.** In *Technical Editing: Principles and Practices.* Ed. Lola M. Zook. Washington, DC: Society for Technical Communication; 1975: pp. 12–16.

The Final Step: Perfecting a Document

Helena Chytil

THE FINAL PRODUCT OF THE TECHNICAL COMMUNICATOR *should be as nearly error free as possible. Eliminating errors is never easy, but here the author describes a systematic approach based on proofreading fundamentals and organized into logical phases that should help all technical writers and editors.*

Typographical errors used to haunt me. Although I have a spell-checker on my word-processing software and a good proofreader in the word-processing center, ultimately, documents are *my* responsibility. The finger points only in one direction—at me.

Word-processing spell-checkers are a must, but they do not catch a "then" that should be a "the" or a "no" that should be a "now." Nor do they catch verbs and subjects which—for some mysterious reason—just don't agree. And even good in-house proofreaders can make errors, especially if they are not familiar with the content.

Ideally, one hires a professional proofreader. Unfortunately, budgets do not always permit this. And some documents, just by their nature, do not lend themselves to outside proofreading. In larger shops, a copy editor helps eliminate many errors and inconsistencies. "In traditional publishing houses," writes Bruce C. Lewenstein, "proofreading is a two-person operation. One person reads the copy aloud to another" [**1**, 23]. In my case, and especially for my freelance work, I am the writer, the editor, the copy editor, and the "lay" proofreader—so I'm on my own.

Having found that the success of a document depends not only on technical accuracy and quality writing, but also on correct grammar, good appearance, stylistic consistency, and ease of use, I began paying a great deal of attention to detail. Sound familiar? Read on.

PROOFREADING FUNDAMENTALS

The following guidelines help minimize errors and take the "dread" out of proofreading while ensuring the best possible impact of a document.

Seek Out a Proof Buddy When I am confident that my draft is "ripe" for proofreading, I find a "proof buddy." Basically I propose that I proof my buddy's work and that my buddy proof mine. After working on a document for so long that I dream about it, I often miss even the most blatant of errors, whereas my buddy easily detects them.

Stick to a Style Guide Valerie Mitchell's article "A Style Guide—Why and What" will convince you that if you do not already have a company "stylebook" in use, you should develop one that " . . . attempts to standardize spelling, punctuation, and word divisions [and] . . . gives preferable word choices and acceptable jargon. In general . . . [a stylebook] sets rules to follow, allows few exceptions, and chooses authoritative dictionaries and grammar books" [**2**]. For writers who have just come on board, style guidelines—coupled with a buddy system for proofreading—are especially helpful in teaching the use of terminology particular to an industry, products, services, and corporate conventions.

Prepare Yourself If I have to proof my own work, I let it "cool off" for a day or two, time permitting. Then I forward my phone, find the quietest place I can (preferably *not* my office),

surround myself with reference books, and start listing and "grouping" the items to check. More on the list and how I go about putting my "grouping" technique to work later. For now, here are just a few more fundamentals. The first is my favorite.

Take Breaks Taking breaks is vital. Since I tend to lose track of time, I set a timer and take a short break every twenty minutes. I call someone. I get coffee. I attend to something I have been postponing. I do anything to clear my mind and "re-sharpen" my detective skills.

Use a Mechanical Aid When working through documents, I use a straightedge such as a ruler or a thick piece of paper to help keep my focus on a single line [**1**, 24]. On longer documents, I prefer using a piece of paper so I can make notes on it and also indicate where I have stopped before taking a break (e.g., "Page 2, Paragraph 3").

Make Changes Neatly When noting changes—especially if the corrections will not be made by me—I write neatly and explicitly. So that corrections will not be missed, I use red ink. If working on camera-ready proofs, I use a light blue pencil which the camera will not pick up.

Although some typesetters use proofreaders' marks which they will make available to you, I find that many typesetters use their own systems of symbols and abbreviations. If in doubt, I check a style manual. Most style manuals include a section on proofreading and proofreaders' marks. Whatever system is used, *clarity* and *consistency* avoid confusion and increase turnaround.

Keep a Copy Before turning over my proofed document to the person making the changes, I *always* keep a

Reprinted with permission from *Technical Communication,* published by the Society for Technical Communication. February 1989, vol. 36, no. 1, pp. 53–56.

copy of it so that I can quickly check the changes made. (Not keeping a copy just one time taught me that lesson.) If working with a typesetter, I request that the original be returned with the corrected copy.

Make the Last Check This is the last step—so tempting to forget when rushing to deliver a document. Once all indicated changes have been made, I check that no change has been omitted and that each change has been made as indicated. Taking the time to find errors and then forgetting to verify their correction is certainly counterproductive—if not plain "silly."

LISTING AND GROUPING THE CHECKS

Keeping the fundamentals in mind, here is how I go about improving the proofreading process:

Make a "Core" List Since there are certain items to check in just about any document, I start with a standard or "core" list which I then customize to a particular document by adding or deleting items. Here's an idea of what my core list includes:

Proper Nouns, Numbers, and Symbols

- Mathematical Operations
- Terminology
- Visual Aids, Charts, Columns
- Acronyms
- Trademarks/Service Marks

Cross-checks

- Cross-references
- Page Numbers
- Table of Contents
- Indexes
- Outlines
- Footnotes
- Bibliographies

Correct and Consistent Punctuation
Typographical Faux Pas, Spelling, and Grammar
Consistency

- Running Heads
- Footers
- Headings/Subheadings
- Type Sizes and Fonts
- Indentation
- Placement of Columns, Visual Aids, etc.

Appearance

- Layout
- Margins
- Packaging

Ease of Use

- Readability
- Ease of Reproduction

The core list above is not all-inclusive, and your list will, no doubt, differ from mine. What is important is being *organized*, i.e., developing a core list, adding detail, and customizing it to each document instead of "keeping it all in your head."

Check Items in Logical Groups One pass through the document to verify "everything" can make the checklist seem endless and the task of proofreading, unmanageable. So I facilitate the process by making several passes. In each pass through the document, (or section if, for example, it is a manual), I look for just the items in a particular "logical" group as noted above in my sample core list. Once I've completed one group, I move on to the next. Here's how the grouping technique works:

- *Proper Nouns, Numbers, and Symbols*. I group proper nouns and numbers together because they are easy to spot as I move through the text. And since verification of names and numbers sometimes requires outside assistance, I tackle these first. When checking names and numbers, I take my *original* document and match each number and name to those appearing on the working draft. Although the original should be correct, if there is any doubt, I never trust the original or my memory. I pick up the phone and put my suspicions to rest. During this pass through the document, I also check any mathematical operations; I never *assume* they are correct.

When I am certain I have accurate information, I check for *consistency*. I start with terminology. For example, does "micro-computer" appear hyphenated throughout the document [1, 24-25]? Have I used more than one term for the same piece of equipment? Using the word "widget" to refer to an object in one place and then using "thingamajig" in another will confuse the reader and should be avoided in technical writing.

During this first pass, I also pay special attention to any visual aids such as diagrams and flow charts and make sure that the text agrees with any terms, names, or numbers used in the artwork. Since the writer is rarely the artist, discrepancies between the copy and the artwork occur frequently and must be checked.

At the same time, I ask the following questions: Have I used the company name correctly throughout the document? Does it always appear in the same way? For example, in upper case? If using an acronym, have I informed the reader what it represents by spelling out the name and placing the acronym in parentheses behind it? Have I identified registered trademarks and/or service marks as required? Have I always spelled out numbers zero through nine? Fern Rook's "About Numbers" article offers tips on using numbers correctly and consistently [3]. More detailed guidelines on the correct use of numbers can be found in a style manual such as *The Chicago Manual of Style*.

- *Cross-checks*. This is my "flip the pages" check. I group cross-reference, outline, footnote, and bibliographical checks together. I start by checking any references such as "See Page 13 for Additional Information" or "Refer to Diagram B-2." These checks are extremely important when revising a document. Sometimes a diagram or a part of one is no longer used, yet the text continues to include the reference. Or the index includes a feature which is no longer available or a part which has been replaced.

In this same pass, I verify that page numbers in the table of contents and those in the index really do lead to the information referenced and that number sequences are, indeed, sequential. Even if you have software which automatically provides a table of contents, indexes, and outlines, these checks are well worth it—especially if typesetting.

Next I make sure that footnotes appear on the correct pages. I check that each note number is placed correctly in the text and that each corresponding footnote appears on the bottom of the appropriate page. I perform the same cross-check for numbered reference lists.

- *Correct and Consistent Punctua-*

tion. Having checked my cross-references, I start, again, at the beginning of the document and get nitpicky, but not too nitpicky for perfecting a document. Are words hyphenated correctly? If a comma is used before the last item in a series, is it always used? If quotes are used, are they all opened and closed correctly? Consistently?

If not typesetting, are there two spaces at the end of each sentence? (I was once amazed when a co-worker told me that only secretaries adhere to this rule!) If ever in doubt, I check my style manual. If still in doubt, I ask my proof buddy or peers.

• *Typographical Faux Pas, Spelling, and Grammar.* The next pass through the document is especially crucial when typesetting since the danger of omitting a letter, a word, a sentence, or even an entire section is increased. Most typesetters are very conscientious, but they do make mistakes, so it is always best to verify all proofs.

To check spelling and better detect typos, I begin by reading each word, one at a time, starting from the *end* of the document. By working backwards, one word at a time, I can focus on each word as a unit, not as a part of a meaningful sentence or phrase.

Then I start from the top again and slowly read, checking for grammatical errors, dropped or duplicated words, verb/subject agreement, etc. As I make this pass, I also try to catch any wording which still—in spite of my initial editing—may not make sense. Dennis E. Hensley recommends reading out loud. "Reading something aloud helps you gauge its rhythm, pace, sound, and degree of difficulty. If you discover that certain passages cause you to be tongue-tied or long-winded, rewrite them in more simplified language" [4].

• *Consistency.* As mentioned, the types and number of checks you will have to perform depend on the nature of the document; however, one of the keys to a professional document—no matter what the content, the format, the purpose, or the audience—is *consistency.* Lack of it is one of the most common problems which can be

avoided if you "keep in mind all that you have already read, or at least an awareness of what you have read. Pay attention, not to content, but to the physical document—the letters, the spacing, the punctuation" [**1**, 25].

When verifying consistency, I check the use of running heads, footers, headings, and subheadings. Are all letters in upper case? Bold? Underlined? Centered? Or left-justified? I also check consistency in type size and fonts, indentation, and the placement of columns and visual aids on pages.

• *Appearance.* Another key to a successful document, related to consistency, is *appearance.* Because appearance *can* be everything. I begin by simply looking at the document. Page by page. Section by section. And ask: Is the document "busy?" Has order been made out of chaos? Is there a page bursting with text while another is half-empty? If the document is to be "comb" bound, will the left margin be too wide? Are the margins balanced? If three-hole punching, will I punch right through the text?

When checking the "look" of a document, I ask my proof buddy and co-workers—especially those in the art department—for their opinions.

• *Ease of Use.* Now that the document "looks good," I ask if using it will be such a bother that no one will want to do so. Are the lines so long (over five and a half inches) that readers strain their eyes unnecessarily? Are folios printed so near the bottom of the page that they might not appear on photocopies? Was the printer or typewriter ribbon on its last leg? (Tucking away an extra ribbon—just for final copies—will help avoid last minute delays.)

CONCLUSION

My sample "grouping" of items is not magical or cast in stone. It, like my list of items to check, is provided as a guide for developing your own list, grouping system, and technique—a technique to follow relentlessly so as to copyedit/proofread *systematically* and therefore *efficiently.*

The final product of whatever you are writing—a letter, a proposal, a

manual, or technical specifications—is a reflection of you and your capabilities. So *make* the time to *make* the document as presentable as possible. Try to enlist someone to share and improve the task of proofreading. Be systematic. Develop your own list of checks to perform. Check items by logical groups. Whenever in doubt, use reference books. Ask the advice of co-workers. Take plenty of breaks. Whatever you do, do not let up once you have completed the research and the writing. Leaving no stone unturned will often make the difference between an effective, professional document and one that collects dust. Ω

REFERENCES

1. Bruce V. Lewenstein, "What You See Is What Your Get: How, Not What, to Proofread," *Technical Communication* 32, no. 1 (First Quarter 1985): 23-25.
2. Valerie Mitchell, "A Style Guide—Why and What," *Technical Communication* 33, no. 4 (Fourth Quarter 1986): 232.
3. Fern Rook, "About Numbers," *Technical Communication* 33, no. 4 (Fourth Quarter 1986): 266.
4. Dennis E. Hensley, "A Way With Words," *Dallas* 65, May 1986, 57-58.

Additional Reading/Reference Material

Butcher, Judith. *Copy-editing: The Cambridge Handbook.* Cambridge: Cambridge University Press, 1975.

The Chicago Manual of Style. 13 ed., rev. and exp. Chicago and London: University of Chicago Press: 1982

Fowler, H.W. *A Dictionary of Modern English Usage.* 2d ed. revised by Sir Ernest Gowers. Oxford: Clarendon Press, 1965.

Mish, Frederick C., ed. *Webster's Ninth New Collegiate Dictionary.* Springfield, MA: Merriam-Webster Inc., 1984.

Pierson, Robert M. "Set A Spell." *Writer's Digest* 65, April 1985: 20-23.

Sadowski, Mary A., "Elements of Composition." *Technical Communication* 34, no. 1 (First Quarter 1987): 29-30.

Smith, Peggy, *Proofreading Manual and Reference Guide.* Alexandria, Virginia: Editorial Experts, Inc., 1981.

Strunk, W., Jr., and E.B. White. *The Elements of Style.* 3d ed. New York: Macmillan, 1979.

Webb, Robert A., comp. and ed. *The Washington Post Deskbook On Style.* New York, NY: McGraw-Hill Book Company, 1978.

HELENA CHYTIL is a freelance technical writer.

Collaborative Writing in the Workplace

CHARLES R. STRATTON

Abstract—More and more, technical experts are teaming up to produce technical documents. Dividing the workload horizontally, with each team member handling a separate chapter or section, doesn't work very well. Stratifying the project vertically, with a project team leader, a data gatherer, a writer, an editor, and a graphics person, is a more efficient and more effective method of collaborative writing. The process is quicker and the product is better because team members get to do what they are best at.

MORE AND MORE often in business, industry, and government agencies, technical experts are teaming up to write reports, proposals, articles, and other technical documents. This is not at all surprising when we consider the widespread application of the project team approach to research, development, and engineering design. Recently, Andrea Lunsford and Lisa Ede surveyed 1200 working professionals in management, engineering, behavioral science, chemistry, and communications to seek data on collaborative writing in on-the-job contexts. Of their 530 respondents, "87 percent reported that they sometimes wrote as part of a team or group." [1] I began writing professionally in 1960 and joined the Society for Technical Communication (then the Society for Technical Writers and Publishers) in 1962. At that time, the collaborative or project team approach to writing, editing, and producing technical documents was well established, so well established, in fact, that it wasn't even a topic for discussion. Corporate authorship was the overwhelming norm.

The days of the single person handling all aspects of product development are long gone, and the days of the solitary author for technical documents are long gone, as well. Perhaps the model of the solitary writer never had any validity in business, industry, and government agencies. This article is an outgrowth of my experiences with multiple-author documents over the past 34 years and offers some suggestions for efficient and effective collaborative writing in the workplace.

THE HORIZONTAL DIVISION MODEL

My first experience with collaborative writing was as a college student. Professor Jones would assign three or four students to research a topic and write a report. The only model we knew was one of horizontal division (figure 1). If we had four students on the team, we made sure we had four chapters or sections in the report. Joe wrote chapter 1, Skip wrote chapter 2, Mary wrote chapter 3, and Harold wrote chapter 4. What could be more natural than equitable division of responsibility? Each of us researched one subtopic, outlined one chapter, wrote one draft, edited one copy, and typed one final draft. Our idea of teamwork was to get together once at the beginning and once toward the end of the project, to compare notes.

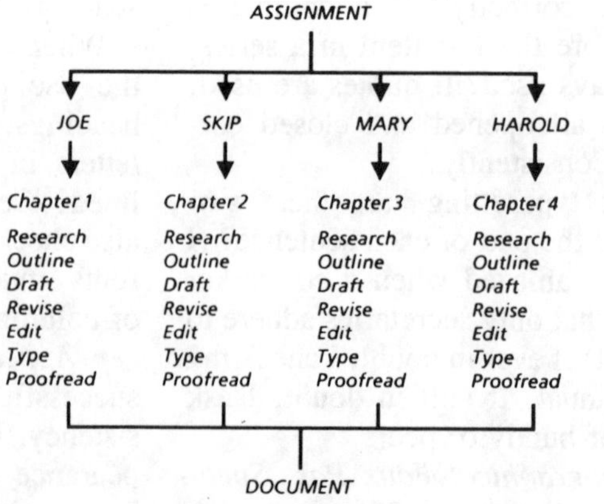

At best, this procedure led to a mediocre report. We would wind up with a patchwork quilt, good swatches of material but of different colors and textures and with very obvious seams. Joe's chapter was well organized and developed but poorly edited. Skip's chapter was lengthy but dull as the hood of an old pickup. Mary's chapter was well crafted but incomplete. Harold's was complete and concise but as pompous as Howard Cosell. Terminology was inconsistent from chapter to chapter. Illustrations abounded here and were absent there. There was lots of duplication, a few holes and, above all, inconsistency. And this was under the best of circumstances; most of the time, it was worse.

Unfortunately, this horizontal division model still seems to be very popular in business, industry, and government agencies—probably because managers aren't aware of any other way to improve on the solitary writer model. Lunsford and Ede found that 72 percent of their respondents used the horizontal division model at least occasionally in their writing; 24 percent, of the respondents used this organizational pattern "often" or "very often." [1]

Figure 1. Collaborative Writing by the Horizontal Division Model

Charles R. Stratton is a Professor of Technical Communications at the University of Idaho.

Reprinted from *IEEE Trans. Prof. Comm.*, vol. 32, no. 3, pp. 178–182, September 1989.

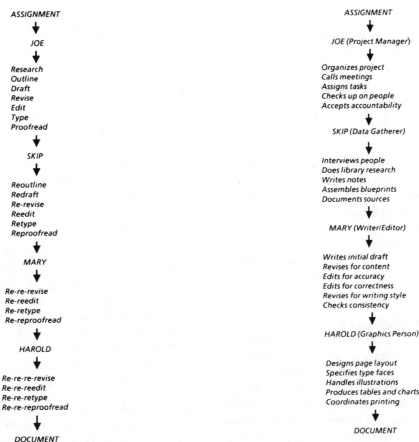

Figure 2. Collaborative Writing by the Sequential Model

Figure 3. Collaborative Writing by the Stratification Model

THE SEQUENTIAL MODEL

Early in my writing career, in the computer center of a big-ten university, I was exposed to the sequential model of collaborative writing (figure 2). In this model Joe started the ball rolling. (By this time, Joe was a technical expert—a software engineer.) Joe got the ball rolling by organizing the material, writing a draft, revising it, editing it, and having it typed up. In other words, Joe prepared what he viewed as a final draft, or pretty close to it. Joe would then give his version of the document to Skip, who by now was a technical editor. Skip regarded Joe's stuff as raw data and would promptly reorganize it, rewrite it, re-revise it, and re-edit it, much to Joe's chagrin. Skip would then pass his version to Mary, who had become a publications manager. She would say, "Ehhh, this isn't quite what I had in mind," and she would re-reorganize, re-re-revise and re-re-edit. (She didn't re-rewrite, because she was a manager, and managers don't do any writing.) Finally, the document got to Harold, the programming manager. He would take one look at it and give it back to Joe, saying "The publications people really screwed this thing up. See if you can straighten it out."

After two or three trips around this circuit, the document finally got printed and distributed. The end product was nearly always better than that produced by the horizontal division model. At least it was consistent. More than that, though, each person in the sequence contributed to the *entire* document, and, as a result, the document communicated more effectively. However, the sequential process was every bit as time consuming and inefficient as the horizontal division process. And it polarized the shop. Joe no longer talked to Skip, Skip was mad at Mary, and Mary muttered a lot about Harold and Joe. Each person in the sequence regarded each other person in the sequence as a barrier to efficient and effective communication—a hurdle that had to be jumped in order to get a document published. The sequential way of doing things was effective but not at all efficient.

THE STRATIFICATION MODEL

Finally, I landed in the service publications division of a Fortune 500 corporation. Here, we produced documents with the vertical division model, or stratification model (figure 3). Joe functioned as project manager; Skip gathered data and drafted the document; Mary revised and edited; and Harold handled format, layout, and production. Each did what he or she was really good at; all worked together, right from the start; everyone had a sense of ownership (or authorship) in the document; and the team was both effective and efficient. Here's how we divided up the duties.

Project Team Leader

Top management at the company was convinced that every project that involves more than two people needed a project manager or project team leader. This included writing projects. I agree. If the project team leader never writes a single word, he still earns his keep. The project team leader organizes the project; calls and chairs team meetings; assigns various data-gathering, writing, and revising tasks; checks up on people; keeps the project on schedule; referees disagreements; and basically accepts accountability for the project.

The critical skills for a good project team leader are ability to see the big picture, effectiveness in working with other people, and knowledge of (which usually means several years' experience in) the organization as a whole. This person need not necessarily have a job title of supervisor or manager. The project team leader can be any member of the team who has decent organizational and managerial skills.

Data Gatherer

The data gatherer or data supplier is the person who assembles all the basic information that is to go into the document. Often, the data gatherer (or data supplier) is the technical expert who knows most about subject matter of the document. In this sense, he functions much as Joe did with the sequential model, only he stops short of organizing, revising, and editing. This person might write notes, outlines, and drafts from his own knowledge, or he may assemble blueprints, specifications, and pertinent copy from other documents. He stops short of writing copy, however.

The requisite skills for a good data gatherer are knowledge of the subject matter, thoroughness, and accuracy. Ability to interview people and care in documenting sources are useful skills as well.

Writer

Following the organizational framework set by the project team leader and using the data supplied by the data gatherer, the writer generates the first draft of the document. The writer's chief job is to get some words down on paper—words that the other team members can work with. The writer is the person who translates data into copy.

A good writer needs to be able to compose text rapidly and have an easy way with words and sentences. A writer also needs a great deal of audience empathy, a sensitivity—preferably gained first hand—to the needs and natures of the people who will read the document and use the information it contains. Above all, the person who writes the draft must like to write.

Editor

The editor is the detail person. The editor is also the consistency person. The editor may do some revising; more often, if the editor sees a need for revision, he sends the draft back to the writer with suggestions or instructions, as the case may be. Mostly, the editor checks writing me-chanics, usage, consistency in terminology, and the like. The editor also views the document with a detachment more like that of a reader, not having the immediate and close involvement with it of either the data gatherer or the writer.

The most important skill for an effective editor is extreme politeness. Other requisite skills include a thorough knowledge of grammar, mechanics, and usage; a sense of effective writing style; a good deal of audience empathy; and an ability to pay close attention to detail. The chief function of the editor is to move the document from good to excellent.

Graphics Person

The graphics person handles all the visual aspects of the document, including type face, spacing, and margins; page layout; headings and subheads; tables, charts, and illustrations; and the cover and title page. Often, too, the graphics person will handle printing and distribution of the document. In some cases, the graphics person will do all the work personally; in others, this individual will coordinate with photographers, illustrators, typesetters, paste-up people, and printers.

A good graphics person needs an artistic sense of theme, balance, and composition. A good graphics person also needs a thorough knowledge of the techniques of publications production and, of course, drafting skills and computer skills.

ADVANTAGES OF STRATIFICATION

The chief advantage of stratifying the collaborative writing process lies in specialization. Each person involved in developing the document gets to work in his or her area of expertise and interest. More important, the entire document benefits from each participant's specialty. The best manager organizes the entire project. The most facile writer drafts the entire document. The most careful editor edits the entire document. And the most artistic team member handles the visual aspects of the entire document. The end product is stronger and more consistent than that produced by individuals working under the horizontal division model.

A strong secondary advantage of stratification, however, lies in the immediate coordination of efforts by the project manager. Each person contributing to the document's production works in the same administrative unit, rather than being in separate sections or even separate divisions of the organization. Each is but one step from the boss. This tends to eliminate the competition between technical people and writing/editing people, or between writing/editing people and publications production people that is inherent with the sequential model. Moreover, with the project team, each participant—manager, data supplier, writer, editor, and graphics person—has a much stronger sense of involvement or authorship with the document than with either of the other methods of collaborative writing.

MIX AND MATCH

Obviously, not every collaborative writing project needs five individuals to function in these five roles. (In fact, in many writing arenas, an individual technical writer has to wear all five hats.) A single person may double up as writer and editor, or project team leader and editor, or data gatherer and graphics person. Similarly, a given person may not fill the same role with every writing project; he or she may be project team leader for one, writer for another, and editor for a third. When I wrote for the advertising department of one of the country's largest manufacturers of heating and airconditioning equipment, I saw many instances where Joe was boss and Skip was worker on one project, while during the same time period, Skip was boss and Joe was worker on another project.

Ideally, each person on the team should perform the function he or she is most skilled at. If all people working on a collaborative writing project are equally skilled at all functions, the choice usually is made on the basis of who wants to do what. (The same is true, I guess, if all people are equally unskilled.) Or sometimes, upper management will designate one team member to be project team leader, and the team leader will assign the other tasks. The important things are to have all the roles filled and to have each remain more or less discrete.

VARIATIONS ON THE THEME

Perhaps the most common variation is to eliminate the overhead and try to get along without a project team leader. I advise against it. For a team with a pretty good writer and a document whose verbal message is more important than the visual, the editor can often double up as project team leader. However, it's important for the editor to keep the two roles distinct. For a document such as an annual report or sales promotion piece, where the visual message is as important as or more important than the verbal, the graphics person might double as project team leader.

Some years ago, I was working as a technical writer with several service engineers who functioned as data suppliers. One of the engineers, however, had a flair for writing and got very testy when I tinkered with his copy. We talked about the problem and decided to shift roles. From time to time, he would fill the role of writer as well as data supplier, and I would then function as editor (which mostly meant suggesting that he make certain changes, rather than going ahead and making them myself.) This worked pretty well, so long as we both kept our roles straight. Once in a while, he would ask me to function as engineer, designing a part or a special tool or developing a service procedure, so that I got to be data supplier now and then.

In another situation some years later, I again switched roles with an engineer. I would gather the data and write the rough draft. He would then edit it, filling in all my technical holes and correcting my technical misconceptions. Although he edited mainly for content, he would take great delight in finding an occasional typographical error or terminological inconsistency. This role switching gave us a more tightly organized and more audience-empathic document and at the same time saved the engineer lots of time and frustration.

In a classroom situation, I had a mediocre student who—among other shortcomings—just couldn't edit the mechanical errors out of his own writing. He missed the project team's organizational meeting and drew, by default, guess what role? Editor. Jerry was a good sport and gave it a try. Much to his surprise (and mine), he found that he could do an excellent job of spotting and correcting the errors in other people's writing. He turned out to be a good editor, and by separating the writing and editing roles he became much better with his own work.

CONCLUDING COMMENTS

Research supports the notion that collaborative writing is both more effective and more efficient than individual efforts. That is, it produces better documents with less time spent. In an article in *Research in the Teaching of Writing* [2], Ann Ruggles Gere and Robert D. Abbott discuss collaborative writing in middle and high school and state that

> Existing research on writing groups supports their efficacy in improving critical thinking, organization, and appropriateness of writing ... in improving usage in writing ... and in increasing the amount of revision done ... .

Gere and Abbott go on to cite some 17 books, articles, and doctoral dissertations that support their observations. Lunsford and Ede's research supports the efficacy of collaborative writing as well. [1] In their survey, they asked, "In general, how productive do you find writing as part of a team or group as compared to writing alone?" Fifty-nine percent of their respondents answered "very productive" or "productive."

My own experience also supports their observations. Of the nearly two million words of scientific and technical information I have published, some were written in isolation, some with prewriting collaboration, some with postwriting collaboration (as with this article), and some with collaboration at all three stages of the writing process. Consistently, the greater the level of collaboration, the better my writing has been accepted and the less it has required revision.

In spite of this strong research and anecdotal evidence, however, I don't see much by way of teaching multiple authorship techniques in college writing classes. In what little I do see, the collaboration is limited pretty much to the horizontal division model. I suspect that the reluctance is simply due to the difficulty of evaluating the contribution of each individual student in a collaborative writing project. (It is possible to do so, however. See my article "A Project Team Approach to Technical Writing in the Classroom." [3]) To be sure, I see a fair amount of prewriting collaboration (group discussion of topics, or brainstorming) and even more postwriting collaboration (peer review, or group editing), but not much in the middle. (See Dene Thomas [4] for a discussion of prewriting collaboration in the classroom and Kenneth A. Bruffee [5] for some suggestions regarding postwriting collaboration. Also, see Bruffee [6, 7] for discussions of some of the epistemological and pedagogical implications of collaborative writing.)

Of the several methods of collaborative writing with which I have had experience, both as a writer and as a teacher of writing, stratification is the most effective and the most efficient. It combines the advantages of specialization with the advantages of project integration. Sequencing of writing tasks without integration is a fairly effective way to collaborate, but it's not very efficient. You can get a decent writing product that way, but it takes a lot of time and effort. Horizontal division of writing tasks is neither efficient nor effective. It takes a lot of time, and the end product is not all that good, because (of course) horizontal division isn't really collaboration at all. It's just individual effort masquerading as a joint effort.

Collaborative writing is fun. Most people enjoy doing what they're good at and welcome respite from chores they're not so good at. Some people like to honcho projects but hate the details. Others like to do the work but escape the pressures of authority and accountability. A person who's not very good with words can contribute substantially in the visual department or as a data gatherer. Job satisfaction is an increasingly important commodity in today's work place.

Collaborative writing is flexible. People can learn to become good writers without having to take on total responsibility for a major document all at once. The project team approach allows for more efficient allocation of human resources. Individuals can fill different roles in different mixes at different times. Workers don't stagnate. People can find out where their strengths and weaknesses are in a matter of weeks rather than years.

In short, collaborative writing makes sense—especially under the stratificational model—and I recommend it as an efficient and effective method of producing technical documents in business, industry, and government agencies.

REFERENCES

1. Lunsford, A., and Ede, L., "Why Write ... Together: A Research Update," *Rhetoric Review 5*, 1 (Fall 1986), pp. 71–81.
2. Gere, A. R., and Abbott, R. D., "Talking About Writing: The Language of Writing Groups," *Research in the Teaching of English 19*, 4 (December 1985), pp. 362–381.
3. Stratton, C. R., "A Project Team Approach to Technical Writing in the Classroom," *Technical Communication 20*, 2 (2nd quarter 1973), pp. 9–11.
4. Thomas, D. K., *A Transition from Speaking to Writing: Small-Group Writing Conferences,* unpublished doctoral dissertation, University of Minnesota, 1984.
5. Bruffee, K. A., *A Short Course in Writing,* 2nd ed., Cambridge, MA: Winthrop, 1980.
6. Bruffee, K. A., "Writing and Reading as Collaborative Social Acts," in *The Writer's Mind: Writing as a Mode of Thinking* (Janice N. Hays, et al., ed.), Urbana, IL; National Council of Teachers of English, 1983, pp. 159–169.
7. Bruffee, K. A., "Collaborative Learning and the 'Conversation of Mankind,'" *College English 46*, 7 (November 1984), pp. 635–652.

Reverse Engineering: The Outline As Document Restructuring Tool

DIETRICH RATHJENS

Abstract—Outlining is usually thought of as an important tool for the initial stage of document development. However, it can also be used to greatly increase the efficiency of restructuring a longer document or developing a composite text from several sources. This essay proposes and details a five-step procedure to effect such a "reverse engineering" of documents.

OUTLINING is a fundamental part of technical writing, part of a process that normally moves from researching and noting ideas, through outlining and writing, to revising. Each of these steps is essential. Outlining, however, exerts the greatest influence on this process because it indicates the degree to which the necessary research and notation have been completed, and it defines the structure and scope of what is to be written.

Although not all engineering writers use outlines, those who do find that arranging concepts (super-, co-, and subordinating them) into an outline serves the following functions:

- It establishes the relative importance of the concepts.
- It ensures continuity of thought.
- It lends proper structure to the document.
- It promotes document completeness.

Assigning ideas to first-, second-, and lower-order headings means defining each section's limits and ensuring document coherence. It is also a *de facto* placement of emphasis.

Consecutively listing coordinate headings is a means of developing continuity, so that the concepts in one section follow from the foundations laid in preceding ones. This helps to create a smooth transition, a logical flow from one concept or subject to another.

Next, the outline gives a good picture of the scope of each section. We can compare each one's length and level of detail with those of the others. This can point up anomalies among (and within) sections, such as greatly varying lengths or single subordinate headings.

Finally, outlining helps us determine the (in)completeness of the presentation. The outline is the document in skeletal form; it is easy to use this structure for checking integrity, to preclude insufficiencies in the completed text.

These points show the importance of outlines within the context of the incipient stage of document development.[1] Yet this is only taking advantage of half their usefulness. Outlines can also be used to excellent effect when revising, or even trying to understand an ineptly structured document. They can save the writer, editor, and reader much time, as well as greatly increase the efficiency of that revision or attempt at understanding. In fact, the longer the document, the more time is saved.

This procedure is based on the concept of reverse engineering, a process of disassembling a (usually complex) device for the purpose of gaining a better understanding of how it functions or what new developments have been incorporated into it. The concept and procedure can be applied analogously to the basic restructuring or analysis of any document, by engineers and technical writers as well as editors. The procedure is simple and highly efficient; it consists of five steps, detailed in the following paragraphs.

Step 1: Make an outline that accurately reflects the structure and content of the original document.

It would be understandable, indeed even seem sensible, to begin a revision by turning to page 1 of the document and starting to edit words, sentences, and paragraphs. Three problems are associated with this intuitive, "brute force" approach.

First, it provides no overview of the task's magnitude, for use in budgeting time. Further, although it may ensure proper word choice, sentence construction, and paragraph development, it has little or no impact on overall coherence and continuity.

---

1. For additional information on outlining at this stage, the following articles are recommended:

   Hall, H.A., "MARCO—A New Technique for Outlining Technical Publications," *STWP Proceedings*, 14th International Technical Communications Conference, May 1967.

   Landreman, D.M., "A New Look at Outlining—The LSN/SN Approach," *STC Proceedings*, 21st International Technical Communications Conference, 1974, p. 88.

Dietrich Rathjens developed the Technical Writing Program for the Department of English at San Jose State University (California), where he teaches courses in technical writing and editing.

Reprinted from *IEEE Trans. Prof. Comm.*, vol. PC-29, no. 3, pp. 19–22, September 1986.

Second, any emendations made at this level may be vitiated later, when it becomes apparent that certain sections should be deleted or vastly restructured. The interrelation of a document's sections may render some already-edited parts superfluous or redundant.

Third, it is difficult to be sure whether you can delete certain sentences or paragraphs, or where they should be repositioned, without memorizing the entire document. Only after completing the initial edit do you know that a matter is discussed repetitiously in several parts of the document. At that point, you still have the job of combining all references to the subject into one section, deleting some, and rearranging others. This approach is like going into a labyrinth blindfolded.

More fundamental than the word-level, sentence-level, or paragraph-level edit is the structure edit. Thus, the first task is to make an outline. Even in the absence of numerical designations for sections or paragraphs, this can be done quickly by using the existing section titles and subheadings. Each numerical designation or head/subhead in the original text must have a corresponding entry in the outline.

At this stage of the edit, it may be tempting to correct in passing certain flaws, errors, or solecisms. However, this is inadvisable for two reasons:

- It can cause confusion later, in trying to correlate the revised headings of the outline with sections of the original text.
- It lessens the efficiency and speed of the outlining task.

The only concern should be the document's structure, not editing its contents or making *ad hoc* changes to the outline being developed.

Step 2: Delete those sections of the outline that do not contribute to the document's unity or purpose.

The finished outline, reflecting the original text, gives a complete perspective of the document, its structure (or lack of structure), and the interrelationship of its sections. It is now easy to ascertain which headings have little or no relation to the document's purpose; these should be deleted. However, do not renumber or change the heading designations (first-order, second-order, and the like) in the outline.

Ineptly structured documents are often carelessly written (and vice versa). Therefore, if a document has been found to require this kind of reverse engineering, it is highly likely that some sections can be deleted. This simplifies the task, because the shorter the outline, the better the overview and the better the editor's ability to manipulate headings.

It is important to remember that this step (and the two following) pertain only to the outline. Put the original text aside; do not try to effect in the document the structural changes you are making in the outline. It is shown below that this is unnecessary and time-wasting.

Step 3: Restructure the outline.

The (reduced) outline will lend itself to quick and efficient manipulation. You should now consider the importance, and, therefore, the placement, of each heading. Is it properly subordinated? Coordinated? Superordinated? If the original text was not well structured, then even after deleting irrelevant or redundant material from the outline, rearrangement—possibly even additions—will be required.

The goals of this rearrangement are the same as those of any careful outline construction: coherence and continuity. Logically grouping the headings of the outline gives the existing ideas coherence. Determining the order of the heading groups, and the headings within each group, gives continuity.

The heading designations of the before restructuring outline merely serve as "tags" that help locate the corresponding paragraph or idea in the document (which you have set aside). You need not be concerned with their rank; a third-order heading, for example, would not necessarily persist as such.

As you restructure the concepts in the original outline, you must give each heading a designation that reflects its place in the new structure. To distinguish between the original and the new designation, place the former in brackets, after the new heading.

It is generally agreed that, when subordinate headings occur within a section or subsection, there must be more than one of these (the rule of double subordination). This rule may be suspended during restructuring; single subordinates are of no consequence in this step. They will, indeed, prove helpful in the next step.

Step 4: Determine what is missing in the outline.

Two phenomena in the outline highlight missing information:

- The relative brevity of some sections
- The occurrence of single subordinates

If one or more sections of the new outline are unusually short (that is, have far fewer subheadings than the other sections), this probably indicates that those topics have not been adequately covered. This assumption is based on morphological considerations: most outlines will have few if any subheadings in the introduction and conclusions, but the central sections will often be similar in length and development. A cursory look at the tables of contents of several books will substantiate this.

Single subordinates in the original outline could have several meanings:

- This is a minor point, which should be deleted (case 1).
- There is at least one other subordinate missing (case 2).
- This subordinate should be incorporated into the next-higher-order heading (case 3).
- The subject's importance is actually commensurate with that of the next-higher-order headings; it should be made one of these (case 4).

Consider the following hypothetical outline:

Section	Contents
1.0	PURPOSE AND SCOPE
1.1	Purpose
1.1.1	Need for Evaluation
1.2	Scope
2.0	CONTRACT AND TECHNICAL DIRECTION
2.1	Contractual Direction
2.2	Technical Direction
3.0	MANAGEMENT TASKS
3.1	Program Plans
3.2	Reviews
3.3	Reports
3.3.1	Milestone Report
3.3.2	Status Report
3.3.3	Red Flag Report
4.0	STANDARDIZATION PLAN
4.1	Nonstandard Parts
5.0	PRODUCTION GOALS
5.1	Design to Unit Production Cost
6.0	SYSTEMS ANALYSIS TASKS
6.1	Design Verification
7.0	SYSTEM EFFECTIVENESS
7.1	Reliability, Maintainability, Producibility
7.2	Integrated Logistics Support

Section 1.1.1 illustrates case 1, above; this heading should be eliminated from the outline. Section 4.1 corresponds to case 2; the subordinate headings should be *4.1, Standard Parts,* and *4.2, Nonstandard Parts.* Section 5.1 corresponds to case 3; designation 5.1 and its heading should be deleted, and the title of section 5.0 should be *COST/ PRODUCTION GOALS.* Section 6.1 illustrates case 4; it should be redesignated section 7.0. (This means that the first-order designation for all subsequent sections will be increased by one; section 7.0 will become 8.0 and so forth.)

When you have eliminated single subordinates of the case 1, case 3, and case 4 types, any single subordinates remaining in the revised outline will reflect case 2. When you have thus identified the topics that require more information, you should try to supply the missing headings. Do not try to write the text to accompany these headings. Either write it later or, if you are coordinating the work of others, tell those responsible of the urgent need for specific information. You should now proceed to the final step.

Step 5: Restructure the original text according to the revised outline.

The revised outline will probably consist of some unchanged (original) headings and some changed headings; each changed heading should be followed by the original designation in brackets. Using the original headings as signposts, find the corresponding text in the original document. Then cut and paste (either literally or on a word processor), arranging the text in the order given by the outline. Where there are new headings without any corresponding text (that is, instances where you have identified the need for more information), leave space after each to call attention to the incompleteness.

It is likely that there will be text left over, corresponding to those sections that were deleted from the original outline because of irrelevance or redundancy. Discard this text.

The cut-and-pasted text can now be edited. Incorporate the new text for the headings created in step 4, either by writing it or editing material supplied by others. The entire editing process should be simple and straightforward because the sections that would have provided the most confusion have already been eliminated. Your editing will now have purpose and efficiency. Also, you will be able to ensure consistency, as well as specify layout and format, in this one pass. This certainly would not have been possible before the basic restructuring of the document.

The procedure described above may be more readily understood if illustrated with two realistic examples:

- *Case 1*—You are under time constraints to revise an unstructured (and haphazardly written), 300-page product description into a technical proposal for a potential customer who is interested only in certain aspects of the product described.
- *Case 2*—You must produce a coherent, encompassing report on a subject that has been covered in several books and numerous articles.

Case 1

The unmanageable 300-page document could be turned into a manageable 10-page outline by following step 1. This step seems to take up valuable time, especially if done thoroughly and meticulously in ensuring parity between the numbering of sections in the original text and that of the outline's headings. It is tempting to think of the number of pages that could be edited (even superficially) in this amount of time. However, it soon becomes apparent that the outlining actions are much more trenchant than any editing could have been, and that this seeming detour is actually the start of a remarkable shortcut.

Deleting inappropriate headings (step 2) is easily effected because of the outline's relative brevity. The 10-page outline suggested above may now be only seven or eight pages long. Also, the next step (in effect, the original task in miniature) is easily managed because of the overview afforded by the outline.

Determining what is missing can be a complex task, depending on your knowledge of the product being offered and the customer's needs. This step often requires the assistance of staff members in engineering or marketing organizations. It should not result in a noticeable lengthening of the outline; rather, the outline should only indicate those concepts that must be presented (or presented in greater detail). A copy of the restructured outline will indicate the desired level of detail to the person charged with writing the new sections.

The restructuring of the original text (step 5) can be done by rote if sufficient care was exercised in step 1. Then, only the editing (for grammar, punctuation, style, and so forth) of the document remains to be done—a simple job, requiring only one pass.

Case 2

Here, the task is more complex, producing a complete report by integrating sections from various sources. Although the course is not as linear as in case 1, above, it is still fairly simple. In step 1, you make an outline of each book, using first-order headings for chapters, second-order headings for sections, and so on. (It is possible that the table of contents of each book will serve this purpose; at least it will facilitate outlining the book.) The articles can be treated as one book, with a first-order heading for each, because articles usually compare with chapters in a book, in both level of detail and length.

Because the original texts all pertain to the topic of the report, step 2 will consist of eliminating redundancy among the outlines. Compare two outlines at a time. If both address the same concept, choose the more detailed discus-

sion; then delete the superfluous section from the other outline. This process is continued until all outlines have been compared and all redundancies deleted.

You can now effect step 3 by integrating the remaining sections of the various outlines into a new, composite outline. Since you chose the more detailed discussion for each concept presented more than once, you may have some inconsistencies in length and level of detail between these discussions those that did not result from such a choice. Step 4 will probably consist of noting those sections that are much briefer than others in the new outline. This will lead either to further research on the concepts they cover, or at least to a notation that no more information is currently available.

Developing the report from the composite outline should not be difficult if step 1 was done methodically and carefully. It comprises cutting and pasting texts in the order prescribed by the new outline. Editing for continuity and for consistency in style should, then, require only one pass.

The advantages of using an outline for the reverse engineering of one or more documents should be apparent. Of prime importance is the order imposed on the task, which provides a perspective of the entire project. This perspective, in turn, allows for the effective budgeting of time. Each step taken toward completion of the project is the most efficient one; the tasks are streamlined, each sharply focused.

The most important advantage, however, is the saving of time. Proof of the time saved, and a full appreciation of it, requires a comparison with the usual (brute force) method of revision. The time saved is in direct proportion to the length of the original document. For documents of several hundred pages or more, therefore, this method becomes imperative.

How Writing Helps R&D Work

HERBERT B. MICHAELSON
LIFE MEMBER, IEEE

Abstract – An engineer author can develop better perspectives and even new technical concepts when writing a report of project work done in an R&D laboratory. These new insights eventually help the engineering work in subtle but powerful ways. An illustrative example is given.

INTRODUCTION

Writing can be surprisingly productive for an author, especially where development work is being reported. The benefits of good writing can indeed be subtle. Although many engineers readily admit acquiring new ideas while writing manuscripts, others are not always aware of the insights born of writing. Regardless, such new understandings eventually are reflected in the engineer's own work and achievements. Composing a report or paper, then, sets up interactions among the author, the manuscript, and the work itself.

The relation between writing and thinking is summarized nicely by a brief comment in the famous book on the elements of style by Strunk and White [1]:

> Fortunately, the act of composition, or creation, disciplines the mind; writing is one way to go about thinking, and the practice and habit of writing not only drain the mind but supply it, too.

This is especially true of any report on a research and development (R&D) project that was, during the development cycle, plagued with changes of pace and new directions of thinking. In preparing such a report, the engineer must retrace the thread of development work to construct an intelligible description of what was done, what was found, and what was contributed to the state of the art. In thinking the whole process through, the author may discover limitations and omissions in the data base.

The author is a consultant in technical communication in Jackson Heights, NY.

This paper shows how creativity in writing can actually help the work. It also cites an illustrative example of how a frustrated attempt to describe completed development work produced instead a new engineering concept.

CREATIVITY IN WRITING

Although the discipline of formal writing can expand a writer's own technical ideas, there are also some negative aspects. All writing, of course, is not creative. Much depends on the author's attitude.

The bored author who considers writing a tiresome but necessary evil will inevitably produce a bored, tired manuscript, mechanically written and lacking in imagination. Even though such a paper is clear, correct, and accurate, it can be deadly dull. The author's negative feelings have quenched any spark of creative effort.

The engineer who wants to be known for contributions to the field adopts a different approach to writing. Such an author carefully reviews the methods and results of the development project, identifies the portions of the work that were novel and innovative, and emphasizes those items in the manuscript. This kind of critical probing of the engineering work can produce not only new patterns of thinking but also a more interesting paper.

The mechanisms of thinking that produce sudden bursts of insight are poorly understood. Regardless of how these creative forces operate, they seem to be at play in an engineer's reporting of R&D work. When writing a paper at the completion of a laboratory project, many an author senses a new interpretation or sees a defect in the results and goes back to the laboratory for additional data, a more thorough analysis, or a modified design.

One theory of the subconscious to explain this kind of spontaneous thinking was favored by several scientists, including Helmholtz and Poincaré: While the mind is occupied with a conscious activity (for instance, composing a manuscript), new ideas and relationships are being generated at the same time in the unconscious. According to this theory, fresh, creative thinking consists of ideas rushing pellmell into the conscious mind from a mental arena below the writer's threshold of awareness.

Reprinted from *IEEE Trans. Prof. Comm.*, vol. PC-30, no. 2, pp. 85–86, June 1987.

Whatever the thought mechanism, an author's attempt to sort out the facts and set them down in a manuscript in logical order does tend to yield a fresh perspective and new understandings.

INTERACTIONS BETWEEN AUTHOR AND PERSONAL COMPUTER

Because of the proliferation of computers in industrial laboratories, many engineers now compose reports and papers on a personal computer (PC) or workstation instead of on a typewriter. The interactions between writer and machine tend to stimulate creativity in writing in several different ways – an unexpected fallout from the new technologies.

The obvious benefits of using a PC are fast corrections and insertions, automatic formatting, spelling checks, and freedom from proofreading the revised papers. Less obvious, however, are the machine's influences on the writer's intellectual output and therefore on the engineering work.

Composing on a PC effects some definite changes in the engineer's writing habits. Writing habits, of course, cannot be separated from thinking habits. The modifications of the writing process are due to the speed of the machine. For example, fleeting ideas are part of every writer's composing process; the mind generates new ideas, in fact, far faster than a writer can inject them into a typewritten or handwritten paper. The computer quickly records such thoughts before they are lost from short-term memory. Simplification of the mechanics of writing allows more attention to the intellectual process. Further, the PC does a rapid, legible job of modifications and additions to the text and does it easily; this capability gives the author more time (and more inclination) to refine the technical content of the manuscript.

The machine affects a writer's thinking in another way. For the PC user, "writer's block" tends to disappear because of a psychological reaction: the machine forces the attention of the user and in effect invites an author to take control of the writing situation. In addition, the computer has certain functions that aid the user's reasoning power. Engineering ideas can be explored and tested with programs that build models and do simulations, to confirm and extend technical concepts. Often the calculation and simulation functions can be carried out on the same computer or computer terminal as the writing function. Programs that facilitate moving back and forth between these functions have facilitated the creative process enormously.

In addition, when connected to a communication network, the PC or workstation permits quick interaction with other people. This, of course, is a decided advantage in any kind of cooperative writing project. Co-authors in separate buildings can trade notes or send revised manuscripts almost instantaneously on the network. Here, again, the speed of transmission becomes an actual stimulus to information exchange and to a free play of thinking.

Another excellent function of the PC is its use for writing a report or paper in increments – a technique described some years ago in these *Transactions* [2]. This method consists of writing a manuscript in segments as the development work on the project proceeds. The writer stores successive increments of the paper on disk or in system memory. The insights derived at various times during the writing are thus applied to engineering work. Instead of first beginning to think of writing a paper at the end of the project, the author can print out a completed draft, ready for revision and editing.

Writing in increments is not a theory. It has been used successfully by hundred of engineers to prepare papers for publication while design or development work was still in process. The writing was part and parcel of the development cycle instead of an afterthought.

ILLUSTRATIVE EXAMPLE

Industrial research and development usually takes place in an atmosphere of change. Even with the best of planning, modifications are necessary to meet new requirements or improved approaches to the work. These may be dictated, for example, by changes in marketing strategies or by news of competitive developments in another company.

Or else, an engineering project may get off to an awkward start for one reason or another and require new thinking. Here is an example of such a situation, in which writing actually helped the work.

Many years ago, when applications of electronic computing were beginning to develop commercially, a programmer at an IBM laboratory designed an ingenious system for solving a traffic problem at a given street corner. The problem dealt with variations of pedestrian traffic throughout the day, a specified number of vehicle right-hand turns per hour, a need for periodically adjusting the traffic signals, and so on. When the program was completed, the programmer's manager asked him to prepare a report describing its operation for the customer. Writing the document was a frustrating job and, as a direct consequence of these frustrations, the author first began to realize the complications of the mathematics on which the program was based.

When the report was half finished, the programmer suddenly had a new idea and abandoned the writing. He started over again from scratch and worked out a new program for the traffic problem. This program generated a graphic printout representing the street corner, with built-in parameters and variables. The program was successful and later became widely used by traffic engineers.

REFERENCES

1. Strunk, W., Jr., and White, E. B., *The Elements of Style*, Third Edition, Macmillan Publishing Co., Inc., New York, 1979, p. 70.

2. Michaelson, H. B., "The Incremental Method of Writing Engineering Papers," *IEEE Transactions on Professional Communications*, PC-17, No. 1, 1974, pp. 21-22.

Part 7
Oral Presentations
Speaking Effectively to Groups

MANY PROFESSIONALS are required to present information orally as well as in writing, and if you are in management or marketing you may speak before groups even more frequently than you write. The articles in this section, therefore, offer ideas on how to prepare and make successful oral presentations. Some topics, such as controlling nervousness, keeping audience attention, and using notes or visual aids are discussed in more than one article, so you will get a variety of help on the primary concerns we all have when giving a presentation.

Watching the Time

How do you make sure you can present your information effectively in a limited amount of time? Robert Wismer discusses this concern of speakers who give technical presentations to management, and shows how to stress key information at the stages in the presentation when the audience is most receptive. His final recommendation that you practice a presentation seven times before delivering it may seem excessive, but the advice is worth following if you are not yet skilled in making oral presentations.

Delivering Your Message

Although Bert Decker focuses on helping managers and executives speak in public, his insights can help all of us who have to do so. In the author's words, when you give a presentation you are "both the delivery system and the message itself." An oral report involves verbal, vocal, and visual elements, not just turning written words into spoken ones. Decker's concluding pointers on maintaining eye contact with your audience and using visual aids, particularly overhead transparencies, are taken up in more depth in some of the remaining articles in this section.

The Hi-tech Talk

Giving technical presentations at professional conferences can both enhance your career and provide valuable publicity for your company. Sara Pritchard shows how to improve such presentations by analyzing the techniques of five of her engineering society's most popular speakers, thus identifying and illustrating those factors that make the difference between a mediocre presentation and a superior one. Ronald Rosen-

burg, on the other hand, concentrates on hi-tech presentations in government and industry, covering the spectrum from audience analysis, planning, and rehearsing, to handling visuals, conveying confidence, and providing handouts.

Presenting the Proposal

Michael Warlum looks at presentations you sometimes have to make in support of proposals, first analyzing why so many proposal presentations are ineffective, and then describing the steps taken by the Boeing Company to remedy problems such as poor audience analysis and a failure to establish a clear goal, with the resultant failure to stress key points. Warlum's comments on borrowing storyboards from the proposal team to ensure an accurate presentation of the proposal show a further use of these aids, discussed in Part 5 of this anthology by Greenly and Barakat.

Practice, Practice, Practice?

Most experts stress the need for practice before making a presentation, but Susan Dressel provides an anecdote illustrating the possible dangers of *over*preparing. You could tie yourself into a straitjacket if you plan every syllable and action in advance, give your talk in a robot-like manner, and leave little flexibility to respond to cues from your audience. Some preparation is unavoidable, however, and Dressel's fellow contributor to this article, Joe Chew, has some good ideas on how to practice your presentation in such a way that your confidence, spontaneity, and personality will all work for you when you get up to talk.

Lively and Emphatic

Morey Stettner follows up on Dressel's concern that you should make facts come alive rather than present them in a monotonous manner. He chides the "it's-okay-to-be-boring-as-long-as-I've-got-the-facts" attitude that some of us might entertain. The trick is to know which facts should be presented and to highlight them with anecdotes and dramatization. Stettner feels your presentation should contain no more than three main points, an opinion worth considering even if you can't always comply with it. His analysis of stage fright and how to deal with it is insightful, while his argument that enthusiasm is one of the most vital components of your presentation may be validated from your own memory of presentations that have most (or least) impressed you.

HANDLING THE MOB

No one wants to face an unfriendly audience, but it can happen, often through no fault of your own. Gilda Carle's narrative of how one speaker handled an awkward situation is instructive and reassuring. Several articles in this section stress eye contact with an audience, and as Carle shows, you can use visual contact or "eye-alog" as an effective tool for softening negative attitudes among listeners. Eye contact alone is not enough, however—responding to hostile listeners also involves appreciating their point of view and talking with them in language they can relate to.

TYING IT ALL TOGETHER

In the final article, John Baird ties together much of what has been covered in this section. He pinpoints ten shortcomings encountered in public speaking, six of which result from poor preparation. Three potential problems—nervousness, poor delivery, and a lack of concern for the audience—are found in the delivery stage, while handling questions is the final challenge. Each problem is discussed and remedies suggested. Since Baird's article contains a lot of detail in an easily accessible format, you might want to use it as a guide or checklist for your own presentations.

MAKING AN EFFECTIVE TECHNICAL PRESENTATION

The most common criticism I hear of engineers and scientists is that they "don't know how to communicate." This is a very damaging perception of professionals who must work with others to accomplish their objective. Communication—"the ability to explain yourself"—is vital to our success.

My formula for successful communication is:

- Know your subject.

- Know your audience.

- Prepare.

A simple formula to be sure, but like the second law of thermodynamics, its application is more complex!

1. *Know Your Subject.*—To make a successful presentation to management you must be *credible*—which means you must exhibit confidence in yourself and your work. That shouldn't be too hard to develop if you think about it. You spent four or more years in a technical university and have spent months or years working on the project you're reporting. If you have been paying attention, you know infinitely more about the technical aspects of the project than management is ever going to know—or wants to know. Thus you should be confident of your results, and you want to exhibit that in your presentation. In addition, you begin your presentation to management with *credibility*—they hired you and continue to pay you all this money, so they must believe you know "your stuff." Therefore, you don't have to

convince them by endless recitation of technical details, punctuated with technical jargon to *prove* your technical competency. They accept that, so get on with conveying the information *they* want to hear. To do that, you must know who *they* are!

2. *Know Your Audience.*—Knowing your management audience means recognizing their values and interests and their knowledge level relative to your subject. Since I define "management" as having a business bias/emphasis, they are going to be interested in the business implications of the results, seeking answers to such questions as:

- What does it do?

- How does it compare to competitive products or processes?

- What is its cost?

- What are its capital, engineering and service requirements?

- What should we do?

These questions, which define management's interests, can be categorized under four headings:

- Technical results.

- Business implications.

- Technical implications.

- Recommendations.

Note that I list *technical results* first. We should not lose sight of the fact that our trade is technology, and that is why we have been given the group's time and attention.

The second item is *business implications*—of the technical result. Now we are translating from our area of *expertise*, "technology," to the province of others, "business." But someone has to start the process and we are elected. Remember, research and development is "the reduction to

practice of scientific principles" and that implies business constraints.

Technical implications include engineering, manufacturing, service and capital requirements of the results, and magnitude of capital and organizational requirements.

Finally, *recommendations* or the answer to the question, "What should we do?" varies with the specific presentation. Sometimes the recommendation may stop with an identification of the alternatives for action. If the presentation is a status report, the recommendation may be implicit—follow the established program. I am personally suspicious of a memo, report or presentation that does not recommend a course of action. Presumably, everything we do as industrial scientists and engineers leads to a decision point—ours or someone else's. If we don't have a recommendation relative to that decision, I question whether we have done our job. We should not overstep our expertise in making a recommendation, but I do believe we need to stand up and be counted.

3. *Prepare.*—Telling "management" what they want to know is not enough. We must tell them in words and ways that they will understand. It is *our* responsibility to be understood by management—that's why they pay us. To do this, we must select words, structures, models, and examples that are consistent with their backgrounds and technical comprehension level. We can't talk over their heads, but we can't talk down to them either. Knowing the group and knowing your subject is your best bet.

Here is my check list for making effective technical presentations to management:

Objective

Whoever initiated the meeting did it expecting some desired outcome. It is

Robert Wismer is director of research and advanced technology at Deere & Company in Moline, Illinois.

Reprinted with permission from *Research•Technology Management,* vol. 32, pp. 9–10, July-August 1989. © 1989 Industrial Research Institute.

incumbent upon the presenter to understand that objective if he or she is to organize an effective presentation. What will be different? What will happen that would not have happened without the meeting and the presentation? If you can't come up with answers to these questions, you are better off not having a meeting.

Time

Conformance to time allotments is important in effective presentations. The organizers of the overall activity have determined that your part of it has an importance of so much time and that's what they have given you. You may not agree with it, but you either have to change it or live with it. The amount of time you are allocated also suggests whether the organizer wants you to tell them "what time it is" or "how to build a watch." You have to make sure you are on the right wavelength, and the time allocated to you is a strong indicator of what is expected.

Audience Makeup

Find out who will be attending the meeting and what their values, interests and backgrounds are. If you are unfamiliar with those people, talk to someone who knows them—your group leader, department manager—anyone who can give you insight into the individuals' responsibilities, background and "sensitivities." The second time you make a presentation to a group or an individual, you will better know how to do it. I encourage you to become a student of individuals or groups, categorizing their characteristics and interests and their understanding so that you can better communicate with them in the future.

Inverted Triangle

Traditionally, as technical individuals, we have followed the "pyramid" structure of communication. We start at the "peak" with items of least importance and proceed to the "base" with items of greatest importance.

Unfortunately, this approach leaves the audience in the dark as to where you are going. And the audience may lose interest and attention by the time you get to the "good stuff." This has led to the "inverted triangle"

approach in which you begin with a brief summary of the conclusions and recommendations—the most important items—and then proceed to establish the context of the effort in the form of an introduction. Then present the details of the work that was done, the results that were obtained and, finally, a more detailed description of the conclusions, implications and recommendations that resulted from the effort. This approach is similar to the approach recommended by Toastmasters International:

● **Tell them** what you're going to tell them.

● **Tell them.**

● **Tell them** what you told them.

This scheme emphasizes two important principles of technical presentations: You don't keep the audience guessing and you tell them more than once what they really need to know. An entertainer often wants to keep the audience guessing; an engineer or scientist never does. Our responsibility is to convey the significant technical information in an understandable context as concisely as we can. Keeping the audience in the dark doesn't help. We also need to help the audience sort out the most important items from those of lesser importance. Restatement helps accomplish this. Tell them what you're going to tell them, tell them, and tell them what you told them.

Appeal To the Senses

A written report appeals only to our sense of sight. An oral presentation with visual illustrations appeals to both sight and sound. But how about touch? The use of models, physical displays, parts, components—anything you can hold, heft, manipulate—can add to the comprehension of a group that is relatively unfamiliar with the subject. For example, our group has made several presentations on a new combustion engine. At every one, we have placed on the conference room table parts from this engine, along with any physical representation of the engine which would help our executives understand the parts we were talking about and how they related to the operation of the engine.

We have begun using videotapes

regularly in our presentations, and find them extremely helpful in conveying information that is difficult to understand without seeing it in action.

Another option is to actually hold the meeting and the presentation in the middle of the subject. If it's a bench test, conduct the meeting in the laboratory. If it's a prototype, gather everybody together indoors or outdoors, whatever is appropriate for the subject and the time. Appeal to as many senses of your audience as you can. It is very difficult for us to comprehend what management people understand and don't understand about a technical subject. The more senses we can appeal to, the more likely we are to convey our message.

Review with Others

No matter how many presentations we make, we are never absolutely sure of the effectiveness of an oral presentation until we try it. So, find a friend or someone whose opinion you respect and who is not that familiar with your project or subject, and try it out.

Practice, practice, practice

I suspect it's possible to overpractice or overprepare for the presentation to the extent that you're bored, but I've never experienced it. Practicing your presentation makes you intimately familiar with the subject, order of presentation and word choice. All of this will increase the clarity and conciseness of your presentation. You are less likely to get things out of order or to struggle for the right choice of words. The speech consultant we've used in the past recommends going through the presentation seven times before you actually deliver it. That seems like a lot, but it works. Even if you don't get to seven, at five or six, or even four, you're going to be more effective than if you only practice once or twice.

Remember—*communication* is the *ability to explain yourself* and explaining yourself involves:

● Knowing your subject.

● Knowing your audience.

● Practice, practice, practice. ▲

A Good Speech Is Worth a Thousand (Written) Words

BERT DECKER

Abstract—Management often thrusts the role of speaker on people untrained in that art. Although they know that printed words are effective for communicating information and data, they seldom realize that spoken words may have little to do with the effectiveness of a presentation. More important to believability are the visual and vocal characteristics projected by the speaker. Several suggestions are given for becoming a more effective speaker.

TO obtain desired results and actions, information processing professionals must be able to communicate. Speaking effectively and delivering plausible presentations can make the difference and is a good place to start.

In any management situation, it's critically important to touch the emotions and the senses. Listeners will not react and won't be moved if they are given nothing but literal information. Unfortunately, managers often ignore the basics. They don't realize how their own nerves help cause those "blahs."

While technology has catapulted organizations into an information maelstrom, the human capability of communicating has moved at a snail's pace. For each technological advance that distances one person from another, there is a corresponding increase in the need for human contact.

The most important skill any manager can have is the ability to verbalize and motivate people—to put ideas into action. Yet fewer than one percent of the business people in America have done anything to improve their skills.

Thomas A. Murphy, former chairman of General Motors Corp., once said, "Few of us are trained in public speaking; it isn't how we got where we are. But when you reach the general-management level, you are thrust into that role."

There is nothing complicated about speaking. It comes naturally to everyone, but too many business people approach it the wrong way.

NOT A WRITTEN REPORT

If there is one critical concept that the majority of business people have not realized, and one essential for verbal effectiveness, it is that a presentation is not a written report.

Reprinted with permission from *Data Management,* September 1983, vol. 21(9), pp. 30–31, 55; copyright 1983 by Data Processing Management Association, Inc., Park Ridge, Illinois.

The author is president of Decker Communications, Inc., 999 Sutter Street, San Francisco, CA 94109; (415) 775-6111.

The printed word is the most effective tool for getting across data and information. People can read four times faster than they can speak. Business traditions are based on the rational, logical approach of the sequential printed word; so is the educational system. But a live, verbal presentation is much more than a linear progression of words delivered through the mouth.

When a person is speaking, many other things are going on. Studies have proven that when nonverbal messages are inconsistent with words—even contradictory—people believe the nonverbal over the words. The speaker is the medium of the message; and to a very large extent, the speaker *is* the message. If a speaker grasps a lectern, gazes up at the chandelier, displays signs of nervousness, and begins a speech with "My associates, my good friends, ... ," no one in the audience will believe he or she is their friend. Actions speak louder than words.

RICH BRAIN, POOR BRAIN

One way to understand speaking is to recognize that it is largely a right brain function. So are seeing and listening to a speaker.

"... And that government of the people, by the people, for the people, shall not perish from the earth."

Reprinted from *IEEE Trans. Prof. Comm.,* vol. PC-27, no. 1, pp. 32–34, March 1984.

Much has been written in the last four years about right brain/left brain processing. What is important for the business executive to understand is that everybody thinks in both modes, but by knowing what is to be accomplished, the speaker can achieve more success by using the strongest mode for the goal at hand.

The right brain handles intuition, emotion, and holistic methods of the thinking and decision-making processes. This is where such inputs as sound, color, movement, and patterns register.

The left brain (usually located in the left side of the cerebral hemisphere) processes information analytically, logically, and linearly. Reasoning and scientific data are stored and categorized by the left brain. The written word—reports, budgets, analyses, arguments, and even written speeches—is processed through the left brain.

Recent studies have indicated that top executives are largely guided by right brain activities—they score high on intuition and prefer oral reports to written reports because they can "get a feel" for the person presenting. "Seeing the whole picture" is a right brain function, and the best managers operate from that perspective.

Speaking is not a ping-pong game of words. People communicate mind-to-mind in person, not word-by-word. It's not just the words, but *how* those words are said. More than that, it's how the person looks, sounds, and everything he or she does. They all register—either positively or negatively. So it's important to know what counts the most for the message to be accepted.

THE THREE V'S

There are only three elements to any spoken communication:

- Verbal—the words or content—what is said
- Vocal—the voice expression, resonance, tempo, and inflections
- Visual—all that is seen by others: How the speaker looks, eye contact, posture, gestures, and facial expressions.

Dr. Albert Mehrabian of UCLA startled the traditional world of communication (in both education and business) with the discoveries published several years ago in his book *Silent Messages*. One of the foremost communication researchers in the country, Mehrabian measured the impact and believability of the spoken message according to these three elements.

His important findings: verbal—only 7 percent; vocal—38 percent; and visual—a whopping 55 percent. These figures contradict what has been taught in the schools and has carried over into business communication.

BECOMING EFFECTIVE

Very few managers are operating at their optimum level of communicative ability. Most managers block their message, get bogged down in the detail of words or facts, and forget that they are both the delivery system and the message itself. Even those managers who are outstanding decision-makers, administrative whizzes, and financial geniuses often fall short when they stand up to speak.

Following is some personal advice on how to go about becoming a more effective speaker, to deliver ideas with impact.

DON'T READ A TALK

Hiding behind the security of a written script may make *you* feel better, but not your audience. More important, it's not very effective.

When you are speaking from notes or extemporaneously, personally making your point, the audience can see that it comes from you. A written speech might be analytically correct, but most people will assume the words aren't yours. Voice takes on a reading-aloud incantation—often a monotone—and your eyes are chained to the page too long.

Don't Think They're Going To Get It Just Because You Said It!

Most managers erroneously think that because they give five major points in a presentation, their audience will remember them all. Not so!

You'll actually be lucky (or skilled and effective) if they remember your point of view, your key statement, and perhaps one or two supporting points. The truth is, they'll remember a good story or an anecdote. They'll remember a vivid litle detail—such as the color of the socks you said you wore on your first job interview. Abstract facts and data will go in one ear and out the other.

Since nobody can go back and reread your statements, you have to make sure the audience gets your main points. Don't try to overload them. Remember *KISS'M*: Keep It Simple, Stupid—and Memorable.

Be The Host, Not The Roast

When you're at the lectern, you're in charge of your audience's well-being. You are momentarily the host—not the main course they want to devour—so don't let your nervousness show.

If you have butterflies (all speakers do) don't assume everyone knows it. As a matter of fact, they don't. If you think your voice is quavering or your knees are shaking, chances are nobody else notices or cares. Simply release your knee lock and let your natural energy carry you through animated gestures and movement.

210

Look at your listeners the same way you would in your office, with "extended eye communication." This goes beyond the traditional notion of eye contact: Look at one person for four to five seconds, then move on to someone else at random, covering all corners of the group. By looking at people with extended eye communication, you show you're interested in communicating personally with them. They feel it, and you feel it.

Get Feedback

Speak as often as you can and use both audiotape and videotape to find out what others are hearing and seeing. Forget about asking friends and associates how you did. Most will say "Wonderful!" and mean it—they're too close to you and your goals to appraise objectively.

And don't tear yourself down in audio or video replay. Most speech teachers are counterproductive in telling people what *not* to do rather than pointing out their strengths as well as their weaknesses. Use the tools available. We are in an electronic age and there is no better learning tool than video, when used constructively by professionals.

Discovering bad habits in yourself does no good unless you know what to do about them, what specific steps to take for your style of speaking. You are your own best teacher.

Use Visuals, Be Visual

Since 85 percent of what we know has come through our eyes, it's obvious that we learn more from what we see than from what we hear or touch; people remember images far longer than facts. They will remember the image of you, too—animated or flat, energized or rigid, enthusiastic or dull.

As visual aids, overheads (transparencies) are probably the most versatile presentation tool, with 35-mm slides and flip charts also used for the right situations.

Keep simplicity in mind: One thought per page, three lines per page (or slide), bold graphics and color. Visual aids should highlight key points, not tell the whole story. Never put yourself in the dark or half-hidden behind equipment. You are the host. Think of your visual aids as a good waiter—there to assist you, not replace you.

Self-Confidence!

These are just a few essential guidelines for speaking effectively.

Tips, techniques, and ideas are great. Anyone can find a million more in books, but if you don't use them, they're about as valuable as a mail-order course on tap dancing if you don't get up to dance.

What counts is experience and self-confidence. Speak at every opportunity, get the kind of expert feedback that's immediately useful for your environment, and keep extending yourself.

More Tips for Technical Speakers

Friends, Romans, Cost Engineers . . . Can We Talk?

by Sara Pritchard
AACE Communications Manager

If you're going to be a speaker at AACE's 34th Annual Meeting in Boston, then June is just around the corner. You've already submitted the final draft of your paper, and hopefully you consulted Jim and Janet Stevenson's article, *How to Write and Present a Technical Paper,* in the November 1988 issue of **Cost Engineering;** watched their video; or attended their workshop at the 1988 Annual Meeting in New York. *How to Write and Present a Technical Paper* tells you how to choose a topic, how to organize your ideas and actually get down to the nitty-gritty of writing your paper, how to plan your visual aids and integrate them with your presentation, and how to prepare for your hour at the podium.

But, what will it take to give a great presentation? What makes some presentations great and some just ho-hum?

When it comes to technical speakers, what separates the cream from the milk, the superior from the good? Is it the topic? Is is the speaker? Is it the visuals? Is it a combination of everything?

AACE asked five of its top technical speakers (based on audience evaluations from the 1986-89 annual meetings) to share their secrets and tips with other speakers and with **Cost Engineering** readers. In this companion article to *How to Write and Present a Technical Paper,* we present their responses: tips from the pros: the inside dope on delivering winning technical presentations.

"He whose face gives no light shall never be a star"

(William Blake)

What do you think our experts said was the most important element of a great presentation? You may be surprised to hear that it's not the topic alone as much as the **ENTHUSIASM** of the speaker for the topic. A hot topic and great title may draw you a big crowd — once, but if you don't do your topic justice, that crowd won't come back next year. On the other hand, no matter what you're talking about, if you acquire the reputation for being a good speaker (as our "experts" have), you'll continue to draw participants to your presentation just on that reputation of consistently giving a good presentation.

The only way to give a superior presentation is to be certain that your material or your approach is superior. The only way to show enthusiasm is to *be* enthusiastic about your presentation. Your choice of topic and your enthusiasm go hand and hand.

Be Enthusiastic About the Topic You Pick — When you choose a topic, you will consider, of course, who your audience will be. More importantly, though, think of yourself. Choose a topic that interests and excites *you.* Certainly you will want the audience to share your interest and enthusiasm, but at this preliminary point, the audience should not be your main concern. How can the interest of anyone be piqued or enthusiasm stirred if you are not "gung ho" on the topic you're presenting? Enthusiasm is con-

tagious, and it is hard not to enjoy a presentation by an ardent speaker, so pick a topic that turns you on.

When preparing your paper or presentation, our experts agree that you should try to (1) present some new and unique aspect of your profession and (2) make a definite contribution to the techniques and practices of your profession. Technical growth would be hampered if we stuck to just popular subjects. A quality paper must have some new or unique material. The topic can actually be old as long as the approach to that topic is unique. A restatement of the work of others or a compilation of statistics is OK *only* if it is a unique perspective or a novel interpretation of data. The material presented should force the audience to look at something in a new light or from a different angle or to think about something a different way. This new viewpoint will hold your audience's attention. As the speaker, you can be confident that you are making a contribution, and the audience will leave with a new knowledge or perspective.

One expert told us he always likes to be iconoclastic — to attack a cherished and unchallenged belief — in an amusing and thought-provoking way. It is important to offer suggestions, opinions, and ideas that are a little different from the conventionally accepted. Neither you nor your audience profit from a simplistic statement of the obvious. Consider how you can generate difference of opinion, controversy, and debate — all of which are fertile ground for discovery. If you are reluctant to nurture a new thought or to

Reprinted with permission from the American Association of Cost Engineers (AACE), *Cost Engineering,* vol. 32, no. 2, February 1990, 17–20.

take a controversial position, remember Christopher Columbus and recall that virtually every great idea was greeted first with skepticism and, often, derision. Audiences tend to relate well to authors who take risks with ideas and approaches. (Many people are conservative, suspicious, and defensive about change but admire those willing to challenge a fixed system). Even if you have an idea, a method, etc, that is not proven, you could present it as speculative.

"Float like a butterfly. Sting like a bee."

(Mohammed Ali)

Style. It may not be everything, but, whether they're willing to admit it or not, our experts have it. Their presentations are lively, informative, and interesting, and they present their material in a relaxed, congenial manner. They are enthusiastic and show it. They are animated. They gesticulate. They enjoy themselves. They may use a little bit of humor. They are *light*.

When you give your presentation, make liberal use of slides, graphs, tables, etc. Draw your audience into your presentation by asking rhetorical questions. Remember also to leave time for questions from the audience. If there aren't any for you, *ask* some. A question is also a good way to start your presentation. Ask a question, and straightaway you involve the audience. And, even if your question is rhetorical, your audience has the immediate sense that you are not talking *at* them but *to* them.

One of our experts told us that he sees the presentation of a technical paper as nothing more than a specialized product that a vendor is trying to sell to a consumer. The vendor in this case is the author/speaker, and the consumer is the audience. Thus, in selecting a subject, the author should ask, "What will sell?" "What title will look inviting?" The paper and the presentation are the two products to be sold. How good a sale you make will depend on (1) the quality of your product (how good your paper is), and (2) your "marketing" techniques (how well you present your paper), which include

your use of visual aids (slides/figures, graphs, etc) and the rapport you establish with the audience through your writing/speaking style.

Remember that although the subject matter of the paper and the presentation are the same, the presentation sequence does not have to be the same. Your presentation may contain more background, examples, etc, and will not necessarily include every point in the paper. When you are at the podium, you can follow the same outline for your presentation as that of your paper, but *be flexible*.

Don't read your paper!
Don't read your paper!
Don't memorize or read your paper!

We cannot say this enough. The only people who get away with standing at a podium and reading their work verbatim are poets. Look at it this way: If you are going to read your paper, you may as well walk into the room where your presentation is scheduled and tell your audience, "Please open your **Transactions** to page xx and follow along while I read aloud my paper on (whatever)." Does this technique ring a bell? Did you ever skip school?

Only a comedian could read a technical paper and keep the audience's attention, and the result would be fine for **Saturday Night Live** but not our annual meeting. So, please. . .

A way to avoid the pitfall of reading or memorizing and reciting your paper is to make yourself cue cards from your outline. Don't write out complete sentences on your note cards, just key words. That way you won't be tempted to read. Practice presenting your paper with the cards as your prompts. If possible, tape yourself with an audio or video tape. Instant replay. Where are the rough spots? Do you need more prompts? Does the order of the presentation feel natural? Are the transitions good? If not, rearrange your cards. Don't think you have to stick to your paper exactly.

Also, slides and other visual aids can be excellent prompts. One of our experts told us that he prepares himself a script of his presentation. This script has miniature pictures of the slides he is us-

ing, and beside each picture are reminders about what points he wants to make or what questions he wants to ask. Any veteran speaker will tell you that the use of slides or any other visual aids lends a relaxed atmosphere to the presentation. It focuses attention away from you directly, and it gives you something to do rather than just stand in front of a lectern.

These tips should help you keep your audience interested and enthusiastic about your topic and make your presentation more enjoyable for them and *you*.

"Say what?"

Don't be too technical. All our top speakers say it. Albert Einstein once advised that nothing should be explained in such a way that it could not be understood by an intelligent 12 year old. When you write and present your paper, remember that 12 year old. Look at your audience and imagine that each one of them is 12 years old.

Yes, sometimes you may *need* to be technical. Don't sacrifice knowledge for simplicity. But, whenever possible, use the simplest analogies to explain concepts and phenomena. We learn well through analogies, by establishing relationships between things, watching patterns emerge. We learn best by analogies that bring together seemingly disparate things. (We remember how peculiar are the striking differences with a common denominator.) Also, we retain information longer if it is presented to us in small packages. So, if you are presenting technical information, think of your audience as a flock of pigeons, not a pride of lions. Don't throw them big slabs of information and expect that information to be swallowed in one piece and digested. Give them little bits of information. Bread crumbs. Popcorn.

Concerning the problem of very technical presentations, one of our experts told us, simply, "be real." Sure, some members of your audience will want to talk theory, but for the most part, people want to learn practical information they can use. Give them something that has a real, working application. There are fewer professional philosophers than practicing cost engineers.

If you have a choice between explaining something with an equation or with an analogy or example, what should you do? Trick question. Do *both*. Remember, there are all kinds of people in your audience. Some are visual learners, some audio; some are "number people," some "word people." Some have difficulty remembering phone numbers; some can never remember a joke. Some are right brain dominant, some left. Your job is to reach them all. If you have to use complicated diagrams or equations, refer the audience to the appropriate page in **Transactions**, or distribute a worksheet. *Never* expect your audience to follow you if they can't see what you are talking about, and never expect them to actually read what is on a slide or overhead transparency. Use whatever you can to enhance your presentation: a video, slides, overheads, handouts, a 3-D model, a demonstration, a short exercise, a pop quiz, a volunteer from the audience. Think of yourself for that one hour as a teacher, and bring as much ingenuity and **enthusiasm** (there it is again) to your class as possible. And, make everything as uncomplicated, simple, and clear as possible.

"What's in a title?"

What's in a title? Plenty. Believe us. Remember taking those reading comprehension tests in high school? Here are the directions: Read each paragraph and then pick the title that best suits the topic. Well, the "paragraphs" were blocks of minuscule, justified type that looked like 8" x 10" bricks; the topics were always something like operating mining equipment, the synthesis and antithesis of polypeptotes, or sewing machine repair; and the titles were always . . . *Operating Mining Equipment, The Synthesis and Antithesis of Polypeptotes,* or *Sewing Machine Repair.*

Listen . . . , you don't have to be so literal. Put a little spunk, a little panache, a little dash, a little chili powder into your title. A little alliteration always draws attention. Too much is corny. Here's a chance to be witty and have some fun. It's like writing a jingle for a commercial or entering a product-naming contest.

Be as creative as you want, but make sure the title conveys what your paper is about, that it targets the audience you want. One case in point: Some years ago a paper, *Don't Raise the Bridge, Lower the River,* was presented at AACE's 12th Annual Meeting. Intriguing title, no? But, what's the paper about? Without the subtitle, *Value Engineering for the Cost Engineer,* no one would have the foggiest idea. So, if you don't want to sully your great title with technical identifiers, please add a subtitle for the sake of meeting participants choosing what sessions to attend and for indexers of **Transactions.**

Picking a title, choosing the *exactly* right words is not such an easy or trite matter as it might sound. Finding the perfect title actually can be one of the most complicated thought processes because what you are automatically doing is analyzing and sorting through *all* of the information that you are going to present. Titles are organic. You'll find that one will change from day to day until you finally get it just right; some will change the whole time you're writing the paper itself as the paper "works itself out." The search for a title can be a very rewarding process that leads you deeper and deeper into the subject. A great title can give you the angle and the impetus to write a better — a great — paper. Don't pick your topic, plod through your paper, then try to come up with a winning title. Find that great title before you ever begin. You may even find that the great title changes your thesis — gives you an angle you never really considered in depth.

Zen and the Synthesis and Antithesis of Polypeptotes?
How To Operate Mining Equipment and Influence People?
B O B B I N.

"Who, me?"

You never studied elocution. You failed Public Speaking 101. When you have to speak in front of a group, you feel like an endangered species. Your legs are tottering skyscrapers of jello. Your throat is coated with a 1-inch patina of Bon Ami.

It's you talking, but that voice . . . it sounds like . . . is it . . . Jiminy Cricket?
So what!

Speaking in front of a group is the *number 1 fear* of many Americans. It's probably not *the number 1* fear of our experts, but, would you believe . . . even these guys, some of whom are at the podium 10 to 30 hours a week, get nervous?

Fear and nervous anxiety come from the same source as all our other emotions. Of themselves, they are neither good nor bad; how we interpret them determines their worth. Many actors and other performers have come to recognize that the feeling they know as stagefright is a positive, energizing force that gets them "geared up" for a performance. It has been described as "battery charging" or as an energizing feeling that makes you superalert. For those of us who don't perform in public often (like most of our annual meeting speakers), this feeling is labeled *"Fear,"* (with a capital "F"), and we all have been conditioned to think: *"Fear is bad."* Thus, we become doubly anxious. But to veteran performers, this feeling is well known and welcomed as an energy that makes them do well. After a while they learn to dissociate the anxiety that unnecessarily accompanies it and to feel only the "high." Welcome your stagefright. Shake hands with it and thank it for helping you get ready for your star performance. Once you begin to speak, it will take a back seat.

(Misery Loves Company, Strength in Numbers, or What to Do if Your Mother Cannot Make It To Your Presentation) — Burns and Allen. Gilbert and Sullivan. Simon and Garfunkle. Great teams. If you want to present a technical paper but don't want to do it alone, why not follow the example of our own Jim and Janet Stevenson and give a joint presentation? There are innumerable possibilities and benefits. You can give a joint presentation with someone who has a totally different opinion of the topic you want to present. That way you can show both sides of the coin. You can play devil's advocate. You can present a formal or informal debate. This kind of presentation is sure to invite interest and comments from the audience. Or, you can give a joint presentation with

someone who shares your point of view. The two of you can "play ball," taking turns "serving" different parts of the paper. This "buddy system" approach is great not only for beginners but could blossom into a beautiful technical speaker relationship. ("Come with me to the annual meeting. We could make beautiful CPM diagrams together.")

Coauthoring is also rewarding because each coauthor has his/her own personal editor in the form of the other coauthor(s). Invite someone you know or think you could work well with and whose work you admire to coauthor and present a paper with you at the next annual meeting.

Bringing It All Back Home

That's it. Five rules of thumb:

1. **Pick a topic that you can be really enthusiastic about and *be* enthusiastic when you give your presentation** — If there's nothing exciting going on, challenge some long-held theory and see where it takes you. Think it through, and make sure it's a workable topic. Be ready for some opposition. You can take it.

2. **Develop a positive attitude and relaxed style.**

3. **Don't be too technical.** Remember Einstein and the 12-year-olds.

4. **Promote your presentation. Lure** those meeting participants into your presentation **with a great title** they can't resist.

5. **Don't worry about being nervous** — Have a good time. Instead of worrying about being nervous, go ahead and be nervous! It's natural. It's good. Remember: Once you start your presentation, the nervousness will subside because you don't need it. Think of these two quotations:
   (a) *"That which has not killed me will strengthen me."* (Fredrich W. Nietzsche). You'll be a better person for giving a presentation. Guaranteed. Ask our experts.
   (b) *"It's just a movie."* (Gil Scott-Heron).

"Why Bother?"

And . . . what should you get out of all this? Satisfaction. The satisfaction of having taught something well and of having learned in the process. As a technical speaker, you have the tremendous opportunity of speaking to a *live* audience, not a camera, not a mirror. This audience is not your enemy; they are your allies. Your friends. People like you and me — fragile, ephemeral beings trying not to combine with the elements on this marbled jaw breaker we call "Earth." Invite these friends to participate. Welcome their questions and comments and their criticisms. Don't alienate them, but draw them closer and let them help you enhance your presentation. And . . . when it's their turn, be attentive, learn from them, and give them all the help you can.

Special thanks to Michael Curran, Richard A. Mazzini, PE CCE; James M. Neil, CCE; Linda Ramme; James J. Stevenson, Jr., CCE; Richard E. Westney, PE; and Charles P. Woodward, PE.

The Engineering Presentation—Some Ideas on How to Approach and Present It

RONALD C. ROSENBURG

Abstract—Achieving a successful presentation begins with the consideration of three things: what to say, how to say it, and how to conduct the presentation to convince an audience that you mean it. Determining what materials, topics, and details to use must be based on an analysis of the potential audience, their sentiments, interests, technical disciplines, and possible responses to what would be presented. Presentation material must be put into a unified, professional, and easily understandable format and thoroughly rehearsed so that it can be delivered in a manner that instills confidence that the subject has been well-researched. The author presents some strategy, guidelines, and pitfalls, based on experience, for consideration toward these ends.

MOST of the time we are given only one chance to present our ideas. If this is not done effectively at that time, the success of any future communication on the same subject could be jeopardized or, even worse, willingness of the audience to ever listen again to a particular speaker who "came across" as ill-prepared, boring, or incompetent could be destroyed.

For communication between speaker and audience to be successful, it must be the culmination of a preparation process started long before, one which included a carefully thought-out "speaker–audience interaction scenario." Such a scenario hypothesizes the type of audience expected, their sentiments, prejudices regarding the subject, orientations (political beliefs or group structure, for example), technical disciplines, personal interests, and questions that might be raised.

Based on the results of this audience analysis, you can determine *what* and *how much* must be said to satisfy both your and the audience's needs. When this is accomplished, organize the information into a suitable format and take it through as many full-scale dress rehearsals or dry-run presentations as necessary to become confident of convincing the audience that you know what you are talking about. To help achieve these goals, some guidelines, pitfalls, and ideas on strategy, based on experience, are presented.

PREPARING THE PRESENTATION

Determining What to Say

1. Thoroughly research what the presentation is to cover, from a general and a contractual viewpoint. Know what your audience *has* to see and hear and what information they *would like* to receive.

2. In researching your contractual documents, *extract all of the categories you are obliged to cover*. You may be surprised at some of the items at a subtler level that are specified in an applicable military specification. For example, in preparing an aircraft avionics presentation, I compiled material on the major and commonly thought-of topics such as displays, signal acquisition techniques, and processing capability but was surprised to find that I also had to address such topics as fungus control and acoustic noise limitations.

3. *Cover all areas*, even if very briefly, and specify them on your agenda. It's better to be complete and include them than to be criticized later for their omission. Your audience may know of the topic, however obscure, but generally they will not know what they are supposed to evaluate or question you on. They will depend on you for that information. If you cover it with details *you* feel are satisfactory, the chances are that no one will want any more.

4. After formulating ideas as to what is expected and what must be covered, contact representatives of your potential audience for further information and confirmation of your own ideas and plans. Again, find out what *they* expect to see and hear and what they would *like* to hear. You can readily gain the appreciation of your audience if you "just happen" to talk about an area they are interested in, or one they may have to cover in a future presentation of their own.

5. *Research your audience's past correspondence* on the subjects you will be covering and be knowledgeable enough to discuss them in depth. Understand their reasons for generating such correspondence so that you will be able to direct any discussion that may arise.

6. It is very important that your presentation *answer your audience's specific questions and concerns*. If possible, gear the material and language of the presentation in that direction. Have viewgraphs presenting *data* that address their specific questions. Try to think about other questions your presentation may raise, and be prepared to answer them.

7. *Thoroughly research all current and past problem areas*. Don't let your audience be first to introduce you to them. Get the latest and complete story on any problem. If it

Received April 7, 1983; revised August 25, 1983.

The author is a senior project staff engineer at the Grumman Aerospace Corporation, C11-40, Bethpage, NY 11714; (516) 752-3651.

Reprinted from *IEEE Trans. Prof. Comm.*, vol. PC-26, no. 4, pp. 191–193, December 1983.

involves hardware, go to the manufacturer for first-hand information on the nature of the problem and on what is being done to correct it. Maybe even include a discussion of those items in your presentation. In this case, however, ensure that for every problem you have either a solution or a direction you will take toward its resolution.

Putting the Information in Presentable Form

At this point in the preparation, even the most well-versed person, with a wealth of information to present, has, at times, fallen flat on his or her face!

1. At the outset, generate a detailed fact sheet covering the subject of the presentation. For example, if a work status is to be presented, outline all efforts to date; if it is a design concept, list design requirements, assumptions, trade-offs, and rationales. Review your material for correctness, clarity, and applicability. Next, extract only the highlights of this information to use as viewgraph points, making a separate (general) viewgraph for each subtopic and ensuring that there are no more than six points on each viewgraph.

2. In general, *have two viewgraphs per subject*—one general and one detailed. You may never use the latter, but it could be a lifesaver should the need for more information suddenly arise. The general viewgraph should contain only terse statements of fact to key your memory on what to say. The detailed viewgraph, on the other hand, should be slightly "overstuffed," especially with pertinent backup data that cannot be easily memorized. For example, the design review requirements for a particular Air Force avionic development contract called for the contractor to summarize the overall design requirements together with his concept for each. In this case, the general viewgraph contained a summary of the eight major requirements and was backed up by eight detailed viewgraphs presenting laboratory and analysis data. Although the speaker was quite knowledgeable, he was not ready for the barrage of inquiries with respect to "hard numbers." He was able to recover, however, with the help of one of the detailed viewgraphs he hadn't really planned to present.

3. *Make good quality viewgraphs.* Audiences tend to lose interest in faintly observable, hand-written viewgraphs. This is followed by loss of interest in the presentation itself. Make sure that viewgraphs are typed, all with the same typewriter.

4. *Prepare viewgraphs well in advance.* It is better to redo one or two of them than to not have any completed on time. Even having to do all of the viewgraphs over (as a result of, say, a dry run) would not cost much, especially compared to the business that might be lost by not getting your point across. Remember, *you only get one chance at the final presentation—have everything perfect for that time.*

5. *Always have an introductory viewgraph* before you get into the main part of your presentation to clearly tell your audience *what you intend to cover.* Never start off right into the "meat" of the subject as I saw one person do in an attempt to review changes made to a receiver design. The all-important *reasons for* the changes were lost, together with most of the audience, in an explanation *of* the changes. Close with a viewgraph telling the audience *what you have told them.*

6. *Have at least one dry run with the final presentation material,* and have as many knowledgeable people as possible present to critique it.

Notwithstanding the fact that the value of the dry run has been proven over and over again, I have found from personal experience, and through discussions with others involved in various presentations, that many organizations appear to be lax in this important area. Dry runs are always planned but often, due to lack of time, are cancelled or reduced to only a quick last-minute review. *Dry runs should be mandatory for all presentations.*

7. In the dry run, go through each viewgraph in *exactly the same order and manner you will formally present it.* Don't skip areas because you think they are of lesser importance. What you might think is insignificant or will be easily understood by your audience may be just the area that causes problems. Don't wait until the actual presentation to find out about it.

CONDUCTING THE PRESENTATION

1. The most important thing to remember is that *you must, from the outset, instill in your audience confidence that you know what you are talking about,* that you will be proceeding in the right direction with your work, and that, ultimately, you will give your customer what he or she wants.

This is extremely important, for the degree of success you achieve will manifest itself in the effectiveness and ease with which you are able to conduct your work in the future.

2. *Know your audience.* If your audience will contain a majority of people from a particular area (flight crews, for example), expect many questions and discussions in that area (displays and "knobology," in this example). Be prepared so that you will have to defer only a minimum of questions. The more you answer during the presentation, the fewer action items you will have to worry about later.

3. *Don't dwell on what may be boring or not universally interesting topics.* People tend to go into unnecessary detail explaining things they are very knowledgeable about. I attended a presentation to the Air Force in which one speaker went into a long and precise technical explanation of an avionic system. His knowledge of the subject was excellent but the details he presented were unnecessary. When he finished, a number of Air Force personnel gave him a standing ovation, partly because of his expertise, but mainly because he finally finished! *Avoid this pitfall.*

4. *Don't engage in arguments or any other confrontation with your audience.* Stay calm and unflustered—postpone (if necessary) touchy discussions to splinter groups.

This advice may sound unnecessary but I was, unfortunately, witness to such a confrontation during an initial engineering design presentation, where the thrust was really to establish the confidence I spoke of earlier. Needless to say, it did not help us toward that goal; nor did it help the people scheduled to speak next.

5. *Speak loudly and actively throughout the presentation.* Too many people start loud only to become inaudible either toward the end of the talk or toward the end of each sentence.

6. *Have handouts* consisting of at least a topic outline of your presentation if not a copy of your entire set of viewgraphs. Handouts of data are also desirable.

7. *Have enough handouts.* Make sure that there are enough to go around. I knew one individual who was insulted and subsequently appeared hardened to the presentation because he didn't receive a copy of the handout. Such a person may be rare but sometimes this may be the voice that counts.

8. *Never allow your presentation to be a reading of your viewgraphs.* A person who does this only proves two things: that he or she can read and that he or she is the wrong person to give the presentation.

9. *Always maintain a "lecturer" stature during your presentation.* You should strive to maintain a speaker–listener relationship with your audience in which *you* are the one giving the information for their review. Do not fall prey to a barrage of questions that puts you on the defensive. This will destroy your presentation and violate item 1. Once you lose the speaker or lecturer role, it is difficult to fully regain.

A FINAL COMMENT

After all is said and done, remember that your audience is human, also. They are just as anxious to learn about the subject you are presenting as you are to teach them. The easier you make it for them, the easier your task will be, and the more favorable an impression you will make. Always try to *put yourself in their place*, both when you prepare and when you deliver your presentation.

Improving Oral Marketing Presentations in the Technology-Based Company

MICHAEL F. WARLUM

Abstract—Oral presentations by a company's representatives affect its reputation and competitive position. Typically, oral presentations exhibit certain shortcomings. These can be overcome by the application of good developmental techniques. Recognizing the importance of good oral presentations, many companies are providing their employees with professional help in developing them. It is recommended that all companies address this issue.

THE NEED FOR ORAL PRESENTATIONS IN BUSINESS DEVELOPMENT AND MARKETING

A factor vital to the health of any corporation is the constant exchange of accurate information. Although traditional, written forms of business communication remain important, the oral exchange of information occupies an increasingly important role in the functioning of today's companies.

The fast-paced business climate demands that this oral communication, whether within the company or with the customer, be clear and concise. Misunderstandings can be extremely expensive in both time and money. Contracts can, and have been, lost at this point.

Technology-based corporations are no exception to the growing emphasis on oral presentations. Management briefings, design reviews, product demonstrations, marketing and business development efforts, and other requirements for oral communication are standard practice. Oral presentations are assuming a pivotal role in government contracting decisions. Today, and for the foreseeable future, the ability to present oral information successfully is integral to corporate and individual success.

ORAL PRESENTATION PROBLEMS IN TECHNICAL MARKET DEVELOPMENT

Perhaps the greatest impediment to oral communication in the technological corporation is that most engineers and other technically educated individuals are not specifically trained to plan, produce, and deliver oral presentations. As a result, many do them poorly. Client interest declines, and proposal scores suffer.

Several barriers to effective oral communication are seen consistently in the technological community. Included among these barriers are failure to analyze the audience, develop a logical outline, use effective visual aids, or employ appropriate delivery techniques. Lack of such careful preparations usually spells poor reception.

Inattention to audience analysis lies at the heart of many ineffective oral presentations. To ensure success, certain basic facts about those in attendance must be determined. What level of detail is appropriate to the particular audience? What aspect of the subject is of most interest or relevance to the members of that audience? Are they likely to be hostile to the topic, to the speaker, to the point of view the speaker represents? Do they need to be persuaded or do they come seeking information eagerly? The answers to these and related questions about the audience have an immense bearing on how any presentation is crafted and delivered. Figure 1, a checklist for marketing speakers, can result from effective audience analysis.

As a result of their training and experience, technology-oriented presenters typically focus on technical content rather than on communication. "We'll blow them away with our technological approach," is a commonly heard statement. There are instances in the technology marketplace where such an attitude may be appropriate, others in which it can lead to disaster.

Put simply, the underlying purpose of audience analysis is to discover common ground between the speaker and the listener. What do members of the audience already know that the presenter can build upon to give new information meaning for them? If this building process does not take place, chances are that communication will not take place either.

Sometimes the presentation is not organized logically. Again, a building process has to occur. The presenter must build a case, keeping in mind that good speaking, like good writing, has a persuasive element. Too often the speaker approaches the platform with a vague idea of the

Michael Warlum is employed by The Boeing Company, Seattle, WA, where he acts as an advisor in oral presentation development.

Reprinted from *IEEE Trans. Prof. Comm.*, vol. 31, no. 2, pp. 84–87, June 1988.

SPEAKER'S CHECKLIST

Number of View Graphs: _____   Manager: _____
Duration (minutes): _____   Video Budget: _____
Presenter: _____   Video Manager: _____

Section	Treatment	Yes	No	Remedy
INTRODUCTION	All messages there? All parts explained?			
CLIENT THEME #1	Our reply			
CLIENT THEME #2	Our reply			
CLIENT THEME #3	Our reply			
CONCLUSIONS	Summation? Positive sales message?			

Next Meeting: Action Items:

Figure 1. Checklist for Marketing Speakers

points to be covered and does so, more or less, treating each item as though it were a discrete subject, not part of a total story. The result is apt to be audience confusion. We simply cannot afford to invest hundreds of thousands of dollars on a proposal only to allow the intended listeners—a Source Selection Board perhaps—to go away with an unclear message about what we plan to achieve.

In an equally ineffective variation of this theme, a plan of organization does exist, but it is obvious *only* to the speaker. Unless care is taken to provide plenty of verbal signposts, such as "I intend to make three points. They are ..." or "To summarize, let me reiterate my major points," the members of the average business audience may lose the thread of the discussion.

In the technology-oriented presentation, visual aids have a tendency to become the presentation rather than to act as a support and enhancement to it. The use of unnecessary or inappropriate visual aids encourages the speaker to actually read the words on the charts to the audience, a task most audience members could perform more efficiently for themselves.

Many presenters pride themselves on having file drawers full of visual aids, the remains of presentations they have

given over a period of several years. When faced with the challenge of a new presentation, these individuals pull a few visual aids from here and a few from there, amassing a conglomeration of disparate images. Once they have selected their visual aids, these presenters improvise some words to go with them, never actually outlining the presentation logically to ensure that the message is appropriate to their particular audience.

All too often, technical presenters neglect the rehearsal phase of presentation development completely. No matter how experienced a presenter may be, rehearsal is nearly always necessary to good performance.

Dependence on old habits, such as failing to analyze the audience, develop a logical outline, create appropriate visual aids, or rehearse and deliver the message adequately keeps presentations from being as effective as possible.

Moreover, because they have relied for so long on oral presentations that are not as successful as they could be in either content or delivery, many people do not see a need for change. Unfortunately, their presentations usually show it, and in today's demanding marketplace they frequently show it on the bottom line.

PROFESSIONAL HELP FOR ORAL PRESENTERS

To enhance their competitive stance, corporations must find ways to improve the quality of oral presentations given by their representatives. A number of companies are moving aggressively to offer oral presentation assistance to their employees. In addition to calling on the consulting services that operate in this field, firms are developing their own assistance programs.

The Boeing Company, for example, has established a group called Oral Presentation Development. Using a combination of group sessions, one-to-one counseling, customized written materials, and brief tipsheets, this group provides assistance to any Boeing employee preparing for an oral presentation, whether it is aimed at an audience within the company or outside of it. The service is based on the belief that, to be truly effective, presentation development assistance must be available every step of the way, from conception and planning to final rehearsal and delivery.

An initial step in assisting presenters is to help them analyze the potential audience. This analysis consists of identifying the individuals who are likely to make up the audience. If it is impossible to obtain actual names, an educated guess is made as to the types of people who will attend. The expectations, biases, knowledge, and level of interest of the members of the audience all have a bearing on the presentation and its thrust. If the presentation includes a request for action, the attitudes of key decision-makers likely to be in attendance are determined. The biases of the potential audience toward the presenters themselves and the firm they represent are also worthy of note, and a study is made of how the presentation will benefit the audience.

The next step is to clarify the purpose of the presentation. Presenters must ascertain why they are talking about this particular topic to this particular audience. Is the primary goal to inform, persuade, answer questions, or some combination of these? Once the purpose is agreed upon, the presenters can concentrate on stating their objective in one succinct sentence.

Out of this exercise comes a statement of the message. It is important that it be phrased in the way it will be stated to the audience, bearing in mind that subtlety is not usually appropriate to the business presentation. Richard J. Kulda of Professional Eloquence says that speakers who fail to state conclusions for their listeners can expect only about 12 percent of them to arrive at those conclusions on their own. Moreover, fully 30 percent will draw some conclusion the speaker did not intend. Conversely, if the presenter states conclusions clearly, 55 percent of the audience will draw the appropriate conclusion.[1]

With the help of oral presentation development advisors, Boeing presenters plot strategy, determining the most ef-

fective method of communicating the message. They pinpoint any shortcomings in the proposed program or product that must be addressed and anticipate probable questions from listeners, including those they hope the audience will not ask. Once these questions have been identified, the presenters can formulate effective responses.

When initial planning is complete, speakers concentrate on articulating the major points they want to express. Developing a storyboard, they organize these points into a logical structure, adding necessary subordinate points and expanding the outline to include explanatory detail. The result is the basis for the body of the presentation. Once the content and flow have been determined, appropriate closing and opening remarks are developed. Often, the actual proposal storyboards are borrowed from the proposal development group to ensure that the oral presentation reflects the technical proposal's themes and messages.

Advisors help presenters tailor their material to the time available for presentation. They treat the outlined points as time modules, link them with logical transitions, and allow ample opportunity for questions from the audience.

Presenters are urged to steer away from developing a full script. The extemporaneous technique, delivery based on an abbreviated version of the outline, promotes a conversational, unmechanical interaction between the speaker and the audience.

Appropriate visual aids are developed. They are finalized only after the content has been agreed upon, saving time, staff effort, and money by avoiding construction of art that ultimately goes unused.

REHEARSALS AND VIDEOTAPE PRESENTATIONS

In the latter stages of preparation, Boeing oral presentation development advisors assume the role of coach and critic. They attend and critique dry runs and rehearsals, videotaping them for review by the presenters when appropriate, and work with individual speakers to improve their delivery and put them at ease with their material.

The Boeing Company is by no means alone in its determination to improve the quality of oral presentations. Other companies have begun or are starting programs of their own to promote adoption by their employees of some or all of the improvement techniques described in this article. The ultimate aim of these efforts is to ensure development of carefully prepared, quality oral presentations that communicate clearly and effectively—presentations that sell.

SUMMARY

With the increased emphasis by both governmental and commercial customers on oral presentations and the growing need within technology-based companies for clarity and conciseness in verbal communication, programs of assistance, whether organized within the company or provided

[1] Kulda, R. L., *Presentation Workshop Notes*, Orange, CA, Professional Eloquence, 1973.

by outside consultants, are clearly called for. It is vital that any corporation seeking to compete effectively develop some method of assuring that oral presentations given by its representatives show the company in the best possible light. Thus we recognize that business development and technical marketing rely on a personal, verbal touch that only quality presentations can establish.

BIBLIOGRAPHY

1. Leech, T., *How to Prepare, Stage, and Deliver Winning Presentations*, New York: American Management Association, 1982.
2. Wilder, L., *Professionally Speaking*, New York: Simon and Schuster, 1986.
3. Woelfle, R. M. (ed), *A Guide for Better Technical Presentations*, New York: IEEE Press, 1975.

Authenticity Beats Eloquence

SUSAN DRESSEL
Associate Editor, *PC Transactions*
with JOE CHEW

Just last month, a new engineer spoke up in a staff meeting to explain the advantages of a design modification he had recommended. He made his point effectively, drawing a few helpful diagrams on the board and providing an example that clarified a technical detail for several managers in the group. The engineer spoke clearly, confidently, spontaneously. His explanation took about $3\frac{1}{2}$ minutes.

A few days later, the project manager asked the engineer to present his explanation as part of a briefing scheduled for some customers in another 2 weeks.

The engineer prepared his drawings more thoroughly and had the audiovisual lab make some nice transparencies for overhead projection. Although his explanation at the briefing took more than 5 minutes, it was not nearly so complete, cohesive, and effective as his extemporaneous effort in the staff meeting.

At the staff meeting, his attention was fixed sharply on helping others understand his point. At the briefing, his attention seemed to be on his own performance as a speaker. Although he spoke well, he had lost an authenticity far more valuable than the studied eloquence that he, like many other speakers, had attempted.

In preparing his more structured presentation, the engineer locked himself into a script and audiovisual material, on the assumption that all listeners would interpret his words just as he intended. Under this assumption, he was unlikely to notice a puzzled look on a listener's face. And being tied to his script, he couldn't offer spontaneous explanations or examples. He was no longer an authentic person communicating with others.

In his levity-laced contribution to this column, Joe Chew advises speakers to rehearse "in the interest of spontaneity." The purpose of rehearsing, don't forget, is to ensure that you have all the main points in the right order and well supported, NOT to ensure that you have all the words committed to memory.

Joe Chew, guest contributor to Dr. Dressel's column, is a Senior Technical Writer with Softcom, Inc., a Hayes Research and Development Company, San Francisco.

Knowing that you are in control of your subject and the message you want your audience to understand, you will have more confidence. That confidence allows you to deliver your message spontaneously. Just as Joe's personal style and voice come through in the following sample of his writing, your own authentic personality should come through to your audience when you speak. Now listen to Joe.

A FEW WORDS OF ADVICE TO SPEAKERS

You may be one of those enviable souls who can stand up at a moment's notice and make an extemporaneous discourse on anything. If you are, congratulations; we look forward to hearing you, and you can look forward to basking in a room full of envy. Maybe you'll even get a promotion.

Most of us, though, get a bit rubber-legged at the thought of giving a talk. Some butterflies in the stomach. A little sweat on the palms. In fact, if that's all you suffer, you're lucky. Willard Scott passed out on the set of The Today Show a few years ago; upon being revived, the old trouper shocked a lot of people by revealing that he'd been having violent attacks of stage fright for years. (A little stress is actually a good thing; it gets your blood up, so to speak, so you can perform at your best. Serious speaker paralysis is something different.)

If you're a hard-case talkaphobic, it will probably take more than encouragement to turn you around. Being convinced that nobody wants to hear you, or even cares about your subject matter, and that it's a moot point because you just couldn't go through with it anyway, is a bad-news problem. But if you just have a little case of opening-night jitters, there are a few simple and effective things you can do.

Rehearsing probably does the most good for the most people. Now don't write out a script and memorize it by saying it over and over. The only memory work needed here is to store and retrieve your main points and supporting points in the order you plan to use them. You won't have trouble finding the right words when you're in control of what you want your listeners to understand. You ought to have this control established at least by the night before your talk. A quick flip through your slides

Reprinted from *IEEE Trans. Prof. Comm.*, vol. PC-30, no. 2, pp. 82–83, June 1987.

just before the talk is fine, but frantic last-minute cramming will just get you even more keyed up without really helping.

When you have control of sequence and support for your points, you'll need to see if you can sustain that control while talking and applying some of the pointers below. For a live audience, why not try your local chapter of IEEE? If you're not quite ready for them, the bathroom mirror is always handy; so is the rear-view mirror of your car if you don't mind some curious stares from fellow commuters. You can give your talk to the television set, where there are always some confident (if lobotomized) models to emulate, or try it out on your Significant Other if constructive sarcasm is more to your taste. Some people get an ego boost from a dog that barks at all the right times; others prefer the noncommittal bubbling of goldfish. Run through it at least once somehow.

Having demonstrated that you can talk while remembering what you're supposed to be talking about, how can you help yourself further? In a word: relax. An internal attitude of self-confident defiance is helpful to some people: "If they don't like it, to heck with 'em." (Of course, being able to say that and make yourself believe it indicates that your problem was not too severe in the first place.) Yoga, TM, and other mental disciplines or relaxation techniques are great if you're into that sort of thing; there are those who swear by a martini.

Probably, though, the best thing to do is to take a deep breath and dive right in. Once you move into the technical material, you'll get yourself squared away.

Relaxed? Check. What now? Believe it or not, the hardest part is over. You're already in control of your material, so the part you will be judged on – the substance – is in place and ready to go. So here are those pointers we promised.

Speak up.

A seemingly basic and trivial requirement, yet often neglected, especially when someone in the front row asks a question. Don't shout; just get a feel for how your voice carries (or, rather, doesn't carry) and crank it up accordingly.

You're probably not going to use the training technique Demosthenes supposedly came up with, which involved a mouthful of pebbles and a beachful of crashing waves to outshout. (Four of five dentists surveyed disapproved of this technique anyway.) But you don't need that much power just to make your voice carry across the room.

Speaking loudly, addressing the room in confidence, is also a self-fulfilling prophesy. You become what you act like. If you're shy, BLUFF shamelessly. What they don't know won't hurt you.

Stand tall.

Let them see you. Stand in a relaxed but upright posture; don't crawl behind your visual aids. And here's an eye-contact trick: look 'em right in the bridge of the nose. They can't tell the difference, and you'll achieve that magic eye contact without paralysis.

Don't hide rhetorically, either.

In his collection of stylistic essays, *On Writing Well*, William Zinsser reminisced about an old editor of his, an individual allergic to equivocation. "Don't go peeing down both legs," he would say. If you have something controversial to bring up, think it through, decide if it's right, and figure out a fair way to put it across. Then say it.

Be informal – engage your audience.

Life is not long enough to sit there listening to some droning nebbish read a paper verbatim – but you only have to think back as far as your last professional conference or convention to come up with some horrible examples. Use crib notes if you have to. But don't read; talk.

Stay in charge.

Balancing this goal with the previous one is quite a trick sometimes, and unfortunately nothing helps except practice and a strong personality. Try gently but firmly to keep the presentation on track even as you briefly explore sidelines that attract people's interest.

SUMMING IT UP

The Three Rules of Public Speaking:

- Be forthright.
- Be brief.
- Be seated.

HOW TO SPEAK SO FACTS COME ALIVE

SUSAN D. COHEN

Morey Stettner
MAS Communications

Recently I was invited to address a group of engineers and offer some pointers for delivering effective presentations. During a question-and-answer period, a burly fellow seated in the back of the room stood up and said, "What you've talked about makes sense, but as an engineer I let my facts speak for themselves. My presentations can get pretty technical, and my colleagues don't expect me to be a dynamic speaker."

Nods of agreement led to a collective sigh of relief. I could tell what the audience was thinking, "Whew, I'm off the hook. No need to worry about my public speaking, 'cause no one expects me to be a great speaker."

Perhaps engineers do not need to radiate confidence when addressing their colleagues or bosses at work. Per-

Don't hide behind a tangled web of technical terms. Above all, make your presentation understandable

haps a mastery of technical details leaves room for a monotone delivery. Perhaps a truckload of fancy graphs makes up for a sloppy, poorly organized presentation. But I doubt it.

Whether you are an accomplished speaker or a novice, the following tools will help you enhance your presentation skills. They apply to one-on-one conversations (as, for example, between you and your boss), staff meetings (with your peers), management meetings (with your supervisors), and technical meetings (with your peers from other companies or organizations). They may even help you win promotions or greater pay. The overall message is: Rid yourself of the it-is-okay-to-be-boring attitude.

The key to a successful presentation is *selectivity*, or knowing what facts are most important for your listeners to remember. You have probably discovered from your experience as a speaker that listeners do not retain everything you say. They listen selectively, recalling statements that are specially relevant or meaningful. Good speakers prepare by identifying the crucial points —the points the audience must remember. Then they make sure and emphasize these points when delivering their speech.

You might be wondering, "How can I be selective when I am describing a complex process. All my facts are equally important." That's what you think, but your audience may not agree.

As your prepare for your next presentation, select the three most critical points that you wish to make. Ask yourself, "If the audience remembers just three ideas from my speech, what should they be?" Organize your talk so that you emphasize *three* main points or *three* chemical processes or *three* vital considerations for a proposal. Force yourself to limit your horizons; although you may be tempted to discuss tangential details or drone on about theoretical possibilities, resist this urge by thinking in threes.

Why choose three?

The choice of three items — rather than two, four or more — is not an arbitrary one. Psychological research has shown that when a list grows beyond three items, most listeners start forgetting some of them. But there are also common-sensical reasons for sticking with three. When there are only one or two ideas expressed, it is easier for a listener to shoot down the entire topic by shooting down one of them; three makes it harder (assuming all three ideas have *some* merit). But the more ideas there are on a list, the more probable it becomes that one of those ideas is something that turns off the listener. In persuasive-speaking (selling) situations, this thought is often expressed as "overselling:" presenting so many ideas that the listener rejects the whole message when one less appealing idea is expressed.

A middle manager at a large aerospace firm, for example, was prepar-

ing a presentation to senior officers about the need for more research funding. He came to me with nine pages of notes filled with charts, statistics and projections. The question: How best to organize this mass of data? When asked to list the three most important ideas he wanted to express, he realized with much excitement that his proposal could be organized in threes: client needs, competition, and staff morale.

When he finally made his way to the lectern, he began riding a wave of confidence:

This company has a tradition of investing in the minds of its people. Now, in this challenging economic environment, we need to fund research projects more than ever to fill our clients' needs, keep pace with the competition, and motivate our talented engineers to take risks and explore!

As you prepare your presentation, pick three chapter headings that represent the three stages of your speech. The above example could be labeled with CLIENTS, COMPETITION, and MORALE. Plan supporting points and illustrations as needed, but make sure the body of your speech fits into threes. Do not burden yourself with excessive data. Do not stretch the attention span of your listeners with too much techni-

cal jargon. Most of all, establish a crisp tempo so that you do not linger on isolated points.

Defining terms

Now that you have a mental roadmap to direct your speech, you need to define your terms. Many technical speakers hide in a tangled web of verbiage without establishing a clear framework for their ideas. When you describe a complicated concept or use words with six or more syllables, you risk alienating your listeners. Even if you are addressing seasoned technicians who speak your language, do not assume that you must sound like a rocket scientist to make yourself heard. Indeed, I have observed brilliant engineers deliver highly technical talks that even a layman could understand. How? They used analogies, personal anecdotes, and simple examples to illustrate their points.

Define your terms within the context of your speech. For example, if you are describing a chain of events or reactions, continually link each event to the larger theme of your talk. If you are discussing a new technology, do not wax philosophical and then sit down. Define the technology itself, why it is important, and its potential uses. Strive for precision; eliminate assumptions by explaining the whys behind your points. (I know one engineer who rehearsed his presentations with his 11-year-old son; his son would ask "Why, Dad?" after almost every statement!)

Recently I met an engineer who learned the hard way to define his terms. A freshly promoted manager, he called his staff of fifteen into the conference room for a goal-setting meeting. He wanted to prove himself and demonstrate his vast knowledge. He told me that he used "highly technical formulas for highly specialized situations." An hour into his speech, he noticed that his staff kept grabbing sodas, coming and going through the door, even whispering among themselves. He understood the problem: His desire to establish credibility and mask his nervousness led to technical overkill. His listeners were not expecting or seeking such information. They arrived hoping for an informal "let's get to know each other" discussion, and they did not understand or care for much of the manager's technical terminology.

Strive for precision. Define your terms. Do not assume the audience knows what you know. Consider the needs and expectations of your listeners. You may be thinking, *Providing too much definition can be condescending and unnecessary. I want to treat people like professionals who know their stuff, not like children.* This view, although reasonable, fails to consider the disastrous effects of misunderstanding. If you assume your audience understands you, and you neglect to test their level of comprehension during your presentation, then you risk speaking in a vacuum. *You may understand what you say, but your listeners may leave the room running their palms through their hair trying to decipher your message.*

Once you organize your thoughts and define your terms, the next step involves making your facts come alive. Listeners pay attention when they know, *What's in it for me?* Your job as speaker is to capture their self-interest from the outset. Phrases such as "You will find...", "As a result of this..." and "You will come away from this discussion with..." motivate your listeners to truly listen.

Using anecdotes

Another way to enliven dry facts is to relate a personal story. Some examples: *When I first learned about this formula, I was up all night with my equations* or *There was a moment when I was in the shower, my magic moment as I call it, when I realized what this could mean to our project.* Such reflective statements lighten the load of your data and make your speech far more interesting. Ask yourself if there are any funny stories that relate to the topic. What previous experience have you had with the topic?

What does the topic remind you of?

Presentations loaded with facts and technical explanations pose a danger. When we recite a string of facts, we may lapse into lecture mode. This means that we start pontificating rather than engaging the audience in our topic. No one wants to be lectured unnecessarily. One of the most interesting attention-grabbers in any speech is a dash of gossip, a story of individual failure or triumph, an admission of your own confusion, or self-deprecating wit. These nuggets of personal experience help humanize your talk.

Equally effective in a fact-laden speech is the use of dialogue to dramatize your topic. You may wish to repeat an exchange that you had with a col-

Resist the temptation to drown yourself in slides, and forget the audience

league or a conversation that you overheard in the elevator. If you can unleash a bit of your personality and capture the original tone of voice, the dialogue becomes even better. A colorful dialogue, if delivered with energy and a bit a theatrics, shows that you are more than a technician — you are a leader with a sense or humor! (Discretion, of course, dictates that your dialogue is tasteful and well-intentioned.)

Overdependence on audiovisuals

You may have noticed that up to this point there is no mention of audiovisual aids. Some speakers waste time playing with viewgraphs and preparing fancy slides with multicolor graphics while

their actual speech goes unrehearsed. Audiovisuals have a small but significant role in most presentations. They illustrate key points and provide additional data that complements what is said. But they do not and cannot replace a speech.

Resist the temptation to drown yourself in slides, thereby shifting the spotlight away from you. Audiovisual aids cannot do your work for you, so do not place too much emphasis on your handouts or your graphs. Use them only when they offer a clear visual expression of your point. The ultimate test of whether to use audiovisuals is to ask yourself, *What does this add to my speech? How would the speech differ without these aides?* If the answers do not pop into mind immediately, then just say no to audiovisuals and concentrate on your preparation and organization instead.

When delivering your talk, do not retreat into the world of slides and forget your audience! Keep your head aligned with your listeners and maintain eye contact throughout your presentation. Glance quickly at a new slide, but do not keep looking at the screen as if you are seeing the slide for the first time. Most of all, do not read from your slide. You have probably been in an audience when a speaker insisted on reading from slides or handouts that you were perfectly capable of reading yourself. Do not insult the listeners' intelligence or waste their time by reading from a prepared text. Use those precious moments at the front of the room to breathe life into your subject matter.

Dealing with fear

"Stage fright" and a host of other terms have been used to describe the simple fear of speaking before other people. This is a very real psychological condition that many people are stricken with, and that could require intensive psychological counseling. For our purposes, however, the fears we run into when speaking are more typical, and, in the final analysis, more manageable.

The fears we usually have before a speech or presentation come from two sources. The first is the *fear of ignorance:* being shown to be not an expert by someone in the audience. This fear causes speakers to ramble on, trying to cover every possible eventuality or interpretation of a set of facts, and even to attempt to explain things beyond one's knowledge. Everything conceivable can get said in order to avoid the dreaded "I don't know" admission.

The second fear is *fear of rejection*: the belief that one's ideas will be unacceptable to the audience. Especially when there is an authority figure (say, one's boss) in the audience, the idea that someone could disagree with the facts and conclusions of a presentation is enough to stop one's words from being expressed.

There aren't many easy answers to confronting these fears. To be afraid of appearing before groups is human and natural; even the most polished speakers admit to some trepidation. Although it is a platitude, it is true that one should "think positive:" An idea expressed confidently is more likely to be accepted that one that is expressed tentatively. Keep your attention focused on the value and worthiness of the ideas you are putting across, not on the reception they may be getting while you speak. That fellow shaking his head in an agitated manner while you speak may be trying to get rid of an insect hovering nearby, and not disagree with your conclusions. A third point is not to take criticism personally, and to explain clearly one's area of expertise. People are more willing to accept you for what you are, than for what you pretend to be.

Your success as a speaker will ultimately depend on your level of enthusiasm. If you genuinely care about your topic, then all the dry facts in the world will not bore your listeners. They will stay tuned because your excitement sparks their interest. They will ask questions because you provoke them to think and probe. And they will leave the room feeling that they really gained something from you. ∎

The author

Morey A. Stettner is president and founder of MAS Communications [320 East 22nd St., New York, NY 10010; tel: (212) 353 0402], a consulting firm for advising on public speaking, sales training and customer service.

Handling a Hostile Audience—With Your Eyes

GILDA CARLE

Abstract—**This paper defines a method for turning confrontation between a speaker and a hostile audience into engagement of common concerns by effective use of eye contact, as well as vocabulary related to the audience's perceptive style.**

DAN WESTON, an engineer with the Clifton Power and Light Company (CPL), stood before the local Kiwanis Club. It was his job this evening to inform the 30-member audience about CPL's new power lines, soon to be constructed in this community. Until yesterday, CPL was looked on as a community asset, providing hundreds of jobs and less expensive electricity. However, just one day before Weston's presentation, the Clifton News published a study linking electromagnetic radiation from power lines with cancer. The Clifton Power and Light Company was the enemy. And, for this audience, Dan Weston was Clifton Power and Light.

As Weston was completing his introductory remarks, he sensed an undercurrent. As an experienced member of the Power Company's speakers bureau, Weston knew that he should discard his prepared outline dealing with the design modifications of the new structures. Instead, his two most pressing objectives were to let his listeners know he understood their concerns, and to get their attention for the information he had come to share. How would he achieve these goals?

According to the *Book of Lists* [1], people consider public speaking their number one fear. Fear is certainly to be expected when a hostile audience can turn public speaking into public persecution. But audience opposition and anger must be confronted as soon as possible [2]. If antagonism is allowed to germinate, additional resentment builds, and with the slightest provocation, an entire auditorium can become unmanageable.

Many public speakers open their presentations with a prepared *monolog*—one that bridges the audience's agenda with their own. Other presenters immediately immerse the audience in participative *dialog* to build enthusiasm and activate involvement. But both *monolog* and *dialog* heavily rely on words to get their meaning across. And because

Dr. Gilda Carle is President of InterChange Communications Training, Yonkers, New York.

as little as 7 percent of our information is communicated through words alone [3], Weston chose to concentrate at first on communicating with facial and body expressions, particularly with his eyes. Before he uttered a single word, Weston conducted *eye-alog*, prolonged empathetic eye contact. More penetrable than monolog, and more disclosing than dialog, eye-alog can help to establish rapport and cooperation—especially among hostile parties.

SECRETS BEHIND EYE-ALOG

Former President Reagan used eye-alog before he began each press conference. He looked reporters in the eyes—and then smiled and greeted them by name. This initial eye-alog told his viewers, "I'm on your side." It promoted Reagan's reputation as "the great communicator [4]." Mikhail Gorbachev is considered by many to be one of the world's best speakers. Using his hands and voice as well, he first dominates a meeting with his eyes.

America focuses much attention on the eyes with flattering eyeglass frames, tinted contact lenses, and colorful eye makeup. The vocabulary substantiates the richness of eye-alog: *gaze*, *glance*, *glare*, *gawk*, and *gape* are just a few synonyms for "look." Because eye-alog should precede dialog and monolog, the wise communicator can use it to choose an appropriate vocabulary. Neurolinguistics teaches us that "visual" communicators look up before telling us they "see" our point of view; "auditory" communicators look from side to side before explaining that they "hear" our side of the argument; and "kinesthetic" communicators look diagonally downward before admitting that they "feel" we are right [5]. Thus, by noting the direction of a listener's glances during eye-alog, a speaker can be alerted to choose words that are derivatives of *see*, *hear*, or *feel* in an effort to connect with the eye cues before him. Based on the principle that "Like likes like," this connecting of speaker's vocabulary to listener's eye-alog enhances subliminal interpersonal communication, even under hostile conditions.

WESTON'S PLAN

After his unceremonial introduction, Weston paused silently before the angry group. He divided the audience into imaginary quadrants, and for three seconds each, 12 seconds in all, he sought friendly eyes among the faces. He had an eye-alog with each—and then smiled. Finally, he

Reprinted from *IEEE Trans. Prof. Comm.*, vol. 32, no. 1, pp. 29–31, March 1989.

took a deep breath and acknowledged the fear: "By now you have read the findings of the Savitz study. You know the information is inconclusive. You also know that Clifton Power is subsidizing further research—which will take time to conduct, analyze, and report. This is a societal issue—and one we will *all* solve together. Tonight, let us openly discuss our mutual concerns. I will be glad to fill you in on the facts we currently have."

Weston had used eye-alog to establish unity and set the stage for believable dialog. With each question from the audience, he could offer a sympathetic monolog in response.

Weston conducted eye-alog for another few seconds, then smiled to some friendly audience eyes. He called upon a receptive face for question 1. With subsequent queries, when possible, he noted the interrogator's eyes and was sure to respond in *visual*, *auditory*, or *kinesthetic* terms: "I *see* what you are saying (visual), "I can *hear* how upset you are" (auditory), or "I *understand* why you *feel* as you do" (kinesthetic). Finally, when a hostile interrogator got carried away, Weston was quick to disengage eye contact with him and enter into eye-alog with other audience members.

Thus, beginning with eye-alog, Weston set the stage for each response. He read each inquirer's eye-alog and responded to questions with the most acceptable language. He used eye-alog effectively with the hostile group before him and allayed their irrational concerns. Eye-alog was the perfect prelude to the subsequent dialog and monolog that contained Weston's real agenda.

The EDM Formula for Hostile Audiences

The EDM Formula comprises the following components:

E: EYE-ALOG

- Divide the audience into imaginary quadrants.
- Pause.
- Hold an *eye-alog* with the friendliest eyes in each quadrant, three seconds each, 12 seconds in all.
- Use the SOFA:
  — *S*: Smile.
  — *O*: Open your posture to demonstrate receptivity to your audience.
  — *F*: Forward lean—into your audience.
  — *A*: Acknowledge your listeners by nodding slightly while smiling "Hello."

D: DIALOG

- Think NOT in terms of making a *presentation to* your audience; instead, think of having a *conversation with* them.
- During your conversation with the audience, especially while answering questions, use the SOFA.
- If you are close enough, note the position of the interrogator's eyes. Also note the verbs he chooses: are they visual, auditory, or kinesthetic?
- Pause.
- Mirror the visual, auditory, or kinesthetic terminology and eye-alog used by your questioner—but sweep your gaze from the questioner to other audience members.
- By disengaging the eyes of the hostile interrogator, you will publicly and gracefully empower your own authority by *conversing* with the people you are there to influence.

M: MONOLOG

- With each response, paraphrase the question to emphasize your own objectives and eliminate negative buzz words [6].
  For example:
  "Why is your company constructing those unsightly power lines by our homes, especially in light of the potential dangers to our public health?"
  Responding Monolog: "The question is ... [Use the SOFA] ... 'What are the safest and most cost-effective means that Clifton Power has at this time to transmit electricity to our homes?'
  The question is deliberately paraphrased to formulate a positive response.
- Do not offer the usual, defensive response of "Power lines are *not* unsightly ..." or "We are *not sure* that there are potential dangers in power lines ..." Instead, reposition positive questions that will prepare you for positive responses.
- Using eye-alog and dialog, you have already primed your audience to support you. Now, through monolog, you can crystallize your position and *sell* the reason you are actually there.
  For example:
  The words of the questioner above included the visually oriented terms "unsightly" and "in light of." Thus, monolog following the paraphrased question should incorporate *visual* terminology: "Clifton Power *views* power lines as a cost-effective means of bringing electricity into our homes ..."
- *Monolog* can therefore come to life in the audience's terms. The subconscious subtlety of familiar terminology and congruent eye-alog set the tone for dialog that will encourage acceptance of the monolog of information.

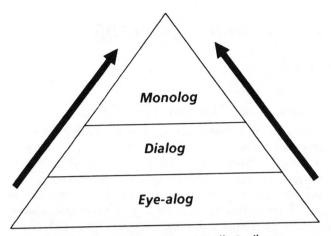

EDM Formula for Handling a Hostile Audience

The EDM formula inverts the usual tendency of speakers to begin with their prepared agenda and to monolog an audience to sleep. Especially with hostile listeners, the EDM formula establishes immediate rapport through eye-alog to be followed by dialog, and finally by monolog. Using the EDM formula, a speaker has the chance to allay audience resistance and achieve his objectives.

REFERENCES

1. Wallechinsky, D., Wallace, I., and Wallace, A., *The Book of Lists*, New York: Bantam Books, 1978.
2. Carle, G., "Coping with Corporate Rage," *Shell World*, internal publication of Shell Oil International, December 1987, pp. 28–29.
3. Mehrabian, A., and Ferris, S. R., "Inference of Attitudes from Nonverbal Communication in Two Channels," *Journal of Consulting Psychology 31* (1967), pp. 248–52.
4. Ailes, R., and Kraushar, J., *You Are the Message: Secrets of the Master Communicators*, Homewood, IL: Dow Jones-Irwin, Inc., 1987.
5. Richardson, J., and Margulis, J., *The Magic of Rapport: How You Can Gain Personal Power in Any Situation*, San Francisco: Harbor Publishing, 1981.
6. Lustberg, A., *Winning at Confrontation*, Washington: U.S. Chamber of Commerce, 1985.

How to Overcome Errors in Public Speaking

JOHN E. BAIRD, JR.

Abstract—Ten common errors cause most public speaking failures. A positive approach in three stages can increase your chance for success. *Preparation:* 1. Plan to have an effect; don't just be speech-giving-oriented. 2. Determine the audience's interests, motives, knowledge, attitudes, and values. 3. Be aware of your credibility with the audience. 4. Capture attention with a confident, motivating introduction. 5. Organize your information for understanding. 6. Plan the conclusion to again place your objective before the audience. *Presentation:* 7. Control stage fright and channel the energy into dynamic speaking. 8. Be physically active and purposeful with gestures; vary speech characteristics; be natural and direct. 9. Sense audience feedback and adjust to it. *Preservation:* 10. Prepare for questions from the audience.

O F THE many types of communication prevalent in business and industry, none is more important than public speaking. Through presentations made to large and small groups, speakers attempt to persuade people, inform them, impress them, and entertain them. Unfortunately, public speaking is also the most frequently mishandled form of communication. Because of inadequate speaker performance, attempts to influence misfire, misunderstandings occur, images suffer, and ceremonies collapse.

While such public speaking failures occur in a variety of settings and produce an array of catastrophes, my experience with speakers in both industrial and college settings suggests that the causes underlying these failures are relatively few in number: ten, to be exact. Moreover, if speakers are aware of these causes, they can usually avoid them and thereby greatly increase their chance for success.

Public speaking involves a sequence of three stages: preparation, presentation, and preservation. The preparation stage includes setting objectives, analyzing your listeners, planning strategies, and gathering and organizing information. Presentation consists of the verbal and visual delivery of the message and your adaptation to audience responses. Preservation, finally, comprises your ability to handle audience questions in a manner which preserves the positive impact of the speech. Because specific causes of communication failure occur during each stage, we consider them separately.

STAGE I: PREPARATION

Most speaking failures occur before a word is even uttered. In fact, committing any of the six preparatory mistakes will virtually doom you to failure.

The author is a consultant with Modern Management Methods, Inc., 2275 Halfway Rd., Bannockburn, IL 60015, (312) 945-7400.

1. Procedure Rather Than Product Orientation

Too often I have found that people view speaking as an end, rather than as a means. When I ask them, "What are you trying to accomplish?" they reply, "I want to give a good speech," or "I want to avoid making a fool of myself." While those may be legitimate concerns, they are not acceptable objectives for speech-making. You shouldn't deliver a speech simply to prove your artistry in speech-making, and you certainly shouldn't view speaking solely as an opportunity to avoid looking foolish. Rather, speeches should be designed to have some impact on audiences—to give then new information, to change their attitudes, to secure their commitment, or, at the very least, to keep them entertained. Communication is a tool for achieving results, not an art to be admired for its own sake.

The problem with a "procedures" orientation is that it distracts you, the speaker, causing you to become concerned with matters that are really of secondary importance. For example, people have often told me, "My main goal is to give this speech without the audience's knowing how scared I am." Their over-concern with stage fright only frightens them further. Eventually this vicious circle reduces them to virtual incoherence. But perhaps more important, their concern for self rather than audience causes them to lose sight of the real objective of public speaking: to have an impact on their listeners. Similarly, speakers who are determined to enunciate clearly while speaking, to use the most colorful visual aids ever seen in the Western Hemisphere, or simply to "get it over with" are preoccupied by the procedures of public speaking, not the product. Rarely do such speakers deliver successful speeches.

To train college students and professionals in public speaking skills, I have developed an approach called "communication by objectives" in which speakers begin their preparation by phrasing a single declarative statement:

I want (who) to (what) (where, when, how) because (why).

This statement, after the blanks have been filled, becomes the central concern of the speaker, and all of his or her preparation is directed toward its achievement. Through this technique, then, the speaker is forced to focus on the desired results, rather than on secondary considerations. Communication by objectives thus helps to ensure a product, rather than a procedure, orientation.

2. Inadequate Audience Analysis

Speeches are given for audiences. Again, their sole reason for existence is, or should be, to have some impact on the audience members—to produce some change in their beliefs, attitudes, or behaviors. As in any analysis of a situation involving change, you should ask three questions: "Where are we now?" "Where do we want to go?" and "How do we get from here to there?"

Reprinted from *IEEE Trans. Prof. Comm.*, vol. PC-24, no. 2, pp. 94–98, June 1981.

Establishing speaking objectives answers only the second question, but from it, speakers often proceed directly to the third, never bothering to answer the "Where are we now" query. Clearly, this is a serious oversight. You cannot effectively plan change without knowing *what* is to be changed. But all too frequently, the sort of audience analysis needed to answer that first question is neglected by prospective speakers.

Although a variety of audience characteristics are relevant to the concerns of public speakers, five are of particular importance: interests, motives, knowlege, attitudes, and values. In preparing your speech, try to answer five analytic questions:

- What things relevant to my topic is the audience probably interested in?
- What things that the audience wants or needs can my speech provide?
- What things relevant to my topic does the audience probably know about?
- How does the audience probably feel about me and my topic?
- What values probably underlie this audience's attitudes?

Certainly, the accuracy with which you can answer these questions is determined by the acquaintance you have with the audience. Nevertheless, my experience suggests that most speakers know enough about their audiences to answer these questions with a reasonable degree of accuracy.

Many sources of failure in public speaking are eliminated by this sort of systematic audience analysis. By assessing audience characteristics, you are far less likely to present uninteresting or irrelevant information, to give information that the audience simply cannot understand, to misjudge their attitudes toward the topic and you, or to offend their value systems. In addition, audience analysis allows you to establish the subgoals that must be accomplished if the objective of the speech is to be met. You can then determine whether it is necessary to interest the audience in the topic, to relate the topic to their needs, to establish more credibility, to provide them with specific information, to change their attitudes, or to deal with their value systems in order to achieve desired changes. By analyzing the audience in this fashion, you can then develop strategies that directly relate to the listeners' characteristics.

3. *Underdeveloped Credibility*

Despite its name, "source credibility" or "speaker credibility" is not a characteristic of speakers at all. Rather it is the amount of believability which the audience attributes to the speaker. "Believability" is obviously crucial to a speaker's success. If an audience is inclined to believe the speaker, he or she will have a good chance of convincing them to accept the information and arguments presented in the speech. If the audience is inclined to disbelieve the speaker, however, then nothing he or she says is likely to have any effect. Yet few speakers bother to assess their own credibility with a particular audience, to develop it, or to use it to best advantage, and often they fail as a consequence.

Research suggests that credibility consists of five dimensions, or five specific judgments which an audience makes about a speaker. Briefly, these include

- *Competence*   The expertise of the speaker with the speech topic
- *Character*   The speaker's honesty and sincerity
- *Intent*   The purposes of the speaker: whether he or she has the audience's best interests at heart
- *Dynamism*   The general level of forcefulness, energy, aggressiveness, or boldness with which the speaker presents the message
- *Personality*   The speaker's general likability.

Overall, audiences believe sources whom they believe to be competent, trustworthy, well-meaning, dynamic, and likable; and they distrust speakers they believe to be incompetent, dishonest, self-oriented, timid, or obnoxious. But credibility also is of two types: initial and elicited. Initial credibility refers to your image before the speech and is determined by the audience's previous experiences with you, by your reputation, by your status or rank in the organization, and by the similarity of your and the audience's attitudes and values. Elicited credibility springs from the speech itself. Generally, credibility is enhanced if you show an understanding of audience needs, appear well-informed and fair-minded, express yourself dynamically, and appear to like the audience. Thus, credibility which initially is rather low can be improved as the speech progresses.

4. *Ineffective Introduction*

Perhaps the most important part of a speech is the introduction—the first few words the audience hears. If the introduction is well done, the audience will be inclined to listen to the information and arguments which follow. If the introduction is poorly done, however, the audience will simply stop listening, and the remainder of the speech will be wasted. But despite the obvious importance of the introduction, some speakers fail to give it adquate attention during their preparation. They carefully gather and arrange their information, but they prepare little or nothing with which to introduce it. Perhaps they expect some sort of inspiration to strike just as they begin to speak. Usually, it does not. All too often, such speakers begin by saying, "Uh, this evening, uh, I'm gonna talk about, um, a couple of things." And just as often, the audience thinks "So what?" and mentally exits.

An introduction must be carefully designed to do two things. First, it must get the attention of the audience. Several devices have proved to be useful attention-getters. A series of rhetorical questions causes the audience to become mentally involved in the speech; a brief narrative arouses their curiosity; a familiar quotation appeals to their past experience; a humorous anecdote (one that is truly funny) entertains them; a reference to the occasion establishes speaker-audience commonality, which involves them; a reference to the speaker's own experience interests them. None of these is sure-fire, but each of them, when presented reasonably well, seems to gain an audience's attention.

Second, the introduction must motivate the audience to listen. Audiences are essentially selfish—they will devote their attention to the speaker only if they perceive some benefit in doing so. Thus, the introduction must relate the topic directly to the audience, answering the question, "What will this information do for me?" Perhaps it will help them achieve something they

ZIGGY drawn by Tom Wilson. Copyright 1978 by Universal Press Syndicate, Inc. Reprinted with permission.

desire; perhaps it will allow them to improve their situation; perhaps it will assist them in avoiding undesirable future consequences. In any case, it falls to you, the speaker, to analyze audience motives and then state specifically in the introduction how the topic relates to them. By so doing, you can virtually be assured of an attentive audience—at least for the next few minutes.

5. Unclear Organization

Of the ten faults considered here, this may be the most common. On too many occasions I have watched highly credible speakers talk to attentive listeners and fail miserably—fail because their information was a garbled, incomprehensible mess. If information is to have impact, the audience must be able to understand it, and if the audience is to understand the information, it must be clearly organized. Far too many speakers overlook this fact. They worry about having enough information, but they neglect to organize the materials they have collected.

A variety of organization patterns may be used:

- *Chronological* Organizing according to time of occurrence: "First, Hitler invaded Poland. A few weeks later, he invaded...."
- *Spatial* Organizing according to physical arrangement: "You walk through the door and turn left. Then you come to a hallway and turn right."
- *Sequential* Organizing according to a series of steps: "First, you address the ball with the club. Second, you begin the backswing."
- *Hierarchical* Organizing according to a series of subdivisions: "At the top is the president. Below her are two vice-presidents. Reporting to each of them...."
- *Classificational* Organizing according to category: "There are two types of golf clubs: woods and irons. We will consider each type in turn."

6. Weak Conclusion

When preparing a speech, it is all too easy to overlook yet another important element of success: the ending. Usually, speakers plan their introductions and presentations with some care, but they neglect to plan the conclusion, apparently assuming that the momentum of the speech will serve to carry them off the platform and on to glory. It does not often happen that way. Instead, the speaker comes to the end of the speech and discovers that he or she still is standing in front of the audience with nothing left to say or do but blush, mumble "thank you," and slink from the platform, perhaps destroying the speech's previous impact.

A strong conclusion does two things. First, it summarizes the speech, touching on the important points covered in the introduction and main body. Second, and more important, it incites. It challenges the audience to take action, and then it tells them specifically what to do and when, where, and how to do it. In effect, the conclusion places your objective before the audience and virtually demands that they carry it out. Preparation of such a conclusion, then, is vital to your talk's achieving its objectives.

STAGE II: PRESENTATION

The six errors just described are committed prior to the delivery of the speech; the next three occur during the presentation. Like the preparatory faults, these inadequacies of presentation also can be fatal, turning even the best prepared address into a monumental disaster.

7. Uncontrolled Stage Fright

Stage fright plagues literally everyone to some degree. Fear of failure produces an all too familiar array of symptoms: loss of memory while speaking, trembling voice, shaking hands, quaking knees, and an almost uncontrollable urge to exit stage left and never speak again. But surprisingly, stage fright can produce another very different set of symptoms: clear thinking, physical energy, and an emotional "high" that makes speaking a truly enjoyable experience. The factor that seems to determine which set of symptoms a speaker will experience is control: unsuccessful speakers allow fear to preoccupy and control them, whereas successful speakers control their fear and use it to their advantage. The key issue with stage fright is not how can it be eliminated, for elimination is impossible. Instead, you must decide how it can be controlled to channel the energy fear produces into more effective speaking.

Several methods for controlling stage fright have, in my experience, proved useful:

- *Try Physical Exercise* By expending energy before speaking, you can release some tension and keep your energy at a manageable level. (Indeed, the stage-fright symptoms of quavering voices, shaking hands, and quaking knees are nothing more than signs of uncontrollable energy.) Thus, if you walk up a flight of stairs on the way to a meeting, or walk a long distance to the room, or use unobtrusive isometric exercises while waiting to speak, you will be better able to control the physical symptoms associated with stage fright and to channel the remaining energy into delivering the speech.

- *"Objectify" the Situation* Think calmly and rationally about the audience, setting, and speech, and realistically assess the latter's potential consequences. I have found that public speaking is most frightening when it is dealt with in abstract terms, and that when speakers begin to rationally analyze the situation, their fears diminish.

- *Establish Realistic Goals for Yourself* Fear of failure decreases as the chances for success grow, and establishing realistic goals ("I am going to deliver this information in a coherent manner"), rather than unreachable ones ("I am going to deliver the greatest speech in the history of the Western world"), does much to bolster your confidence and reduce your apprehension.

- *Distract Yourself* If you focus on the speech, repeatedly rehearsing it mentally, rather than on the situation, your fear will probably be greatly reduced.

- *Get Help* Therapies (such as systematic desensitization), which are too complex to describe here, are available through professional trainers to speakers who still are unable to cope with their fears. Usually, however, using one or more of the first four techniques reduces fear to manageable levels and greatly enhances your odds for success.

8. Ineffective Delivery

Although I firmly believe that content is more important than delivery, the fact remains that many speeches are destroyed by speaking techniques that are boring, offensive, inappropriate, or distracting. The list of delivery faults is endless, but among the more common are overly tense or overly relaxed posture, lack of physical movement or aimless wandering, absence of gestures or meaningless flailing, lack of eye contact with the audience, monotonous voice, too fast or too slow rate of speech, nonfluent speech (that is, speech containing a high proportion of "ums," "uhs," "you knows," and the like), and too loud or too soft vocal volume. Each of these faults hampers the effectiveness of the speech's message and increases the likelihood of failure.

The following guidelines will do much to promote effective delivery:

- *Be Active* Successful speakers make a conscious effort to move as they talk, using gestures and general bodily positions to add meaning to their words.

- *Be Purposeful* While successful speakers are active, they are not uncontrolled. Their gestures, posture, and so on all have a purpose and are added to the presentation to produce a specific effect.

- *Be Varied* Repeating the same behavior time after time becomes distracting. Effective speakers vary their physical and vocal behavior to maintain audience interest. Of particular importance is vocal variety (changes in pitch, volume, and rate), which makes listening to the speech much easier and more pleasant for the audience.

- *Be Natural* Effective delivery springs from a curious mix of planning and spontaneity. Good speakers plan to use gestures, vocal inflections, and so on actively, purposefully, and with variety. They do not, however, plan specific behavior for particular places in the speech. Rather, they do what feels natural, so that their observable behavior springs from the meaning of their words. Thus, good delivery involves a conscious decision to be active during the speech, coupled with a natural flow of behavior as the speech progresses.

- *Be Direct* Effective speakers talk to their audiences; ineffective speakers merely talk in front of them. Particularly important here is eye contact: To talk to an audience, one must look at them. Speakers who simply read their speeches or who study their notes rather than the audience might as well be all alone—and if they perform in that fashion long enough, they will be. Maintaining direct eye contact with the audience is crucial to good delivery.

9. Insensitivity to the Audience

Even when speakers look at their audiences, they may not "see" them. A good speaker adjusts to the audience in both preparing and presenting the message. He or she tries to judge them ahead of time and construct the speech accordingly, but the speaker must also try to "read" the audience feedback and adjust to it while talking. Unfortunately, many speakers are insensitive to or unconcerned with their listeners; they say whatever they want whenever they want, regardless of the audience's nature and responses. On one occasion, I attended a convention at which the after-dinner speaker began his presentation with a joke so off-color that it would empty most locker rooms. The audience gasped, the speaker was embarrassed, and the remainder of the speech was a catastrophe. The speaker should have known better. With just a little forethought, he would have realized that the audience was mixed, sophisticated, and conservative, and that this sort of humor would be highly inappropriate. At still another conference, I watched an after-dinner speaker talk for an hour and a half when he had been scheduled to speak for 15 minutes. After 30 minutes, people began to fidget; after 45 they began checking their watches; after 60 they began shaking their watches in full view; after 75, they began talking to each other audibly, passing notes to the speaker telling him to quit, and leaving. Finally, after 90 minutes and with half the audience gone, the master of ceremonies interrupted the speaker and ended the ordeal. Clearly, this speaker was insensitive to his audience, and a lot of embarrassment and hard feelings resulted.

We have already looked at preparatory audience analysis, so I simply repeat this principle: Carefully judge audience interests, motives, knowledge, attitudes, and values, and then select materials appropriate to those characteristics. But remember, too, that audience analysis does not cease when preparation has been completed—you must analyze and adapt to audience feedback while you are actually speaking.

Feedback from an audience is primarily nonverbal, as the listeners express their reactions through a variety of behavior. Four cues are particularly revealing: audience eye contact, facial expressions, head movement, and posture. Audiences tend to look at you if they are interested in your message, if they understand the message, and if they agree with it; they tend to look away if they are bored, confused, or in disagreement.

Facial frowns appear when the audience is confused or disagrees; pleasant expressions usually indicate agreement. Head nods demonstrate interest (except when a listener is nodding off to sleep), understanding, and agreement, while head shakes usually indicate confusion, disagreement, or even disgust. Finally, an upright, erect posture is usually indicative of interest, while a slouched position suggests boredom. By "reading" all these cues, you can gain insight into how your speech is being received.

The next issue, then, is how to adapt to feedback. If the audience agrees, move on to the next point. There is no sense in continuing to fight a battle already won. If they disagree, explain the point carefully, offer new evidence and arguments to support the point and refer back to things with which they previously agreed—matters from which the present issue logically follows. If they seem to understand, proceed ahead. However, if they seem confused, repeat the point (they simply may not have heard it), and if confusion still reigns, clarify it with an explanation, details, analogies, examples, or definitions. Finally, if they are interested, everything is fine. But if they seem bored, try to enliven the presentation with a story, a joke, increased activity in delivery, rhetorical questions, or some other interest-gaining device. If no device works, get the speech over with as quickly as possible, and begin thinking about how to do better next time.

STAGE III: PRESERVATION

10. Mishandling Questions

Even when a speech has been carefully prepared and beautifully delivered, the battle may not yet be won. Many situations allow time for the audience to question the speaker, and if the question-and-answer session goes badly, the positive impact of the speech is lost. I have seen this very thing occur many times: A speaker becomes flustered, angry, confused, or upset while trying to handle audience questions and ruins the impression made by the speech. Thus, to preserve the effect of the message, you must prepare as carefully for audience questions as for the speech itself.

Two steps are involved in preparing for the questioning period:

- *Know the Subject and Audience Thoroughly* If a speaker's knowledge is limited to only the content of the speech,

audience questions will quickly reveal his or her vast regions of ignorance.
- *Try to Predict Question Areas* What aspects of the speech is the audience likely to find particularly intriguing? What subject areas are covered only briefly? What additional facts is the audience likely to want for their own use?

Dealing with audience questions also requires the careful application of several techniques. Generally you should react to all questions, even if the reaction is simply to postpone consideration of the question until a later time. You should consider the entire audience, not allowing any one person to dominate the interaction. If necessary, admit ignorance, rather than trying to "fake it" on questions to which you have no real answer. Finally, avoid confrontations with hostile questioners and respond to questions in a friendly, encouraging manner so that other people with inquiries will feel more inclined to participate. In actually handling questions, you should repeat the question to be sure that everyone has heard it, rephrase the question if it is unclear (being careful to preserve the original meaning), postpone it if it is potentially troublesome (perhaps by offering to speak with the questioner after the meeting), and analyze it for the audience if it is a "loaded" question asked to trap or trick. By preparing for and handling questions in this manner, you will be able to preserve the positive effects your speech has produced.

DRAMATIC RESULTS

In my contacts with college students and business people, I have witnessed an enormous number of public speaking disasters. Virtually all of these calamities were attributable to one or more of the errors in this "ten least wanted" list. Happily, I also have witnessed an enormous number of improvements among speakers trained to apply the principles explained here. By establishing objectives, analyzing the audience, developing an effective introduction, enhancing credibility, organizing the information, strengthening the conclusion, controlling stage fright, improving delivery, adapting to the audience, and managing audience questions, you can virtually guarantee yourself a successful speech. None of these principles is easy to apply, and all of them demand time, thought, and effort, but the results they produce are dramatic.

Part 8

Meeting, Disputing, Listening
Working with Others to Get Results

A LOT OF communication takes place beyond the written word or standard on-the-job presentation. The articles in this section will help you organize and participate in successful meetings and seminars, recognize when and when not to argue, and listen to others to receive their messages accurately.

WHAT CAN GO WRONG—AND RIGHT

First, two brief selections identify what makes a good or bad meeting. Although written for the paper production industry, Rodney Beary's observations apply to any meetings. He recognizes that people often react negatively to meetings and warns of 17 factors that can cause such attitudes. His companion article gives 11 prerequisites for an effective meeting, suggests three organizational roles to ensure productivity, and provides ten follow-up questions you might ask to gauge a meeting's success.

PREPARING THE EFFECTIVE MEETING

T. F. Gautschi categorizes meetings into three types: information-sharing, instruction-giving, and problem-solving. The focus of a meeting becomes clearer when you establish its "type" beforehand. Then questions of agendas, time schedules, attendance, and decision-making can be considered. You may or may not like the idea of someone playing devil's advocate at every meeting, yet such a person might prevent your meetings from lapsing into groupthink, five symptoms of which Gautschi provides.

If you fail to plan ahead, your meeting will be unsuccessful. Thus Eugene Raudsepp deals with specific preparatory considerations, such as the relationship between the number of attendees and the physical set-up for a meeting. Circulating an agenda beforehand can pinpoint topics and save you time. Your responsibility as the leader to establish bounds, steer the meeting, and handle "problem participants" is unavoidable. Raudsepp's final ten guidelines for successful meetings encapsulate much that has been said so far in this section's articles.

COORDINATING THE SEMINAR

A technical seminar is a highly specialized meeting where you and other participants present technical material. The coordinator may be one of the speakers, but additional responsibilities fall on his or her shoulders. Thomas Ealey discusses the work involved in conducting such a seminar, and by now you won't be surprised that success depends on preseminar planning. Although your participants will be eager for information, many things can distract them (you will learn more about this in the articles on listening). They will need comfortable facilities, supplies, frequent breaks, a dependable schedule, and reliable address systems. Ealey warns against permitting discussion during presentations and, like Beary, stresses the value of following up with an evaluation questionnaire.

WHEN ARGUMENTS ARISE

Disputes can arise in meetings—and anywhere else when humans communicate. W. A. Delaney's use of "argue" in his title might conjure up images of heated words and unpleasant exchanges, but as he points out, there are ways to present conflicting views and ways not to. There are some situations where you should argue, but plenty where you should not. Knowing when and how to argue is essential, and Delaney's insights and examples, together with eight tactics to help others see your point, will be useful in any work situation where discussion takes place.

WITH EARS AND MIND

Listening plays an essential role in meetings, discussions, and arguments—yet listening is perhaps the hardest communication skill to acquire. Thus, the final two articles turn to this complex art. Most of us have no training in listening yet must do a great deal of it, on and off the job. Marion Haynes analyzes the process of listening and the barriers to doing it well, and suggests techniques for focusing attention while you listen. In the final article, Blake Emery and Karen Klamm describe how you can listen for clarification of a speaker's meaning, and suggest ways to handle conflicting ideas, evaluate information, and ask questions. Their seven-point checklist enables you to assess your skill as a listener and their informal exercise will help you sharpen your listening skills.

What's wrong with your meetings?

Rodney P. Beary *PIMA Training Institute*

In a recent training session on conducting meetings, the participants were asked to list their first thought upon hearing the word "meeting." In the group of ten people, all the responses were negative. The reactions included "Oh oh! Here's trouble!" "They're a waste of time." "Boring!" and "Not another meeting!"

These responses are not a scientific study, but they do highlight the fact that meetings are the source of many complaints from employees at all levels.

Meeting attendees often complain about too many meetings, inconveniently scheduled meetings or meetings that last too long.

Shift workers seem to be especially sensitive to meeting length, schedule and content. Some union representatives have even taken up the cause for their beleaguered constituents by insisting that the only meetings an employee can be forced to attend, outside of normal working hours, is a safety meeting, where only safety-related topics may be discussed.

In their book "We've Got to Start Meeting Like This," published by Scott, Foresman and Co., authors Mosvick and Nelson discuss the most frequent meeting problems, reported in Mosvick's surveys of 1982 and 1986. The listing, in order of the most frequently to the least frequently reported problem, is:

- Getting off the subject
- No goals or agenda
- Too long
- Poor or inadequate preparation
- Inconclusive results
- Disorganized
- Ineffective leadership/lack of control
- Irrelevance of information discussed
- Time wasted during meetings
- Starting late
- Not effective for making decisions
- Interruptions from within and without
- Individuals dominate discussion
- Rambling, redundant, or digressive discussion
- No published results or follow-up action
- No pre-meeting orientation
- Cancelled or postponed meetings

With all the dissatisfaction and complaints about meetings, there is a temptation to stop having them. In some cases, plants have all but stopped having meetings.

Not having meetings may sound like a good idea at first, but it is very difficult to use a participative management style and build a work crew into a team without getting the players in the same room at the same time — a meeting. However, many managers do conduct excellent meetings. The best approach to take regarding meetings is to assume that a well-conducted meeting will yield more effective results and fewer complaints.

Accept the fact that some people will always be critical of any given meeting; then accept the challenge as a manager to conduct the best meeting possible under the circumstances.

Reprinted with permission from *PIMA Magazine,* p. 8, December 1988. © 1988, Paper Industry Management Association.

Great meetings:
How to make them happen

Rodney Beary *PIMA Training Institute*

In last month's column we discussed what can go wrong in meetings. A bad meeting is not only a source of frustration, it reduces productivity through lost work hours and leads to poor morale and bad decision making. If a person leaves an unsuccessful meeting feeling poorly about themselves or the people they work with, their job performance will suffer.

What goes into a good meeting? What does it take for people to leave a meeting charged up and ready to work? Participants in meetings have said that meetings are successful when:

- Time constraints are adhered to
- Schedules are considered when planning the meeting
- The topic is important and interesting
- There is adequate time for open discussion
- Mutual respect prevails
- Clear goals and an agenda are spelled out
- Meetings follow a structure or format
- Something specific is accomplished
- The leader is effective
- There are no distractions or interruptions
- A relaxed atmosphere prevails, with a distinct lack of fear or intimidation

The main lesson from this feedback is that a meeting should be a planned event that happens on purpose! A meeting should *not* happen unless someone takes active responsibility for making it a success. The people who want a successful meeting have to effectively perform three separate roles: Champion, Facilitator and Evaluator. The same person or different people may play these roles.

The **Champion** sponsors the meeting and is the driving force behind it. He or she schedules the meeting, invites and solicits input from the right participants, gathers information and prepares the agenda. When the meeting is held, the Champion starts the meeting with introductory remarks and a guiding statement.

The **Facilitator** makes sure the meeting is conducted in a competent manner. The Facilitator makes presentations, leads discussion, creates a climate conducive to discussion, coordinates other speakers, enforces time limits and makes sure everyone participates. The Facilitator makes sure the meeting's goals are met and makes specific work assignments.

The **Evaluator** follows-up with attendees to see that agreed on tasks are accomplished and then communicates the results to interested parties. The Evaluator also solicits reactions to the meeting so that future meetings can be better. A followup survey and a discussion of the results at the next meeting can be helpful. Feel free to use the following questions in your survey. Answers should be given on a scale of one to five, with one being "not at all" and five being "very much."

1. Do you feel your ideas were listened to?
2. Did the meeting stick to the agenda?
3. Did individuals avoid monopolizing the discussion?
4. Did you express your real feelings?
5. Are you satisfied with the results of this meeting?
6. Did you know what to do in this meeting?
7. Were you prepared for this meeting?
8. Was your presentation of value to the meeting?
9. Did the facilitator perform his/her job well?
10. Open ended questions:
    - What did you like about the meeting?
    - What about the meeting should be improved?

Surveys like the one above are part of the commitment someone in your mill needs to make to improve meetings. If your typical mill meeting starts late, gets off the subject and goes downhill from there, then your mill management has to **collectively** make the commitment to hold more professional meetings.

Reprinted with permission from *PIMA Magazine*, p. 32, January 1989. © 1989, Paper Industry Management Association.

Meetings: How to Make Them Better

Dr. T.F. Gautschi, P.E.
Professor & Consultant
Bryant College
Smithfield, RI

When you take a hard look at it, managers are really in the communications business. Duties may include planning, organizing, staffing, decision-making, controlling, and budgeting, but these actions are useless without communication.

The sad truth is that few managers receive any kind of formal communications training—or even give it much thought. Since we learn to talk, write, and interact with others at an early age, we may feel that we can handle all our communications activities intuitively. We must be wrong, because communications consistently remains a management weakness—especially as it relates to meetings.

Meetings are a fact of organizational life. There are 20 million business meetings each day in this country. Managers and executives spend about half their time running or attending them. When properly conducted, meetings are an effective tool for sharing information, giving instructions, or solving problems among several people. But often they are too frequent, too long, ineffective, and indecisive. The tool is okay—it's the manner in which we use meetings that causes problems. Condemning all meetings would be like my throwing out my tennis racket because it doesn't volley well. It's not the racket's problem; it's my problem.

Since a meeting can involve a considerable investment of time, it should be approached in a professional manner. First, the meeting caller must establish specific purposes for the meeting, preferably in writing. This sounds obvious, but it is surprising how many meetings are held without agreement on the purpose.

The leader should determine if there's a better, more time-effective alternative to holding a meeting. Perhaps a telephone call or a memo would suffice. Instead of holding weekly staff meetings, a flexible meeting schedule might focus more on priorities and the communication needs of the organization.

Meetings are only as good as their leaders

There are three general types of meetings: those for sharing information, giving instructions, and solving problems. The leader should establish the type of meeting and plan its details. These include an agenda, a time schedule, a proper attendance list, and providing necessary prework. All should be sent to the attendees in time for them to prepare for the meeting.

The meeting caller should start at the designated time and remain on schedule. Don't delay the start of the meeting to accommodate late arrivals—if you do, soon everyone will arrive late. Try to avoid digressions, or wandering into subjects or discussions that are not appropriate to the meeting's purpose. For example, if a problem involving only two of the participants is identified, assign its solution as an action item to those two and go on with the meeting.

To reach a decision, the group should first concentrate on defining the problem, go on to establish a common data base, then explore alternatives, and lastly discuss the solution. The natural tendency for most groups seems to be to jump to a discussion of the solution before going through the prerequisite steps. This encourages non-productive "groupthink," especially if a high-status person expresses his opinion early.

Be a catalyst to open discussion

The group leader should encourage the participants to feel free to disagree with him or her, or any other high-status person present. He should encourage minority viewpoints as well as those of the majority. The group should be structured and operated to assure that the various points of view will be considered.

Since there's a tendency to evaluate new suggestions or ideas as they appear, and because such evaluation tends to shut off the flow of unique and innovative contributions, evaluation should be withheld until all ideas have been brought out.

Someone should play devil's advocate and challenge the majority position. New perspectives regarding the problem are often helpful. If a new product offering is discussed, for example, a group could try to view it as if a competitor were planning the product, instead of them.

The group process itself should be examined from time to time. Is someone dominating the group? Are some people not making any contributions? Were the disadvantages as well as the advantages considered? Did the group exhibit groupthink symptoms?

The natural tendency for most groups is to jump to a discussion of the solution before going through the proper steps.

Symptoms of when meetings lapse into groupthink include:
- Limiting discussion of alternatives to a minimum.
- Not fully examining alternatives.
- Not obtaining outside, expert opinions.
- Dwelling on "facts" that support a chosen position.
- Leaving out or not considering contingency plans.

For important matters, perhaps a "second chance" meeting should be held after a preliminary decision has been reached, but before the final decision is made. This would give everyone the opportunity to mull over the various alternatives and facts and discuss them with others.

Be sure to review the results of the meeting (decisions made, and action items assigned/accepted) before adjourning. It is also a good idea to document the results in a memo and send copies to the attendees after the meeting.

It's important to assure that only the right people attend the meeting, those who understand what is being discussed, have the power to make a decision, will be responsible for implementing a decision, or will be affected by the decisions made. Meetings tend to be shorter when they encounter a time constraint such as lunch or the end of the day.

What goes on in meetings tends to be a reflection of the organization's culture. For example, if the culture does not tolerate mistakes, participants will be reluctant to admit to a mistake in an open meeting of colleagues and superiors. In fact, the things that cause poor meetings—problems in communication, internal coordination, handling conflict, critique, and feedback—are the same factors that can cause an entire firm's operation to go sour. □

Reprinted with permission from *Design News*, vol. 45, no. 16, p. 210, August 21, 1989. © 1989 by Cahner's Publications, Newton, MA.

Toward Better Meetings: A Psychologist's View

EUGENE RAUDSEPP

Abstract—Meetings are potentially the best means of communication between managers and the people who report to them. Often, however, meeting time is misused and unproductive. The author suggests that in planning your next meeting you consider group size, check the meeting place before the meeting convenes, and prepare and distribute an agenda. Tips are given on guiding the discussions, handling behavior problems, establishing rules of order, and voting.

THE RIGHT NUMBER OF COOKS

GROUP size significantly influences problem-solving and decision-making. In general, experiments have shown that as the size of the group is enlarged, members are able to participate less and become increasingly dissatisfied. Also, in larger groups a consensus is harder to achieve.

Eight participants, for example, are more likely to reach a consensus than are fifteen. In addition, the leader in the group of eight will influence decisions more than the leader in the larger group.

One obvious reason for this is that, in a large group, there is not enough time for all members to participate in the discussion, resulting in dissatisfaction with the entire meeting.

With many employees, lack of opportunity to participate and interact with others leads to the feeling that their opinions are not important, and therefore not worth presenting. Large groups also tend to disintegrate or break into factions, and spokesmen for each faction then take over the discussion.

SETTING THE SCENE

Although the importance of the meeting place's physical condition should be obvious, this factor is often overlooked. The discussion leader should personally check the conference room before anybody arrives, and see to it that

- Participants can sit sufficiently close together to develop the necessary feeling of camaraderie. The attendees should be able to see each other and the discussion leader without having to turn around.
- A blackboard and a space for the display of diagrams, charts, and maps are provided.
- The lighting and air conditioning are working well.
- The room is free from external noise and potential interruptions.

PLANNING AHEAD

The discussion leader should prepare a carefully conceived, written plan or agenda. He or she must be sure of the objectives of the meeting and should list a number of problems or

The author is president of Princeton Creative Research, Inc., P. O. Box 122, Princeton, NJ 08540, (609) 924-3215.

questions that are to be considered. Wording of the questions is important because this determines the success of the meeting to a large degree. Effective leaders often spend as long as eight hours preparing three or four discussion areas for an important meeting!

The following sequence of steps should be the minimum for a meeting plan:

1. Statement of the problem (or problems).
2. Statement of the facts.
3. Consideration of objectives.
4. Examination of proposals in the light of the facts and objectives agreed upon.
5. Summary of the discussion.
6. Consideration of means to implement decisions.

A detailed plan or agenda guides the discussion leader and serves as a reminder of the points to bring up so that the group will gain full value. A group without either a plan or an agenda may have an interesting discussion, but may seldom reach any destination. Despite a lively exchange of ideas, the group may fail to consider the only phases of the problem that can lead to a sound conclusion.

Discussion leaders should take brief notes during the meeting or, preferably, assign someone else to do this.

Summarizing the discussion is a vital step in the meeting procedure. It should not be a rehash of all the arguments presented, but a brief review of the *results* of the discussion—a statement of points upon which agreement was reached or a solution found, as well as a listing of issues left undecided due to lack of information or unresolved disagreements. A statement is in order as to how the discussion leader proposes to supply needed information and how he or she plans to resolve the conflicts.

Disagreements need not be considered negatively. A meeting without any disagreements indicates either lack of involvement on the part of the members or a constricting atmosphere. As a rule, resolving existing disagreements is easier than breathing new life into an over-conforming, uninterested group.

The most frequent shortcoming in preparing an agenda is failure to plan carefully. Most agendas are thrown together with little attention to organizing and highlighting different aspects of the problems to be discussed. As a result, those attending the meeting are not alerted to the important parts of scheduled topics. Problems should be listed on the agenda by priority.

ADVANCE NOTICE

A question that discussion leaders frequently ask is, "Should the agenda be circulated, with supporting documents, before a scheduled meeting?" In general, this is advisable. Circulating an agenda before a meeting enhances understanding, generates thinking, and primes the whole group with the best information available. Group members who come to meetings prepared are

Reprinted from *IEEE Trans. Prof. Comm.*, vol. PC-24, no. 3, pp. 136–138, September 1981.

better motivated and derive greater satisfaction from the sessions.

On the other hand, the major shortcoming of prior preparation is that opinions can crystallize before the general discussion takes place. This may hinder group decision-making.

The advance agenda should consist of three parts:

1. Items of information that require routine or no action.
2. Items with recommended courses of action. (Members of the group should read these items carefully before the meeting so that—if recommendations are acceptable—quick action may be taken.)
3. Items that are classified as "guidance needed," "solution needed," or "discussion indicated."

Some leaders encourage the group to prepare discussion questions, which are submitted in writing before an agenda is drawn up. This gives the members an increased sense of involvement in the meeting.

STEERING THE MEETING

Having many individuals in a group means many ideas, and many ideas can produce a sounder conclusion than a few ideas. However, this situation can also cause confusion. Here is where the discussion leader's effective presence is important. He or she maintains the group's sense of direction and keeps the meeting moving.

When the discussion is underway, the leader should regulate—not lecture. He limits his own participation to starting things and asking thought-stimulating questions (and follow-up questions) that provoke and guide the discussion. The longer a discussion leader talks at the opening of a meeting, the more passive the group becomes, and the more difficult the task of initiating and arousing the discussion becomes.

During the discussion, the group members, although naturally interested in the leader's ideas and opinions, resent being *controlled* by the leader's ideas—they want conclusions to emerge from the body of the group. The leader should draw out all the members of the group and keep their attention on essential problems so that decisions can be reached.

The leader also has the task of clarifying and interpreting the points made. Some participants state their points in a confusing and vague fashion. The leader must clarify obscure contributions by asking the confused member a few pointed questions, by uncovering the hidden premises in the speaker's statement, or by rephrasing the statement. Occasionally, clarification requires stopping the discussion and taking stock. This will make plain which issues have already been discussed and with what conclusions, what aspect of the problem is under discussion at the moment, and what points of conflict (if any) should receive attention next during the meeting.

The leader must be impartial toward the ideas and contributions offered during discussion. It is proper to acknowledge a contribution, indicating approval that the member has participated. But the leader should not indicate in any way whether he or she is in sympathy with the ideas expressed in the contribution, before pros and cons have been heard.

A leader can sharpen a group's thinking and help the members arrive at significant and effective solutions to problems if he or she poses pertinent questions at vital points of the discussion.

The high point of interest in the discussion should occur near the close of each meeting. This ensures effective decisions and provides a carryover of involvement to the next session.

How effective a group is in carrying out a decision depends on several factors. One of the most important is how clearly the goal has been defined in the discussion, and the degree to which the group directs its energies toward that goal. A responsible leader keeps a mental "goal-achievement index" in order to state at any time how much progress has been made toward a goal. This index is the chief guide in formulating an agenda for each meeting.

HANDLING MISBEHAVIOR

Three common types of problem participants are

The Commentator The discussion leader may have to deal with a flurry of whispered side-conversations, especially when a large number of people participate in a meeting. These conversations can distract and split the group. One of the most effective ways of dealing with this is to interrupt the general discussion abruptly and ask for silence. Frequently, of course, a whispered remark can be highly relevant to the discussion.

The Monopolizer Some participants tend to monopolize the discussion. This slows the pace of the meeting and also tends to divide the group. Sometimes the verbose member is incapable of submitting a contribution in a few short, clear sentences, or may present several ideas at once.

A more serious case of monopolization occurs when the speaker talks not only too long but too often. Such people also usually interrupt other speakers, confusing the discussion and annoying people.

As a rule, the monopolizer is also a bad listener; thus, there is much irrelevancy in his or her arguments. The leader should politely interrupt the speaker either by asking a question or by stating the point himself.

The Clam Diametrically opposed to the overly talkative members of the group are the sphinx-like ones who do not contribute anything. If they are hostile, or if their number exceeds two or three, their presence has a decidedly adverse effect. Such people should be invited to express an opinion. After they have heard themselves speak a couple of times, they may overcome their reticence and even prove to be among the more informed and thoughtful members of the group.

RULES OF ORDER

Establishing correct meeting procedures is the best way to minimize disruptive behavior. When a series of meetings is planned, the discussion leader should devote 15 minutes of the first session to some carefully worked-out rules of order. A copy of the rules should be distributed to all members of the group and be discussed by the group leader before the regular business of the meeting.

An effective meeting guide might suggest the following rules:

• Reach all decisions by majority vote (or consensus). The discussion leader should have no preestablished decision on the problems scheduled.

• Urge everyone to participate. Voluntary expression of ideas facilitates group thinking and contributes to a lively discussion.
• Speak out one at a time and do not interrupt another speaker. Show courtesy to the speaker—refrain from private conversation.
• Be brief; present only one point at a time.
• Avoid expressions that arouse hostility, even though you will encounter many differences of opinion. Be frank and honest, but avoid emotionally tinged words and expressions that create unneccessary opposition.

When obstructive behavior occurs, calling attention to the rules should restore order.

UP FOR A VOTE

The discussion should produce something approaching unanimity; frequently, however, a vote is necessary. When the leader is reasonably sure that everything of value has been said by the group, he or she should (1) make a last summation of the discussion, highlighting the principal ideas developed, and (2) put the issue to a vote for decision, after which no further discussion on that matter is allowed.

Some rules that will further help you reach your goals follow.

TEN RULES FOR PROFITABLE MEETINGS

1. Don't call a meeting to decide something you could and should decide yourself.

2. Never get people together if a series of phone calls to individuals would serve your purpose.

3. Never invite anyone who is not essential, but make sure that everyone who would be of value to the meeting is included.

4. Insist on punctuality. If you're two minutes late for a 20-person meeting, you waste 40 person-minutes.

5. Keep the purpose of your meeting firmly in mind, and be sure it can be achieved.

6. Draft an agenda that breaks down all subjects into their simplest constituents. If well constructed, even a lengthy agenda can result in a short meeting.

7. Before sending out your agenda, read it through and examine the points that could be misunderstood. In most meetings, disagreements occur because people are not talking about the same thing. If the issues are crystal clear, the muddlers will have less chance of confusing them.

8. Circulate the agenda in time for people to read it before they come but not so far ahead that they will have forgotten about it by the time they arrive.

9. Set time limits for each section of the discussion. Make sure everyone can see a clock. A discussion, like work, expands to fill the time available.

10. See that the person in "the chair" actually acts as chairperson, i.e., that he or she states the issues, keeps to the agenda, lets everyone have a fair crack at the subject, cuts them short if they wander and sums up succinctly as soon as all have had their say.

Presenting the Successful Technical Seminar

THOMAS EALEY

Abstract—Conducting a successful technical seminar requires careful preparation. Pre-seminar planning includes (1) outlining your material, (2) developing much more material than you think you'll need, and (3) arranging comfortable and appropriate facilities. At the beginning of the seminar you should provide a topical outline and announce plans to take a break between topics or at least hourly. After the presentation, time should be allowed not only for questions and answers but also for an evaluation to help you prepare for future seminars.

THE knowledge explosion and the desire of most business people to enhance their careers has led to a boom in continuing education classes, usually in seminar form. The demand for highly technical seminars has created many opportunities for knowledgeable people with good communication skills, but with these opportunities come considerable challenge and usually unforeseen problems.

Whether your field is law, medicine, accounting, engineering, or computer science, you may at some point be requested to present or coordinate such a seminar, or you may even take the initiative and present one yourself. The seminar can be an opportunity to make a great impression on colleagues, your employer, potential employers, and persons who will be hiring instructors for future seminars. Unhappily, you can also cause yourself considerable embarrassment and discomfort by doing only an adequate job when expectations are high or, worse, by doing a poor job.

The people who attend technical seminars are, as a group, much different from the participants in general interest seminars, which deal with such broad subjects as motivation, basic speechcraft, or general leadership skills. Technical seminar participants have a specific purpose and a sharp focus. They are, for the most part, demanding and serious. They value their time and take great pride in their skills and talents. Most have limited time budgeted during the year for continuing education. They either have paid for the seminar personally or know what their employer has paid, and they expect their money's worth. If they feel their time is being wasted, they will certainly let you know.

The key to presenting a good technical seminar is preparation. People attend technical seminars because they feel

Reprinted with permission from the June 1982 issue of *The Toastmaster*, published by Toastmasters International in Mission Viejo, Calif. (714) 858-8255. A nonprofit organization with 7,400 clubs worldwide, Toastmasters International teaches its members skills in public speaking and leadership.

The author is an instructor of business at Findlay College, Findlay, OH 45840, (419) 422-8313 X346.

that each hour in class has more value than many hours of independent reading and analysis. They expect you to have done the homework, the long hours of reading and analysis, so they can receive a concise presentation of the information they need. They don't expect you to cover each and every fine point within the subject area but they do want to leave feeling they can use something from the seminar in their daily work.

BASIC OUTLINE

The first steps in preparing the technical seminar are to outline the material and plan topical segments of digestible size.

The exact nature of your subject usually suggests your topical organization. A tax seminar may proceed on a numerical sequence of revenue code sections whereas an architectural seminar may proceed from foundation to roof. Don't discount an outline that seems too obvious and straightforward; your group will be able to distinguish between an erudite leader and one who tries to appear intelligent by complicating the material. If you cannot fit your material into an outline of eight or fewer topics for an all-day seminar, you need further planning and revision.

A segment should rarely run longer than one hour and you should never conduct more than two segments without taking a break. One hour is usually as long as the audience's attention span can be stretched. If you do plan to move from

Reprinted from *IEEE Trans. Prof. Comm.*, vol. PC-26, no. 1, p. 35–37, March 1983.

segment to segment without taking a break, schedule a short question-and-answer period and a 30-second stretch-and-yawn session to clearly mark the transition.

If you are truly prepared, you will probably develop 12 hours of material to fit your eight-hour outline. (Remember, however, you're preparing to talk to experts, so there is no such thing as being too prepared!) If this is the case, you must edit the material so your presentation fits the time schedule and is clear and concise. Any edited-out material is still useful, though—it enables you to fill any extra time you may suddenly have in your actual presentation, and it gives you the background to handle questions.

The presentation itself should be tailored to the needs of the audience and the demands made by the material. There's no one correct style for a presentation; rather, your style and format depend on each other and deserve as much thought as the material itself.

Obviously, some materials demand a lecture format, usually supplemented with handouts and projected transparencies. Remember the comments about efficiency and your group's desire to assimilate as much information as possible in the limited time available—a crisp, concise lecture can be both enlightening and efficient.

The lecture format, however, doesn't work in all instances. You may be teaching one of the newer theories of management that are heavily oriented toward democratic participation, enlightened thinking, and improved communication skills. What kind of example would you be creating if your presentation format ignored or violated your own lesson? This type of seminar calls for a combination of lecture, small group discussion, role playing, and films. No matter what your format, though, always provide your group with a topical outline at the beginning of the session. This defines the path you are following and prevents distracting conjecture on what is going to happen next.

Never insult your group by reading material to them. You may need to read specific passages or statistics, but your group will start napping or become aggravated if you do little else but read to them. Add flesh to whatever concepts or techniques you're teaching by providing emphasis, examples, anecdotes, and enthusiasm.

Always include time for questions from your audience, either by allowing questions at any time during the presentation or by calling for them at the end of your presentation. Keep in mind that taking questions throughout the seminar can be disastrous if you allow yourself to be led off the track or if you devote so much time to questions that you cannot finish your presentation in the allotted time. The best approach is to set aside specific times for questions, perhaps as you finish each major topic within your outline. This way you'll keep in touch with the group and still control the flow and continuity of the presentation.

COMFORTABLE CONDITIONS

Motivational speakers and tent preachers may enjoy "packin' 'em in," but crowded conditions and uncomfortable room temperatures aren't appreciated by people interested in capital gains taxes or the load-bearing properties of concrete. Arrange for a room that is large enough to allow each person to have an adequate table top area and comfortable seating. If necessary, place a strict limit on the number of people attending—wouldn't you rather have 40 satisfied participants than 50 disgruntled ones?

Provide a break area that is roomy and well-ventilated for smokers and has access to restrooms. Preferably this should be an area completely separate from the classroom so there'll be a true break from work.

Sound and audiovisual systems are critical and deserve a great deal of your attention. Talk with your host and the facility manager as far in advance as possible, being very specific about the type and quality of equipment you need.

When using a microphone, arrive early enough to test the mike and adjust the sound level before your audience arrives. Protest if the equipment is not delivered as promised and inquire about back-up equipment and access to amplifier controls and power sources. Someone should have immediate access to the amplifier controls during your entire presentation.

If an overhead projector or a film projector is part of your program, have the appropriate spare bulb on hand. If a film is the cornerstone of your presentation, have a splicer or a spare print ready. A good supply of grease pencils and blank transparencies allows spontaneous additions to projected displays. Whatever your setup, no one will ever complain because you are over-prepared.

FEEDBACK MECHANISM

Many seminar leaders include evaluation mechanisms in their programs. They need to know whether their presentations are accomplishing their purposes and satisfying their audiences. If your're planning to conduct more seminars, this information is invaluable.

The most commonly used evaluation technique is the questionnaire. Brief and to the point, it contains just enough questions to evaluate the major areas of presentation, organization, and setting. Designing a questionaire is not an easy job, so seek help if you are not confident in your ability to design a meaningful evaluation tool.

Remember, too, that a questionnaire isn't the only evaluation technique you can use. Consider choosing a few members of the group and asking them for a brief oral evaluation immediately after the seminar. You could ask for a written evaluation from them as a follow-up. Whatever the evaluation tool, if you have the mature ability to accept and use constructive criticism, the information will be helpful as you prepare future seminars.

When the work is done and the evaluations are in, sit back and give your seminar the ultimate test of quality. Would you want to sit through the seminar you just presented? If you can honestly pass this test, you can be sure your technical seminar was well organized and professionally presented. You can take considerable pride in having satisfied a tough audience!

When and How to Argue

W. A. DELANEY

Abstract—Arguing is presenting an opposite, alternative, or different viewpoint and the reasons for your opinion. The author warns us to consider the consequences of arguing by deciding what can be gained or lost. We often waste our time (and the time of others) by not being aware of the difference between *what* to argue and *how* to argue. What to argue calls for considering the problem's permanence, its seriousness, and the reality of your winning. How to argue involves remaining impersonal and gracious, stating what you want, sticking to the issue, listening to and seeing the other person's point of view, and not harping on the same issue.

As in any other form of human relationship, people at work get upset, angry, frustrated, and depressed. During such times, many of us have damaged our careers by inadvertent or deliberate remarks made to superiors, peers, or subordinates—responses that we regretted later on.

If such blunders occur frequently, serious personal attitude and behavior problems are present that far transcend the job. Professional guidance may be called for. But let's consider those situations that occur occasionally, or even rarely, when you want to (or have to) argue for one reason or another.

In such cases, irreparable damage can be done to your career for the momentary satisfaction of putting down someone who has been bothering you for years. Before you argue, consider the consequences and decide what you have to gain or lose.

RULES TO FOLLOW

Even in warfare many countries attempt, through the Geneva Convention, to make the whole terrible affair more humane by setting certain rules and regulations. Procedures have been agreed to, right down to whether to allow a soldier to use a silencer on his rifle, or what type of bullet could explode on impact and what type could not. It's O.K. to execute spies, but it's not O.K. if the person is in uniform when captured. You can shoot him dead while he is shooting at you, but if he drops his weapon and raises his hands, you are supposed to capture him alive, feed him, and protect his life, as you would your own comrades.

As senseless and barbarous as warfare is, people try to make and observe rules on how and how not to fight.

In everyday work, fortunately, no one is ever deliberately killed or injured. So it should be even easier for us to make up a set of guidelines to follow before we fight or argue.

It is best to think out clearly and objectively before you argue what you plan to say. With emotions aroused, we say too much, or say the wrong things, to the wrong person, at the wrong time, and injure a career or hurt someone else needlessly.

Three conditions should exist before you argue at work:

1. The problem should be permanent.
2. The problem should be serious.
3. You should have a real chance to win.

Some examples are in order here to prove each point.

1. If the problem is temporary, don't argue. Save yourself for more important matters. If your secretary, who is your "right hand," goes on vacation and is replaced for two or three weeks by someone who comes in late, leaves early, types sloppy letters, and has a bad attitude, it may be better to grit your teeth and await the return of your pride and joy than to have an argument with the new person over a problem that time will solve.

If on occasion the boss is "out of sorts" and difficult to approach, it's best just to avoid him or her for a few days. If it happens more often or does not go away, then that's another matter. It must then be faced and resolved.

2. The problem must be serious. No one should argue or make noise over trivia. If the coffee machine doesn't work, forget it. Bring in your own pot; don't argue and fuss over an insignificant point. Even if you win, what do you win? You get better coffee and a reputation as a chronic complainer. It's just not worth it, is it? Arguments at work should concern only such matters as promotions, raises, job assignments, opportunities, and the like.

At one of my former places of employment, one senior manager didn't like the food in the company cafeteria. He complained repeatedly to higher management about it and he even formed a committee and went around getting signatures to back up his claim. He eventually won. The cafeteria was closed. Cooks, cleaners, and food servers lost their jobs, and we had to go outside to get lunch. Most of us had found the food to be acceptable, but one chronic complainer got his way. He won nothing and gained a reputation that persists because he really was, and still is, a chronic complainer. He could have eaten elsewhere, but he chose to argue instead.

3. Finally, and most importantly, you must see an opportunity to win your case before your start the argument. If you can't see a reasonable opportunity to win, then don't argue. It's pointless.

If you complain to the manager that his son or daughter who is working for the company during school vacation is a problem and not working hard enough, you won't win. Even

Reprinted with permission from *Supervisory Management*, vol. 24, no. 12, p. 16, December 1979; copyright 1979 by AMACOM, a division of American Management Associations, New York, NY 10020.

The author is president of Analysis and Computing Systems, Inc., 109 Terrace Hall Ave., Burlington, MA 01803, (617) 272-5000.

Reprinted from *IEEE Trans. Prof. Comm.*, vol. PC-24, no. 4, pp. 186–188, December 1981.

if the boss secretly agrees with you and removes his son or daughter, the boss will remember that you complained about it. Parents love their children. You do, too; remember that. You can't win such an argument, ever.

If you are passed over for a promotion, it is reasonable to ask why. When you are given a reason, if you don't like what you've heard, don't argue; remember you are not going to change the boss's mind at this point. To argue or complain that it wasn't fair won't gain you anything and may cost you your job. The time to argue this point is *before*, not after, the promotion is announced. You can't win by arguing or complaining in such situations. You can't change the score after the game ends.

If you consider the three conditions that should exist before you argue at work (or anywhere else for that matter), you will find that the majority of issues most people argue about do not pass the test. Especially at work, be more careful about when and with whom you argue. At home, you have people who love you, know you very well, and may not mind listening to you, periodically, when you want to free yourself of your frustration, anger, or depression by a long talk, discussion, or argument. This is not necessarily true at work. If one tries this sort of behavior at the office or factory, it can, and generally will, adversely affect one's career. Higher management does not, in general, promote people from the ranks of those who "have problems" of one sort or another. A history or record of having a hot temper, or of arguing a great deal about small or unimportant items, can, and generally will, slow down or stop a budding career.

It's a question of judgment, as it is with most things in life. The human "door mat" who never speaks up, no matter what happens, will be taken advantage of sooner or later and will be judged by superiors as one who lacks the moral fiber to assert himself or herself, even when justified. Leaders and top managers are never "door mats." But one must be selective about picking arguments. Also, an argument doesn't have to be a shouting match. This generally accomplishes nothing except to arouse emotions, create resentment, embarrass, or humiliate the initiator and recipient. What is meant here by arguing is presenting an opposite, alternative, or different viewpoint to whatever situation you are in at the time and the reasons for your opinion.

THE ART OF ARGUING

Now that what we have discussed when to argue, the next item is *how*. Again we can list some concerns to consider in advance when you find yourself in an argument and emotions are starting to get "hot." It's much like training in sports or in the military. The theory is (and it does work) if you train and rehearse well enough, certain actions become almost automatic. You then react properly even when under stress later on—in real-life situations of high emotion, strain, anger, or even fear.

Assuming, again, that the three conditions for having an argument exist, then how do we argue our point? Below are tactics or methods to use to win one's point. Some of these may appear obvious, but they bear repeating because many people don't use them, or use them improperly.

• Don't get personal; that is, don't criticize or hurt someone to gain your point. Whether you win or not, by using personal attacks you gain an enemy for life if you hurt someone else or prove him or her wrong before others. It is better to say "I have a different conclusion," than to say "You are wrong."
• Always give your opponent an honorable way to retreat or admit you are right. Don't overkill. Be gracious if you are winning; you can afford it. Try to let the other person off the hook, unless he or she is a repeater.
• Stick to the issue at hand. Don't go back into the past and bring up other problems you want to argue about. Don't save them up for occasional outbursts all at once.
• Be sure you are talking to the person who is able to give you what you want. This is a common mistake many people make. They tell everybody what the problem is and how they feel, but they don't talk to the one person who could do something about it. That person is usually your boss, and he or she does not like to hear about your arguments or complaints second hand. Tell your boss first and no one else.
• Try to see the other person's situation before you argue. You may not agree with it, but in trying to put yourself in his or her position, you can, many times, be better able to present your argument in a manner more acceptable to that individual.

Remember that you have to convince the person to do it your way; anger, noise, and name calling won't do that for you. Calm, reasonable, logical presentations are much more likely to help gain your point or win your argument.
• During your argument, be sure you state exactly what you want. Many times people present a case, yet fail to state what they want done. It may be obvious to them, but it may not be to others—they may need to be told in plain words. Don't assume that others will come to the same conclusion or solution that you have reached; they may not. You can't go back again if you win your argument but disagree with the solution to the problem because you left that to others to decide.

Think your case through before you start. In the past I have listened to arguments or problems and when I asked the person what he wanted me to do about it he didn't know. He just wanted to go on with his "presentation" with no recommended solution. He just wanted to gripe. That is pointless.

What most managers do in these situations is to let the person talk himself out and then quietly say, "Is that all?" or "Do you feel better now?" If any boss has ever said anything like that to you, be careful. That's what psychiatrists do with disturbed patients. It's a form of therapy. And bosses don't promote people who need therapy, do they?
• Be prepared to listen as well as talk. Rehearse your presentation and give time for your opponent to respond. Rushing in, having your say, then rushing out accomplishes nothing, except to convince your boss that you get upset, sound off, then run away. Such people are hardly considered for promotions.

• Don't keep going back over the same issue. Remember, there are two possible outcomes to an argument: You win or you lose. If you lose, either accept it or leave; don't harp on it. You can't win that way, and you may be removed involuntarily, as a result. You may be right in your argument, but it becomes a moot point if you lose your job over it.

Don't argue unless the three conditions for having an argument exist. If they do, then prepare your tactics and present your argument in the best way to win it. Don't "crow" when you win; be gracious and accept the decision if you lose. It's easy to be a good winner but losers tend to moan and gripe. A famous football coach once said, "Winners win. Losers quote statistics."

Remember your long-range goals. Don't win the battle and lose the war. There are times in every person's career when one has to argue, or be ignored and lose by default. A well-thought-out and well-presented argument can do wonders for one's career but the opposite can ruin it, so be careful about how you argue, with whom, and about what issues. Harsh words said can never be withdrawn. You can only apologize. The pain may leave, but the scar remains forever. Those who inflict scars on others, either deliberately or thoughtlessly, will, in general, receive the same in return and very probably from someone else. Think carefully before you sound off. In 90 percent of the situations it's not worth the trouble for what you can possibly gain in return.

Becoming an Effective Listener

MARION E. HAYNES

Abstract—Listening is the process of taking in information and synthesizing it into an understandable message. Clues to understanding the speaker include how things are said, what is not said, and nonverbal behavior. The difference between speaking and thinking speeds should be utilized to review, summarize, and reflect upon the speaker's points rather than to daydream. Physical distractions should be eliminated. Emotional and psychological distractions originating with the speaker or the listener should be discounted or compensated for. First impressions should be subjugated and judgment delayed until the speaker is finished. In conversation, give the other person a chance.

IT IS estimated that the average adult spends about a third of the time listening. Yet it is amazing how inefficient most people are at the process. Many reasons account for this. Among them: They don't like the speaker, they find the person boring, they feel threatened by what is being said, or they feel physically or emotionally tired.

All of these problems can be overcome by learning the techniques of effective listening. This means learning to pay attention not only to the speaker's words but also to their context, to note what's not said, to listen with a purpose, to minimize distractions, and to interpret nonverbal behavior and tone of voice.

ACT LIKE A LISTENER

Listening may seem passive when you observe someone listening to another person, but it is actually a very active, engaging process. So, the starting point for becoming a good listener is to sit in a relaxed but alert posture. This means no slouching. When you get too comfortable, you may become drowsy. Next, maintain eye contact with the speaker. This says you are interested in what's going on and you are paying attention. Finally, make some verbal response such as "Uh-huh" and "I see," restate the last one or two significant words spoken, or summarize a part of what was said.

SEEK UNDERSTANDING

Once you've begun acting like a listener, the next step is to try to understand the message conveyed. This involves three distinct processes—hearing, listening, and understanding.

Hearing is a physical process. If you are not otherwise physically impaired, you hear sounds above a certain threshold of loudness within a certain frequency range. This is hearing.

Listening is an emotional and intellectual process. It is concerned with what you do with the information you obtain from both hearing and observing. Good listening also depends upon how you interpret what you don't hear—silences and omissions in the message. What someone doesn't say or avoids saying is often more important than what is actually said.

Understanding grows from good listening. You understand when the message takes on meaning within your own frame of reference. For example, you can understand how someone feels about an experience through empathy. Or you can understand a concept by relating it to something you already know.

Hearing is not listening. Listening is not understanding. Confusion about these concepts gets in the way of effective listening skills. Understand listening and place it in its proper perspective. Listening depends upon hearing and leads to understanding. It is the process of taking in information and synthesizing it into something that you can understand.

Good listening involves seeking the answer to three basic questions:

1. What does the speaker mean?

People take it for granted that the speaker means the same thing they would mean if they were saying the same words. But the same words mean different things to different people, in different contexts, under different circumstances. A skillful listener starts with the attitude that what the speaker's words mean to the speaker is not known and that, in order to find out, the context and circumstances must be scrutinized. A skillful listener also knows that any interpretation of the speaker's words are the listener's and are therefore his or her responsibility.

Because of different meanings, the listener demands a great deal from the speaker. It falls upon the speaker to take special pains to make meanings as clear as possible, to choose words well—to distrust them, to suspect them of vagueness and ambiguity. It also falls upon the speaker to accept his or her own interpretations of words as being personal and in need of careful clarification.

2. How does the speaker know?

After a good listener finds out what the speaker means, the next thing to work on is what reliable, factual observations have been made, or could be made, to verify the speaker's statements. The good listener wants to see data, pictures, diagrams, or demonstrations. This does not rule out abstractions and generalizations. It does, however, require that evidence supporting them be presented.

3. What is being left out?

The third basic question encourages the art of listening for what a speaker does not say. This, as a rule, refers to omitting important factual details, not drawing certain possible conclusions, or not developing certain implications of the conclusions drawn.

Reprinted from *Supervisory Management*, vol. 24, no. 8, p. 21, August 1979 with permission. Copyright 1979 by M. E. Haynes.

The author is an Employee Relations Associate with Shell Oil Co., P.O. Box 2463, Houston, TX 77001, (713) 241-6855.

Reprinted from *IEEE Trans. Prof. Comm.*, vol. PC-23, no. 2, pp. 91–94, June 1980.

Often what is not said points out that what is said is inadequate, irrelevant, or misleading. Did the speaker overstress issues in his or her favor? Was the full story presented?

Use Spare Time Effectively

When people talk, they speak at a rate of 100 to 200 words per minute. However, people think much faster, perhaps somewhere between 600 and 800 words per minute. What is done with this speaking-thinking speed differential is the key to effective listening. It can be used to become a better listener or to interfere with effective listening. Most people do not use this differential wisely, letting their minds take side trips to worry about something or to think about what they are going to do that evening or about what they will say when they get a turn to talk.

There is plenty of time for these side trips. And most people have been taking them so long that the habit is well ingrained.

But instead of taking side trips, try using this time effectively. Think about what the speaker has been saying or try to figure out where the speaker is going and what the next point may be. This will probably be easy with a well organized speaker who presents points clearly. You could also summarize what is being said or break it down into main points and supporting points. Periodically play back your summary so that it can be verified by the speaker. If you have missed some main points, they can be filled in for you.

Listen for feelings as well as content. Almost every message has a feelings dimension. Many have as a main theme the feelings being expressed and the content is simply a legitimate way of expressing them. By hearing the feelings and exploring the reasons for them, you can understand what led up to the statement. Also look for consistency or inconsistency in what is being said.

Listening is facilitated when you have a purpose for listening and concentrate on achieving it. The next time you are in a conversation with someone, take a few seconds and make a simple declarative statement of purpose. It will focus your attention on what you can get out of the conversation. When tempted to take a mental side trip, your attention can be brought back by a simple reminder of purpose.

Minimize Distractions

Distractions may come from three different sources—the environment, the listener, or the speaker.

• *Environmental distractions* include noise, people passing by, uncomfortable temperature, and poor ventilation. To improve listening, close the door, reposition seating, or, where applicable, turn off the radio. If you are a manager, have your secretary answer your phone and prevent drop-in visitors. Most calls can be returned later when you can give them your undivided attention.

• *Distractions from within* sometimes compete for attention. Instead of allowing your mind to wander, concentrate on what is taking place now—the conversation in which you are engaged—and most distractions will be shut out. You simply can't do two things at the same time, so by forcing attention to focus on the topic of conversation, other thoughts will be excluded.

When listening, you may hear something that causes an emotional reaction and distracts you from listening to all the other person has to say. Watch out for the following three reactions:

1. Defensiveness

In the course of a conversation something may be said that threatens your view of yourself. In response to that, you shift the conversation to a defense of your self-image or you allow hurt feelings to fester instead of diverting the conversation. Attention becomes focused on these feelings and because attention is somewhere else, your listening capacity suffers.

2. Resentment to opposition

It is always easier to listen to ideas similar to your own than to opposing points of view. The resentment that builds when the one to whom you are talking does not accept your ideas may get in the way of listening. Usually when opposition is encountered, the time not spent talking is spent developing a strategy and reply to overcome the opposition. Rather than listening, the mind is busy planning what will be said and how it will be said in order to make a point more clearly and destroy the opposing point of view.

3. Reactions to individuals

Sometimes a general reaction to a person will interfere with effective listening. For example, if you don't like someone, you will have difficulty listening to that person. Likewise, if you resent someone or feel threatened, these feelings will color your views of what that person may say.

There are positive filters as well as negative ones. You may so admire someone that you accept whatever is said without question. Again, a reaction to the individual gets in the way of effective listening. To improve listening ability, recognize these filters and work to set them aside.

• *Distractions from the speaker* can be caused by his or her accent, mannerisms, dress or grooming habits, language usage, delivery style, and so forth. Guard against letting a speaker's entertaining or challenging style become the focus of attention to the exclusion of content. Remember, valuable ideas can be eloquently presented and they can also be presented with a stutter, lisp, accent, or in such a lumbering style that you must fight to keep from losing interest. Probe beyond style to discover the value in the content of the message.

Delay Judgment

Most people have a habit of forming first impressions about what they are listening to. These early judgments serve as filters that color the way the remainder of the message is received. If you decide a message is good or bad, you select evidence from all that follows to back that opinion. Statements in support of a contrary viewpoint are never heard.

To overcome this problem, force yourself to delay judgment until you have heard the other person out. Often this will improve the climate of the conversation because it reflects an attitude of acceptance. Even though they may not say so, people can sense when their statements are being discounted.

SHUT UP AND LISTEN

An obvious point about listening—but one often over-looked—is this: If you want to listen to someone, you must provide the opportunity for the other person to talk. Think of someone you know who chatters constantly and how frustrated you feel because you do not get a chance to say anything. Check yourself on this issue. Don't be guilty of hogging the show. Be sensitive to the other person's need for air time. The person who seems never to have anything to say may simply need an opportunity to speak or a little encouragement. If so, the best encouragement is an attentive, nonjudgmental listener.

Another common problem stems from the listener's jumping in to help the speaker. Unless the speaker actually wants help, this will usually cause some anger to develop that will have a negative impact on further communication. Why does this happen? Some listeners become impatient with speakers who are slow and ponderous. Others feel that all blank spaces in any conversation should be filled with words. So, when the speaker pauses to catch a breath, organize a response, or recall a fact, the listener jumps into the conversation. There are also some people who like to show how smart they are by jumping in to provide information or the punch line to a joke. Begin now to break the habit; eliminate this lack of courtesy. Make a pledge to be patient and to give the speaker time to speak.

LISTENING TO HOW THINGS ARE SAID

Words are never neutral. They are affected by tone of voice, which accounts for 37 percent of a message's impact. Tone of voice is one of the most obvious clues to the speaker's feelings about the topic. Through tone, excitement, anger, joy, frustration, disinterest, resignation, concern, and so forth are communicated to the listener if he or she listens well.

Tone of voice can be broken down into four components—emphasis, speed, pitch, and volume.

• *The emphasis* placed on different words in a sentence—inflection in the speaker's voice—adds important meaning.

• *Speed of delivery* is another thing to observe. A speaker's delivery is slow- or fast-paced in relation to the individual's normal rate of speaking. Variations in either direction, slower or faster than normal, are significant clues to the speaker's feelings and should be noted.

An increase in speed of delivery usually signals an increase in emotional intensity. When people are excited, angry, or frustrated, they tend to talk more rapidly. A slowdown in speech delivery usually signals a resistance to address the topic. This happens in response to some perceived threat, which may be physical (such as the threat of punishment) or emotional (such as a threat to one's self-concept, ego, or sense of values). Or a slowdown in speech delivery may indicate a need to think in order to recall information and formulate responses. So, when you notice a slowdown in speech delivery, look for other data to help you draw appropriate inferences about what the other person is actually experiencing.

• *Pitch* can range from very high to very low. When experiencing stress or anxiety, the throat muscles tighten causing the voice to be at a higher pitch. When engaging in conversation, pay attention to the pitch of your own as well as the other person's voice. Allow a few minutes of related but less significant conversation at the beginning so anxiety can dissipate and both of you can relax and become comfortable. Use voice pitch as one gauge of when someone else is at ease.

• *Volume* is the final tone-of-voice dimension to observe. Some people naturally speak loudly but others use a loud, commanding voice as a mechanism of dominance. Still others, who may prefer to be less conspicuous, speak in a softer, quieter voice. It helps to know the person with whom you are conversing in order to accurately establish a reference base. However, you will be able to observe voice volume during the course of any particular conversation. If volume varies, make note of it. If volume seems louder or quieter than appropriate for the setting, make note of that. Then look for other clues that aid in forming inferences about what is being experienced.

An increase in speech volume usually indicates an increase in emotional intensity. When excited or angry, people tend to speak more loudly. Attempt to determine the reason. Then respond to the cause—not the voice volume. If loudness is in response to joy, delight, or elation, don't attempt to dampen an appropriate expression of these feelings. If loudness is in response to anger, fright, or frustration, you may choose to attempt to calm the person if you feel unqualified to deal with these feelings.

A decrease in volume usually signals feelings of perceived threat. It's as though the person talking were trying to prevent the other person from hearing what was being said or attempting to avoid saying it. Or sometimes the decrease in volume is simply caused by the speaker's not wanting to be overheard by those for whom the message was not intended. A soft voice, therefore, is only an indicator. Look for others. When coupled with slow speech delivery, it probably does reflect some feelings of threat. When coupled with fast speech delivery, it may simply mean an attempt not to be overheard. In any case, also consider other nonverbal clues such as gestures and posture before drawing inferences about meaning.

OBSERVE NONVERBAL BEHAVIOR

Most people are constantly moving. These movements reveal feelings, emotions, and reactions. Sometimes we are very much aware of body movements, as when we smile at a friend, frown, raise an eyebrow in surprise, or wink. Other times we are unaware of them.

Gestures and body movements come in clusters—the way people move toward or away from each other; the way they sit, whether tense, relaxed, on the edge, or slouched. People usually lean forward when they are involved and interested. They tend to lean back when they are not. What people do with their hands, arms, legs, and facial expressions is all part of a cluster.

Try to be particularly sensitive to any lack of congruency among the various verbal and nonverbal messages. A cluster of gestures and movements can be compared to a spoken or written sentence, while a single gesture or movement can be compared to a single word. A word, in isolation, often lacks meaning. It requires the context of the sentence for the full meaning to come through.

A word of caution about gestures. They must be viewed as indicators to be verified through other observations, either verbal or nonverbal, before they become fact. The reason is that some gestures are repeated merely because of habit while others are in response to a physical stimulus. (For example, a person might rub his or her nose from habit, because it itches, or because of feelings of deep concern.)

Let's look at some specific nonverbal behavior.

• *Eye contact* How one person looks at another is a major part of nonverbal communication. You can see interest, excitement, belligerence, warmth, and skepticism if you pay attention to the way others look at you. There are some unwritten rules about eye contact. One says that when talking, look at the person to whom you are speaking. Likewise, when being spoken to, look at the person speaking. People generally feel uncomfortable when the person to whom they are speaking does not look at them; they feel a lack of interest in what is being said. People also are uncomfortable when someone speaking to them does not maintain eye contact. A feeling of insincerity is communicated.

Direct eye contact indicates readiness to engage in the business at hand. Frequent eye contact indicates confidence. These observations let you know that the other person is interested and eager. However, when you observe your conversation partner squinting, it may indicate suspicion or doubt. Stop and check it out. Likewise, should you observe a blank, zombie-like stare, it probably indicates boredom, in which case a change of pace is called for to regain interest or the conversation should be terminated.

• *Use of hands* Next to the eyes, hands present the most expressive nonverbal communications. Open, smooth gestures are indicative of an open attitude. Covering the mouth or tugging at an ear often indicates nervousness, as does jingling keys or coins. Suspicion and doubt are communicated by touching or rubbing the nose. Running the fingers through the hair or rubbing the back of the neck suggests frustration, while thumping the fingers on something indicates boredom or impatience. Steepling, that is, bringing the hands together so they touch only at the finger tips, communicates confidence.

Touching is an important means of communication. A pat on the back; shaking, clasping, or holding a hand; or an arm around the shoulder can communicate more than many speeches.

• *Posture* How one sits during a conversation displays one's feelings about the other person as well as about the topic of conversation. Observe your partner. Is the posture tense and rigid or relaxed? This will indicate the amount of stress being experienced. If it is high, take some time to put the person at ease by talking about a nonthreatening subject.

Sitting on the edge of the chair and leaning forward indicates a readiness to move on and an attitude of cooperation. Acceptance is indicated by moving closer. Watch for these signs as indicators of agreement and conclude the conversation. Sitting erect, but not tense, indicates confidence while slouching suggests defensiveness. Turning the body away from the speaker and tilting the head forward indicates suspicion. Use these observations to know where you are with your conversation partner and alter your style or topic to utilize or overcome what you see.

CONCLUSION

All these components of effective listening show that it is a complex process that involves being alert to all that is heard, all that may be left out, and all that's observed. What's more, these perceptions must be thoughtfully integrated. Only then can a full understanding of the speaker's message be achieved.

EFFECTIVE LISTENING

Blake Emery

Senior Instructor
Boeing Computer Services
P. O. Box 24346 MS 9A-90
Seattle, Wa 98124
(206) 575-7462

Karen Klamm

Senior Technical Writer
Boeing Computer Services
P. O. Box 24346 MS 7R-04
Seattle, Wa 98124
(206) 644-6118

This paper presents several important listening tools that can be practiced and used immediately. These include step-by-step instructions on (1) how to show someone that you are really paying attention, (2) the use of empathy, and (3) how to use different approaches to verbally clarify what the speaker has said, especially when there is a potential disagreement. The use of restatement is discussed and a model for assessment and evaluation of ideas is presented.

INTRODUCTION

Listening is the most used—and least understood—communication skill. Most adults have had no formal training in listening skills, even though we should be typically using this skill constantly—about 36% of our waking hours.

In the U.S., prior to 1978 the basic determinants of literacy were reading, writing, and arithmetic. Then in 1978, the U.S. Primary-Secondary Education Act was amended to include speaking and listening. Listening skills have now been incorporated into the curriculums of elementary and secondary schools nationwide.

For adults, courses in effective listening are now being taught in the workplace as well as by universities and consultants. Some companies concentrate on training their managers or their sales force, since the sales and management arenas are widely identified as requiring topnotch communication skills. Yet, the reality is that any worker in any position can either listen effectively or cause costly errors due to faulty listening. Certainly professional technical communicators have a key leadership role as adults attempt to upgrade and expand their communication skills.

This paper will not present any additional background, but will concentrate on some tools of effective listening that can be practiced and used immediately.

HOW TO SHOW SOMEONE THAT YOU ARE LISTENING

"Attentiveness" is the term used for the process of sending signals that say you are really aware of and receiving communication cues. To put it simply, good listeners really make you feel like they are listening. By their behavior, they make you feel that, for the moment, you are the only thing on their minds.

Here are some tips on looking attentive, which will make the person talking to you feel more relaxed and confident.

o   Concentrate on the speaker, allowing your natural animation to come through (facial expressions, body language, nodding, etc.).

o   Stop all unrelated activities such as opening mail or rearranging your desk. While this is a difficult habit to break, it's a key point. Little distractions such as these detract from your ability to listen and also tell the speaker that you aren't concentrating solely on the conversation.

o   Face the speaker, even if you have to shift yourself around a bit (such as in a meeting).

o   Comment every now and then, as appropriate. You can use "openers" ("Well maybe we should discuss this more thoroughly..."), simple, non-directive acknowledgements ("Okay, Yes, Oh, Really, Uh-huh"), or reflective listening ("You feel that there is too much time being wasted...").

o   Balance comments with silence. Total silence is a negative signal, but people do need enough silence to let them express their ideas completely.

o   Maintain natural eye contact. Don't stare into space for an entire conversation, but don't feel like you have to look at the speaker every second either. There is a natural rhythm of making eye contact, then looking away to gather your thoughts before looking at the speaker again. To some extent, this rhythm is cultural, so you may have to adjust your rhythm.

o   Smile occasionally where appropriate. Again, this helps the speaker to feel relaxed and worthwhile.

HOW TO CLARIFY PEOPLE INTO A BETTER IDEA

When you listen to someone, you pick up their verbal and nonverbal clues and translate those cues into a "message" that seems meaningful to you. But how can you know if the message you constructed is the same message the speaker intended? You must *open your mouth and interact* with the other person until you both agree on a meaning. In a sense, you negotiate until you reach a consensus.

The ideas discussed in this section can lead to better ideas in two ways: (1) listeners who use these techniques will achieve a better idea of what the speaker is trying to say, and (2) the ensuing interaction will itself produce "better ideas" in a creative synthesis. The purpose of this kind of listening is to simply increase the level of understanding between people.

Clarifying By Asking

One way to get a better idea of the speaker's meaning is to ask questions. Here are four kinds of asking:

1.   Ask for clarification of key words

Ask the speaker to explain key words, terms or phrases that are unfamiliar to you, may have multiple meanings, or are integral to the message. Don't be afraid you'll look dumb, because usually people respect you more if you ask them to explain something. They know you respect the communication process enough to want to prevent misunderstanding.

Example:

Attorney: "People are getting more litigious these days."

You: "How exactly are you using that word "litigious?"

2.   Ask for an example

This is one of the most powerful listening tools, primarily because when speakers are trying to make a point, they almost always have an example somewhere in their minds, even though they may not cite it explicitly. Your job as listener is to find out what their examples are.

When you get an example from a speaker, not only will you recognize more exactly what he means, but the example will often provide ideas for further probing. And most people will appreciate the opportunity to provide an example.

3.   Ask for the specific information you need

Reprinted with permission from *Proc. 33rd Internat. Tech. Comm. Conf.*, published by the Society for Technical Communication, pp. 129–131, Detroit, 1986.

If you want opinions, ask for them. If you want more facts, major concerns, specific details, or the speaker's priorities, ask for them. A speaker can't provide you with additional information without getting some direction from you.

Your question should focus on information that is important to you, to getting the job done right, and to the situation. You will save time, money and, frustration if you ask for the necessary information as soon as you need it.

4.   Ask for verification of your translation

Every now and then, verbally summarize what you think the speaker is saying. This gives the speaker a chance to hear what you think is happening, and remember that only the speaker can say "Yes, that's what I meant".

Your summary can be brief, and it may take various forms: major ideas presented, your interpretation of a specific idea, or perhaps the points that are important to remember or require action.

You may be tempted to feel embarrassed if your summary isn't quite accurate. But remember, finding out where the message is breaking down is the whole point. It is costly and unproductive to find out later that your original understanding was inaccurate.

This verification process we have been describing has been popularly called "reflective listening", "active listening", or "restatement." Regardless of the terminology used, the process is the same: the listener restates in his *own words* how he has interpreted the speaker's statement. This provides an information exchange that allows the speaker to assess if he has been understood. This is in direct contrast to non-informational exchange such as a "Yes" answer to the question "Do you understand?" "Yes" provides no *information*, only an opinion.

What If I Disagree?

Handling disagreement is an important listening skill. Let's look at how disagreement typically occurs in American conversation:

STATEMENT: I doubt if we can expect much progress in the first week of the project.

DISAGREEMENT: Oh, I disagree. The last project I was on...

CLARIFICATION: But that's not what I mean. I'm talking about...

It is frequently helpful to temporarily set aside your feeling of disagreement and ask for clarification, using the approaches just discussed. Fight the urge to debate until you are sure you fully understand.   The two approaches can be modeled as follows:

"Typical" sequence:

STATEMENT — DISAGREEMENT — CLARIFICATION

Better sequence:

STATEMENT — CLARIFICATION — DISAGREEMENT

If you find you still disagree, first say something constructive about what the other person is expressing—then state your disagreement.

Example:

"The good point about that is _____. However, I have a concern about ...".

Be sure to describe how you feel and think, not how you assume or guess the other person is feeling or thinking. One way to do this is to use an "I-statement" describing what action, behavior, or idea is bothering you, how you feel about it, and the effects of the situation.

Example:

"Well, my reservation [how you are feeling] is that this new

enhancement to the system [idea being discussed] will slide the schedule more than the customer will allow [potential results]."

NOT:

"Well, you seem to be willing to run the risk of a schedule slide."

For a helpful exercise on the restatement process, see the exercise suggested at the end of this paper."

How and When Not to Ask Questions!

It isn't always useful or appropriate to ask questions. Also, *how* we ask can affect the quality of communication. Here are a few "don't"s:

o   Don't ask unrelated questions. Stick to the topic.

o   Try not to project a judgemental attitude while asking. Watch your tone of voice, body language, and facial expression.

o   Don't probe during small talk, or when people are just letting off steam. ("Boy, am I tired!" "What do you mean 'tired'?")

o   Don't play one-upsmanship, going for a better story or line than what the speaker has just said.

o   Avoid asking "why?". "Why" is probably the most defense-provoking word in the English language. Use "what"s or "how"s instead ("What do you think we should do?")

o   Don't mirror words like a parrot. ("I'm so excited about this new contract." "You're really excited, huh?")

EVALUATING IDEAS

As listeners, we must frequently discern useful or appropriate ideas from unproductive or inappropriate ideas. This requires analytical listening skills. Analyzing and evaluating ideas are assuming greater importance than ever in the workplace, as employers place more emphasis on group decision-making processes such as quality circles, special project teams, and participatory management styles.

The Three-Step Assessment Model

This model shows the process that occurs when you take someone through the development of an idea. One of its greatest benefits is that it allows you to evaluate ideas without evaluating the personalities associated with the ideas.

IDEA	OPINIONS	FACTS
STEP 1: Clarify	STEP 2: Ask for advantages & disadvantages	STEP 3: Ask for evidence, factual examples or cost studies

Probe
* Ask for
  Examples
* Clarify
  Key Words
* Summarize
	Advantages:
	
	Disadvantages:
		Statistics:
		
		Testimony:

Here's how to use the Assessment Model:

1.   Clarify. Make sure you understand the idea clearly. Ask for questions, ask for examples, clarify key words, and summarize what you think the speaker is saying. Make no assessment or evaluation before you are sure you understand what is being said. This can spare you the awkwardness of disagreeing with someone only to find out later that you really agreed once you understood.

2.   Ask for advantages and disadvantages. If you focus on advantages first, you'll help keep the atmosphere positive. If you begin with disadvantages you may provoke defensiveness.

Notice that so far no personal issues have entered into the evaluation process, just viewpoints on the pros and cons of the solution.

3.   Ask for specific evidence. Ask for evidence of each advantage and disadvantage. You can ask for factual examples, statistics, or the testimony of an authority (someone who has used the product under discussion, for example).

The Assessment Model keeps you from making premature evaluations and helps promote positive interaction.

HOW TO EMPATHIZE

Empathizing is a way for listeners to establish and maintain productive relationships with speakers. When we empathize we recognize another person's emotion as a genuine part of the communication. We don't attempt to argue or sweet talk them out of it, or to top someone else's war story with an even worse anecdote.

Empathy is often a part of keeping the communication lines open, because when we empathize we are projecting acceptance and respect for the other person. And let's face it, whether at work or at home, we can't ask that people speak to us only when they are happy or in a good mood!

Here are a few suggestions on how to empathize:

o   Focus on the speaker. Concentrate and let the speaker know you are paying close attention.

o   Allow the emotion to be expressed. Be quiet and don't interrupt unnecessarily.

o   Delegate the emotion back to the speaker by getting him to focus on an action or need. Rather than you "owning the problem" by making a suggestion to the speaker, you can make statements such as these:

"What are you doing about the situation?" [Focus on action]

"What are you going to do about the situation?" [Focus on future action]

"How can I help?" or "What specifically would you like me to do?" [Focus on specific perceived need]

Asking questions such as these helps the speaker concentrate on a coping strategy.

o   Match the emotional tone set by the speaker. This does not mean you have to feel the same way! But let your behavior indicate that you acknowledge the seriousness (or possibly lightheartedness) of the situation.

o   Don't make the speaker feel guilty, try to change his mind ("I'm sure things will get better..."), or try to top his story.

o   Recognize that you don't have to empathize with everyone you meet.

CONCLUSION

Here is a checklist you can review periodically to brush up your listening skills:

1.   I create time to listen when I am needed.

2.   I accept my responsibility as a listener to clarify verbally my understanding of what the speaker has said.

3.   My nonverbal behavior shows that I am a good listener: I maintain eye contact and shut out other distractions.

4.   I balance silence and my own questions or responses.

5.   I evaluate ideas, not people.

6.   I keep evaluations tentative until thorough discussion has taken place.

7.   I can use feedback to verbalize understanding of someone else's point of view, even when I do not agree

EXERCISE

To practice clarification skills, get a friend or group of friends together. Each person should write down a very strong opinion—the stronger the better—on an index card. Then mix up the cards and distribute them. The person who gets the card must interact with the person who wrote the statement, using restatement to clarify the statement on the card. The restatement goes on until the person who wrote the statement can say, "Yes, you understand me." No debating or argumentation is allowed! The point is to understand someone's position, not to change it.

Author Index

Subject Index

Editor's Biography

David F. Beer (M'90) attended Bideford School in Devon, England, before coming to the United States, where he received the B.A. in secondary education at the University of Arizona. He received the M.A. in English linguistics from Arizona State University and the Ph.D. in English from the University of New Mexico.

He taught at the University of Colorado and at Haile Sellassie I University in Ethiopia before joining the English Department at the University of Texas as an assistant professor in 1978. In 1985 he was invited to develop a technical communication program in the Department of Electrical and Computer Engineering, where he currently directs the program. Besides teaching technical communication, he has worked at Texas Instruments in Austin and has consulted and edited for a number of engineering professionals.

Dr. Beer has presented papers on technical communication at IEEE and ASEE conferences and has published articles in several journals. He is a member of the IEEE Professional Communication Society and a senior member and past president of the Austin chapter of the Society for Technical Communication. His hobbies include walking his dog and amateur radio (call sign: KB5HZN).